HYDROGEOLOGY

HYDROGEOLOGY

Editor

Marcel Lafrenz

scitus
academics

Hydrogeology

Edited by **Marcel Lafrenz**

This Book is Printed on Acid-Free Paper. ♾

ISBN: 9781681179698

Library of Congress Control Number: 2017949100

SCITUS Academics LLC,

616, Corporate Way, Suite 2, 4766,

Valley Cottage, NY 10989

www.scitusacademics.com

Notice

Printed in United States of America.

Preface

The field of hydrogeology began to grow during the 1960s and expanded rapidly during the 1970s with the recognition of serious environmental problems from hazardous chemicals and the subsequent evolution of stringent environmental regulations. Hydrogeology is concerned with the study of rocks and the structures that are formed over past periods of time. The way that sediments have been deposited to create diverse layers of rock beneath the surface, or the way that rocks have been heated and folded over millions of years to create complex structures are the subjects of geology. It deals with how water interacts with geological systems. But there is more to hydrogeology than wet rocks. Some parts of the world are blessed with frequent rainfall and plentiful surface water resources, but most countries need to use the water that is stored underground to supplement their needs. Hydrogeologists can help by locating suitable reserves and by assessing how much water it is possible to extract without permanently damaging underground aquifers or surrounding ecosystems. Water is an essential natural resource for people all around the world - whether it is piped to homes or drawn out of wells. Understanding where it is and how it moves under the ground is essential in protecting this resource. The arena of groundwater hydrology and the discipline of hydrogeology have attracted a lot of attention during the past few decades. This is mostly because of the increasing need for high quality water, especially groundwater.

This volume 'Hydrogeology' clearly demonstrates the extensive range of issues that are dealt with in the field of hydrogeology. Karst hydrogeology and deposition processes, soil hydraulic properties as a factor affecting groundwater recharge processes, relevant conceptual models, and geophysical exploration for groundwater. It explores theoretical and applied aspects of hydrogeologic science, including studies ranging from local areas and short time periods to global problems and geologic time; innovative instrumentation; water-resource and mineral-resource evaluations; and overviews of hydrogeologic systems of interest in various regions.

Table of Contents

CHAPTER 1

Groundwater Flow Model for a Tannery Belt in Southern India

Nepal C. Mondal[1, 2], V. P. Singh[2], S. Sankaran[1]

[1]*Environmental Geophysics Division, National Geophysical Research Institute (Council of Scientific & Industrial Research), Uppal Road, Hyderabad, India*

[2]*Department of Biological & Agricultural Engineering and Department of Civil & Environmental Engineering, Texas A & M University, USA*

ABSTRACT

The objective of this article is to develop a groundwater flow model for a tannery belt using Visual MODFLOW Premium 4.4 for analyzing groundwater velocity and its response to various pumping strategies in two stages, viz., steady and transient conditions. The steady state model was calibrated for April 2001, whereas the transient model was employed to forecast groundwater flow under various pumping strategies. The results showed that the total groundwater abstraction was about 80.43% of the groundwater recharge, but 10.25% was used up by evapotranspiration. The groundwater velocity, which is important for contaminant migration, varied from 0.21 to 0.52 m/d in the tannery cluster. The model was more sensitive to recharge from rainfall, hydraulic conductivity and specific yield. Finally, the model showed that the aquifer could sustain a pumping rate of 24892 m^3/day without further decline in water level.

Keywords: Groundwater, Flow Model, Groundwater Velocity, Tannery Belt, Southern India

1. INTRODUCTION

The study area, a drought prone and hard rock region in Southern India, covering about 207.5 $km^2$is characterized by poor soil, scarce vegetation, erratic rainfall, heavy runoff and lack of soil moisture during most parts of the year [1].Recurring droughts, coupled with increasing exploitation of groundwater, have resulted in the decline of groundwater levels by more than 10 m at some places in the last three decades [2]. Existing shallow wells reflect high drawdowns during the dry season (mainly January through to September) each year. The large drawdown or outright failure of wells results in poor availability of groundwater for drinking water purposes. Untreated effluents from 80 functioning tanneries, forming a tannery belt, have also considerably deteriorated groundwater quality [3, 4].

In order to meet the water demand of communities it is necessary to investigate the groundwater resource potential in the tannery belt. Thus, the main objective of the present study is to determine groundwater velocity, which is vital for mass transport modeling, and to assess the aquifer response under different input and output stresses using Visual MODFLOW Premium 4.4, which was initially documented by McDonald and Harbaugh [5]. In order to achieve the above objective, the following tasks are to be carried out: 1) characterization of geological formations through the interpretation of geophysical data; 2) analysis of hydrogeological data for aquifer characteristics; 3) estimation of natural recharge by using well water level corresponding to rainfall; and 4) conceptualization of 2-D groundwater flow model for shallow aquifer making use of the available data and simulation for prognostication.

2. MATERIALS AND METHODS

2.1. Background of the Study Area

The study area is a drought prone hard rock terrain, and is located about 400 km southwest of Chennai, the capital city of Tamil Nadu, India. It lies between $10^{0}13^{/}44^{//}$ -$10^{0}26^{/}47^{//}$ N latitudes and $77^{0}53^{/}08^{//}$-78^{0} $01^{/}$ $24^{//}$ E longitudes (see Figure 1), and encompasses an area of about 207.5 km^2, covering parts of Dindigul, Attur, Reddiarchattram and Sanarpatti blocks. The area is characterized by undulating topography with hills located in southern parts, sloping towards north and northeast [6]. The highest elevation (altitude) in the hilly area (Sirumalai Hill) is of the order of 1350 m (amsl), whereas in plains it ranges from 360 m (amsl) in the southern part to 240 m in the northern part. No perennial streams exist in the area, except for short distance streams encompassing 2nd and 3rd order drainage. Runoff from rainfall within the area ends in small streams flowing towards the main Kodaganar River. From a period of 1971-2007 the average annual rainfall is of the order of 905.3 mm. Normally, subtropical climate prevails over the area without any sharp variation. The temperature increases slowly to a maximum in summer months up to May and after which it drops slowly. The mean of the maximum temperature in plains ranges from 36.5°C to 41.8°C and in hills, it ranges from 7.9°C to 21.8°C. The mean of the minimum temperature in plains varies from 17.4°C to 24°C and in hills varies from 6°C to 8.5°C [1].

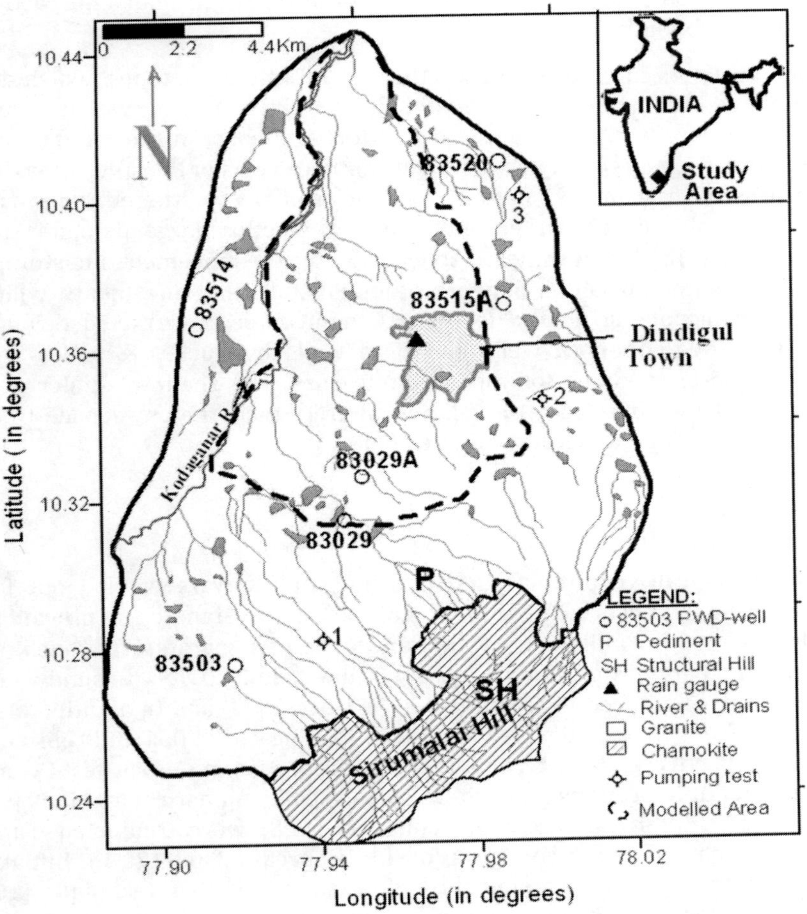

Figure 1. Location map representing the drainage pattern and PWD wells.

2.2. Geological and Hydrogeological Setting

Geologically the area is occupied with Archaean granites and gneisses, intruded by dykes [7]. These formations, including granite, granodiorities, gneissic granite and gneisses, are the most widespread groups of rocks which are mainly composed of gray and pink feldspar with quartz grains, biotite and hornblende [8]. These formations are crossed by sets of joints and fractures, which have also caused weathering of coarser rocks. Weathering occurs due to mechanical and mostly chemical processes that take place, while water in the fractures interacts with the formation. The shallow hard and massive rocks are exposed mostly in the southern part. Another most dominant formation is charnokite, which is found in the extreme southern and southeastern parts of the area (Sirumalai hill) acting as a no flow boundary.

Groundwater occurs mostly in weathered and fractured zones, which are unconfined, semi-confined or confined. The thickness of weather zone varies from 3.1 to 26.6 m but black cotton soil exists in the middle part whereas red sandy soil in northern and southern parts of the study area. It's thickness varies from 0.52 m to 5.35 m. Such shallow weathered zones may not be stable sources of groundwater for meeting large demands for groundwater [2]. The weathered zone facilitates the movement and storage of groundwater through a network of joints, faults and lineaments, which form conspicuous structural features. Groundwater is extracted through dug well, dug-cum-bore wells and bore wells by bucket & pulley and electric motor methods for different purposes. The shallow aquifer gets both direct recharge from rainfall and indirect recharge as seepage from about 93 irrigation tanks and irrigated fields.

2.3. Methods

Hydrogeological, geophysical and entropy studies were carried out for deciphering subsurface litho zones, understanding prevailing hydrogeological conditions, evaluation of aquifer parameters, such as natural recharge, storativity, and hydraulic conductivity. Groundwater levels were monitored in 6 wells by PWD since 1971, and in addition also collected from 45 and 30 observation wells during April 2001 and February 2009, respectively. To determine aquifer properties, pumping tests were carried out at 3 locations and data was analyzed by numerical methods [9]. A total of 37 Vertical Electrical Soundings (VES) were conducted using Schlumberger electrode configuration for a spread of 60-120 m. Initially VES data had been interpreted through a curve matching technique [10] and then interpreted by computer programme [11]. 6-PWD dug wells were selected for the determination of average natural recharge by the entropy theory [12]. The generated data was utilized in the development of the groundwater flow model. The various steps involved in the modeling in the study are shown in Figure 2.

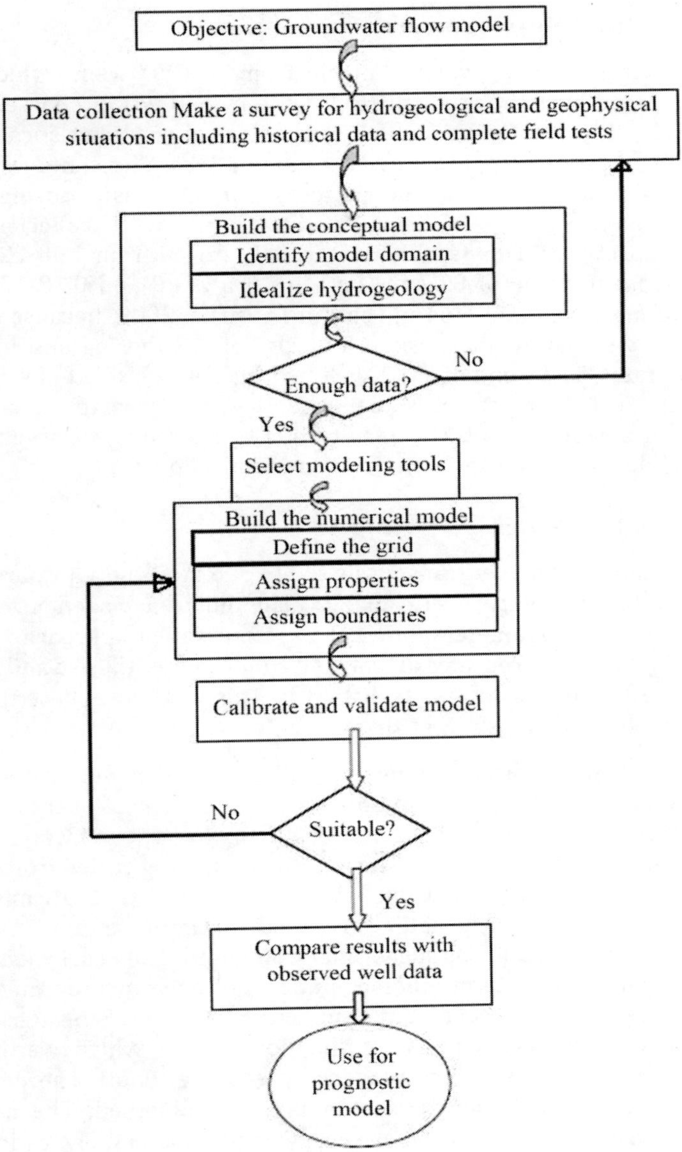

Figure 2. Flow chart for groundwater flow model.

3. RESULTS AND DISCUSSION

3.1. Hydrogeological Investigations

3.1.1. Water Level Measurements

Monthly water level data were available from 6 PWD wells, which are uniformly spread over the study area. A contour map was prepared, from which it was difficult to identify the hydrogeological characteristics of the aquifer with respect to the groundwater flow pattern. The contours don't follow the topography and drainage patterns. For this reason during April 2001 and February 2009, 45 and 30 water levels were collected and analyzed, respectively. The depth of water table varied from 1.00-22.80 m (bgl) with a mean value of 8.07 m (bgl) in April 2009 and 0.70-17.00 m (bgl) with a mean value of 5.64 m (bgl) in February 2009, because of the variations in weathered thicknesses, intensity of weathering and uneven withdrawal rates. The trend of PWD-well hydrographs closely followed the rainfall trend [6]. In most cases the water level returns to its original position after a good rainfall. This may be due to the rapid recharge taking place due to heavy rainfall and also irrigation return flow [2].

3.1.2. Aquifer properties

Aquifer parameters, namely transmissivity (T) and storativity (S), are vital for groundwater modeling. Several analytical methods have been developed to determine these parameters, however, the numerical approach has an advantage in that it incorporates actual field conditions with ease and hence parameters estimated are more realistic [9]. This method is described in detail by Rushton and Redshaw [13].

During field investigations, 3 existing large diameter open wells fitted with pumps (see Figure 1) were selected for pumping test. Most of these wells had been kept without pumping prior to beginning the test and water levels had been continuously monitored. The period of pumping varied from 45 to 126 minutes, whereas recovery times varied from 170 to 1313 minutes. Both pumping and recovery data were used for interpretation using a forward modeling technique as suggested by Singh [9]. The nearby features, such as water body or lateral inhomogenities had been incorporated into individual interpretations. Initial parameter values were considered to generate time drawdown curves for individual tests, which were then compared with observed timedrawdown/recovery data. The aquifer parameters were varied until a close match was obtained. The best-fit match was considered as representative aquifer parameters. The estimated T and S values varied from 15 to 200 m^2/day and 1×10^{-5} to 3.5×10^{-4}, respectively. It was found within the ranges of T and S values from 4 to

1166 m^2/day and 1×10^{-5} to 9×10^{-3}, respectively, obtained in Kodaganar River Basin, Dindigul and Karur Districts, Tamil Nadu [2].

3.2. Geophysical Investigations

The resistivity sounding technique [10,14,15] was employed to identify the aquifer geometry. In total 37 Vertical Electrical Soundings (VES) were carried out with current electrode separation of 60-120 m. The curves obtained are classified as A and H-types, which describe the variation in the resistivity of progressive layers below the ground surface. The A and H-type sounding curves reflect the pattern of resistivity distribution with depth. If ρ_1, ρ_2 and ρ_3 are the resistivities of three subsurface layers beginning with ρ_1 at the top, then $\rho_1 > \rho_2 < \rho_3$ is defined as H-type and $\rho_1 < \rho_2 < \rho_2$ as A-type.

The observed field curves were matched with theoretical master curves to obtain initial parameter values and finally these were used as input in the interpretation of resistivity data through software, namely, RESIST [11]. The interpreted results of VES, when compared with the existing 9 litho-log data [1] and cross-sections of nearby open wells, confirmed the resistivity ranges of different subsurface geo-electrical layers as:

- 2-95 Ω-m: Top soil cover/ clay with kankar
- 6-100 Ω-m: Weathered formation/saturated or saline aquifers
- 100-300 Ω-m: Semi-weathered/fractured granite and gneissic granite
- >300 Ω-m: Hard rock (gneissic granite and gneisses)

The shallow aquifer resistivity in weathered zone ranged from 6.08 to 264.27 Ω-m. The estimated thickness of the weathered zone varied from 5.30 to 26.62 m. It was confirmed that its value ranged from 15.00 to 26.62 m in the western part of Dindigul town. The soil thickness ranged from 0.52 to 5.35 m, whereas the depth of bedrock with weathered thickness: 11.00-26.62 m ranged from 12.00 to 27.67 m (bgl) in western and southwestern parts of the town, which are potentially good groundwater zones.

3.3. Natural Recharge Estimation

For modeling of groundwater resources in the semi-arid area, it is essential to determine natural groundwater recharge. There are several methods for determining groundwater recharge, such as groundwater balance [16]; lysimeters [17]; piston-flow model [18]; RS and GIS techniques [19]; photogeological [20], hydrogeological [21], geophysical methods [22], and ^{14}C-age dating [23]; and regional groundwater models [24]. Among these methods, the tracer technique is one of the best direct methods for estimation of groundwater recharge [18]. This technique estimates recharge on the basis of piston flow model, and has been found useful [25-26]. Other methods are time consuming and sometimes even uneconomical in developing countries, particularly when one has to deal with a large area.

Therefore, an entropy-based approach [12] is developed for assessing natural recharge in this study area. Entropy of a random variable is a measure of the information or uncertainty associated with it. Measures of information include marginal entropy, joint entropy and transinformation. For a random variable x, the marginal entropy, H(x) can be defined as the potential information of the variable. For two random variables x and y, the joint entropy H(x, y) is the total information content contained in both x and y. The mutual entropy (information) between x and y, also called transinformation, T(x, y), is interpreted as the reduction in uncertainly in x, due to the knowledge of the random variable y. It can also be defined as the information content of x that is contained in y. Entropy measures can be expressed using both discrete and analytical approaches [27]. Discrete forms of these entropies can be expressed as:

$$H(x) = -\sum_{i=1}^{n} p(x_i) \ln p(x_i) \tag{1}$$

$$H(y) = -\sum_{j=1}^{m} p(y_j) \ln p(y_j) \tag{2}$$

$$H(x,y) = -\sum_{i=1}^{n} \sum_{j=1}^{m} p(x_i, y_j) \ln p(x_i, y_j) \tag{3}$$

$$T(x,y) = \sum_{i=1}^{n} \sum_{j=1}^{m} p(x_i, y_j) \ln \left[\frac{p(x_i, y_j)}{p(x_i)\, p(y_j)} \right] \tag{4}$$

where x and y are two discrete variables with values x_i, $i = 1, 2, \cdots, n$; y_j, $j = 1, 2, \cdots, m$, defined in the same probability space, each of which has a discrete probability of occurrence $p(x_i)$ and/or $p(y_j)$ and $p(x_i, y_j)$ is the joint probability of x_i, y_j. Note that $H(x, y) = H(y, x)$.

Rainfall is considered as independent random variable (x) and the depth to water table for individual wells as the dependent variable (y). Then, transinformation, $T(x, y)$, is interpreted as the reduction in the original uncertainty of depth to water table due to the knowledge of rainfall. It can also be defined as the information content of water table which is also contained in rainfall. In other words, it is the difference between the total entropy and the sum of marginal entropies of these two variables. This is the information repeated in both water table and rainfall, and defines the amount of uncertainty that can be reduced in one of the variables when the other variable is known. On the other hand, marginal entropy, $H(x)$, is defined as the potential information of rainfall. Then, the ratio of $T(x,y)$ to $H(x)$ is simply a fraction of recharge due to rainfall. Therefore, the percentage of rainfall, R_e (%), contributing to the natural groundwater recharge of an unconfined aquifer is given as [12]:

$$R_e(\%) = \frac{T(x,y)}{H(x)} \times 100 \tag{5}$$

To determine the fractional amount of rainfall (monitored for 8 years from January 2000 to December 2007 at Dindigul rain gauge station), called natural recharge, marginal entropies and transinformation of rainfall and depth to the water table (collected from 6 PWD wells for the same period, see Figure 1) were calculated. Then a ratio of transinformation to marginal entropy of rainfall was used as a measure for evaluating natural recharge.

Rangarajan and Athavale [28] estimated an average natural recharge ratio 10.11% of seasonal rainfall for 15 different granitic and gneiss areas in varying climatic and hydrogeological provinces of India. Out of them, the average recharge ratio was about 9.90% for 4-granitic and gneiss areas in Tamil Nadu state (Southern India). The estimated average natural recharge using entropy varied from 0.63 to 12.53% in the entire period, but it varied from 2.00 to 23.71% only for NE monsoons. The mean of the estimated recharge values was 14.56% of rainfall in NE monsoon (see Table 1).

3.4. Aquifer Modeling

An aquifer model was constructed based upon a conceptual approach. In order to construct the model, boundary condition, grid and time increments were decided, and applied stresses and hydraulic properties were also estimated. Finally, these parameters were tested and adjusted during the calibration procedure, with the intention of reproducing a set of historically observed data. Finally the model was evaluated, considering how reasonably it could represent the actual system.

Table 1. Estimated natural recharge using entropy for the entire monthly and NE monsoon data during January 2000 to December 2007.

For the entire monthly rainfall (in mm) and depth to water level data (m, bgl) from January 2000 to December 2007						
PWD wells	*H* (*x*)	*H* (*y*)	*H* (*x*, *y*)	*T* (*x*, *y*)	*Natural Recharge* (%)	*Average* (%)
83520	0.790	1.378	2.100	0.068	8.61	8.46
83029A	0.790	1.055	1.771	0.074	9.37	
83503	0.790	1.737	2.451	0.076	9.62	
83514	0.790	1.034	1.745	0.079	10.00	
83029	0.790	2.005	2.696	0.099	12.53	
83515A	0.790	0.146	0.931	0.005	0.63	

For monthly rainfall (in mm) in the months of October, November and December and depth to water level data (m, bgl) during NE-monsoon from January 2000 to December 2007						
PWD wells	*H* (*x*)	*H* (*y*)	*H* (*x*, *y*)	*T* (*x*, *y*)	*Natural Recharge* (%)	*Average* (%)
83520	1.497	1.528	2.757	0.268	17.90	14.56
83029A	1.497	2.209	3.320	0.286	19.10	
83503	1.497	1.692	2.887	0.202	13.49	
83514	1.497	0.785	2.115	0.167	11.16	
83029	1.497	1.892	3.034	0.355	23.71	
83515A	1.497	0.095	1.562	0.030	2.00	

H (*x*): Marginal entropy for rainfall; *H* (*y*): Marginal entropy for water level; H (*x*, *y*): Joint entropy; *T* (*x*, *y*): Transformation; All entropies calculated with base 2; and Unit: bits

3.4.1. Conceptualization

The aquifer in the proposed area (see Figure 1) consists mainly of two layers: one-weathered and two-fractured zones of the granite-gneiss formation. The weathered zone overlies the fractured zone (in bedrock), but its thickness varies from place to place. Bedrocks are at a depth of 12.00 to 27.67 m (bgl). The two layers have different hydraulic characteristics and especially the fractured zone has a lower storage coefficient than does the weathered zone. The weathered part of the aquifer was considered as equivalent to a porous zone.

Recharge from rainfall takes place between October and December. The percentage of rainfall that becomes recharge was determined based on the estimated recharge values by entropy [12]. Then it was adjusted during the model calibration. The reason why no recharge takes place between January and September is low rainfall in conjunction high temperature and evaporation. There is no surface water interaction with neighboring sub-watersheds. Groundwater interaction with adjacent area was also considered. The main recharge areas are in the south and southeast, and the main discharge area in the northern part. Thus, groundwater flows southwest, north and northeast. Kodaganar River acts as the main drainage route and flows north. However, it flows only during the wettest months of the year (October-December). The aquifer seems to be in interaction with the river and probably the water table is higher than the river stage. In order to satisfy irrigation, domestic and industrial needs, increasing groundwater abstractions take place in the area. Industrial and domestic abstractions are about 20% of the total abstracted volume. Irrigation abstractions most likely occur during the dry period (January to September).

3.4.2. Model Design Approach

A single layer model of the tannery belt (see Figure 1) was selected and a groundwater flow regime was prepared for the following reasons: 1) geometry of fractured zone was unknown, 2) no data existed about the actual amount of abstractions that takes place in this fractured zone, and 3) the number of abstracting wells reaching the fractured zone was unknown. The geometry and boundary conditions are generally complex. Analytical methods are rarely applicable for determining a closed form solution of the partial differential equations of 2-D groundwater flow equation [13] as given below:

$$\frac{\partial}{\partial x}\left(K_{xx}\frac{\partial h}{\partial x}\right)+\frac{\partial}{\partial y}\left(K_{yy}\frac{\partial h}{\partial y}\right)=S_s\frac{\partial h}{\partial t}\pm W \quad (5)$$

where K_{xx} & K_{yy} are the hydraulic conductivities along the x & y directions, h is the hydraulic head, S_s is the specific storativity, W is the groundwater volume flux per unit area (positive for outflow and negative for inflow), and x and y are the Cartesian coordinates.

Equation (5) was solved using a finite difference approximation technique. The starting point for the application of this method was discretization of modeled area into small grid form. This leads to set of simultaneous algebraic equations, which was solved using the Visual MODFLOW Premium 4.4 modeling code. This code has been widely used and is accepted to produce numerically stable solutions. The method selected for the numerical solution of the algebraic equation set is WHS.

3.4.3. Grid Design

In order to set up the model in MODFLOW set of codes, the area of interest (an area: 75.56 km^2) was divided into a series of grid blocks or cells (see Figure 3), a grid size of 250 m × 250 m (total grids = 1209) and it would be used for mass transport modeling as the hydrochemical data is available in this part. The elevation of the top of shallow aquifer varied from 232.00 to 314.00 m (amsl), while the bottom from 220.00 to 306.00 m (amsl). Both the top and bottom elevations of the aquifer were defined with respect to above mean sea level (amsl) in grid form with the use of SURFER v.8.0 using the kriging method and converted into ASC-format, which was imported into the model under the MODFLOW environment. The variation of aquifer thickness for a typical row number 44 is shown in Figure 4. The layer had the confined/ unconfined condition and corresponded to a layer type 2 in Visual MODFLOW Premium 4.4.

3.4.4. Boundary Conditions

Generalized head boundary: The southern and eastern boundaries of the modeled area were simulated through a generalized head boundary (GHB) in order to represent the groundwater inflow and outflow through nearby sub-watersheds [29]. The values that were used in the steady state simulation in this model were obtained by calibration and were kept constant throughout the transient simulation, because there existed no data at all for this boundary. The values, that were finally used, were boundary conductance of 5.2×10^3 m^2/d and head 4 m (bgl) at or beyond the boundary.

River: Kodaganar River is a special case of a headdependent boundary. It was simulated through a river boundary condition. The properties were set for each river cell boundary as: the river stage (5 m below ground surface), riverbed bottom (4 m below the river stage), riverbed thickness (1.0 m), riverbed vertical hydraulic conductivity (K_z = 2.0 m/d) and river width (10 m). Because there were no data at all for Kodaganar River, the properties were set through the conditional process.

3.4.5. Hydraulic Parameters

Initially the results from the only available aquifer test were used as the hydraulic properties of the aquifer. However, these proved to be inadequate to represent the whole system, which was complicated and heterogeneous due to the presence of weathered zone. Hence various property zones were assigned, and their values were adjusted during calibration.

3.4.5.1. Hydraulic Conductivity

The final distribution of hydraulic conductivity zones is illustrated in Figure 3. The southern part was assigned a conductivity value of 4.5 m/d, in the central part 6 m/d and northern part 9 m/d.

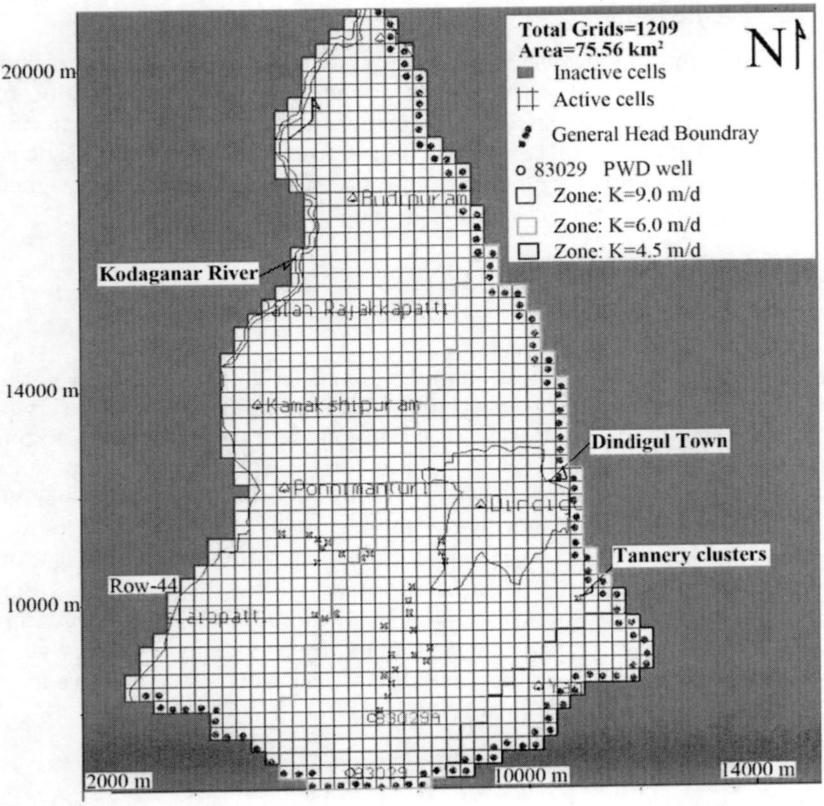

Figure 3. Grid map and hydraulic conductivity distribution of the tannery belt.

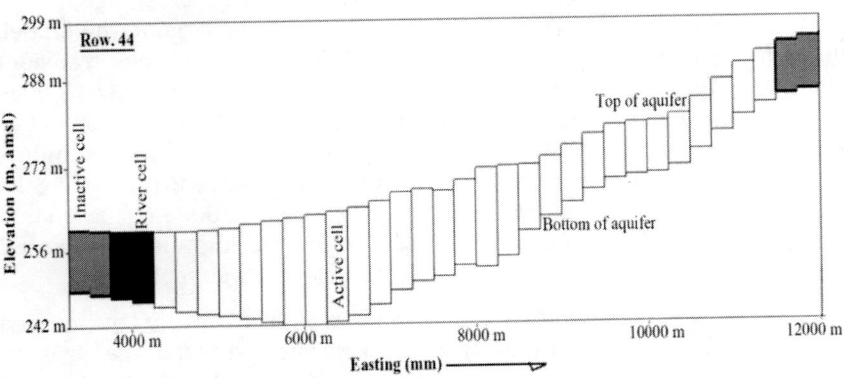

Figure 4. Simulated vertical cross section along row number 44.

3.4.5.2. Specific Yield

The final distribution of specific yield, as indicated by the model calibration, was assigned. The southern part was assigned a specific yield value of 0.015, the central part 0.02, and the northern part 0.03. Similarly specific storages of 0.00005, 0.0012 and 0.0002 were assigned to these zones, respectively. The effective porosity and total porosities were also assigned uniformly throughout the area as 2.75% and 3.00%, respectively [30-31].

3.4.6. Applied Stresses

3.4.6.1. Abstractions

The main groundwater application in the area is irrigation using 80% of the total abstracted volume. The remaining 20% of the abstraction is reserved for domestic and industrial uses [1]. Irrigation abstractions are spread out in the area (except on the hills), while the domestic/industrial abstractions are assumed to be concentrated around towns and the villages. However, all well locations were unknown in the tannery belt. Therefore, the known location of 96 wells, which were used for domestic, industrial and irrigation abstractions, were used to simulate abstractions, because the total abstracted volume was about 23, 588 m^3/day in 2000 [1]. The abstractions during subsequent years had to be estimated by increasing 3% every year. Abstractions from each irrigation and domestic/industrial wells were used for steady, transient and prognostic models.

3.4.6.2. Recharge

Recharge was separated into recharge from rainfall and recharge from irrigation return flows. Recharge from rainfall and irrigation return flows occurs during the wet period (October-December) in the study area, whereas recharge only from irrigation return flows occurs during the dry period (January-September). Initially as recharge, a percentage (2.00-23.71%) of the rainfall was used as a direct rainfall recharge in the model, although this approach proved to underestimate during transient calibration due to its spatial variability and the computed water level values were comparatively lower than the observed ones. Therefore, a higher percentage of rainfall (12-37%) was used (see Figure 5) for understanding the combined effect of natural recharge and irrigation return flows in the study area. Then the evapotranspiration package was added to the system for the dry period (January-September) only. During the wet period evapotranspiration was insignificant, and therefore it was set to zero.

Evaporation was estimated to a maximum rate of 52.5 mm/ year for a depth of up to 3 m based on the type of soil, crops and plants that grow in the area. This way, evapotranspiration occurred when the water table was within 3 m from the ground surface, i.e., the evapotranspiration rate was zero 3 m below the ground surface and 52.5 mm/year at the ground surface. Finally, irrigation return flows were estimated at about 4-38% of the abstracted volume and added to the aquifer system. Based on hydrogeological and climate conditions prevailing in the area, the recharge values of 80, 120, 160, 180 and 250 mm/year, respectively, were assigned to zones 1-5, as shown in Figure 5.

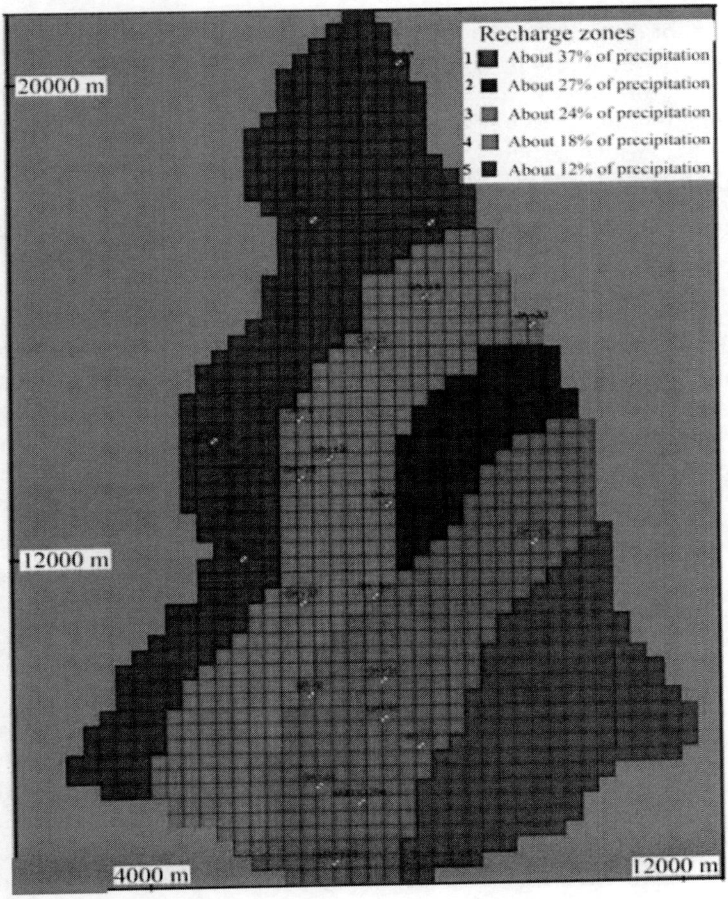

Figure 5. Recharge distribution in the groundwater flow model.

3.4.7. Model Calibration

The purpose of model calibration was to estimate the acute parameters for the given boundary conditions and stresses within a certain established range of error (calibration targets). In this study a trial and error calibration was used. Parameters were initially assigned to each node in the grid. Then these parameter values were adjusted sequentially to match the calibration targets. The calibration parameters, set in the model, were the generalized head boundary, recharge, evapotranspiration, hydraulic conductivity, and specific yield.

3.4.7.1. Calibration Stages

Model calibration was performed in three stages: 1) steady state, 2) transient state, and finally 3) prognostic model under transient state. The purpose of the steady state calibration was to provide a set of initial conditions for the transient model and initial estimates of the calibration parameters, which would be further calibrated during the transient calibration.

The data available for model calibration existed for the period April 2001-February 2009. Since the estimated abstraction for the year 2001 was not zero, no steady data existed. Therefore, the calibration of a dynamic steady state model was considered in order to obtain an initial head distribution for April 2001. The calibrated parameter values were imported into the transient model for the period April 2001 to February 2009. The calibration parameters were further adjusted in the transient model and they were imported back in the steady state model, in order to reproduce the corresponding new initial conditions. Finally the new initial conditions were imported once more in the transient model as well as the prognostic model, and this calibration was at least improved by the execution of sensitivity analysis.

3.4.7.2. Calibration Criteria

Each grid cell had a 5 m difference at the top of the aquifer as well as approximate 5 m difference in the head from the adjacent ones. Each monitoring well could be anywhere within the same cell. Therefore the error due to space discrimination allowed in the steady state calibration targets ±2.5 m [29]. Similarly, because it was not known when within a month the observation measurements were taken, the error due to time discrimination was ±0.5 m. However, in transient calibration errors larger than ±2.5 m were allowed, considering the size of the area and grid, and in fact abstractions were completely unknown, especially for early simulation years. During transient calibration the main idea was to match the trends in the hydrographs reproduced at two PWD wells. But the numerical round off error was evaluated by examining the water mass balance calculated by the model. This error should ideally be less than 0.1% [32] and was tested for both the steady state and transient models.

3.4.7.3. Steady State Calibration

The steady state simulation had been performed for averaged rainfall and abstraction values for the period April 2001 to March 2002. The model was run for 365 days. The estimated abstractions for these months were added and divided by 365 to get an average constant daily rate. Similarly, rainfall during the wet months was also added and divided by 365 to get an average daily value for this period. This model was calibrated against the general head boundary, recharge, evapotranspiration and hydraulic conductivity through a sequence of sensitivity analysis runs, starting with the parameters for which the least data were known. The values were adjusted during trial and error runs, aiming at the smallest sum of squares of residual errors in the targets.

The steady state model was calibrated against 22 head values observed in April 2001. The calibrated water level at the targets varied from 225.91 to 277.96 m (amsl) with a mean of 255.66 m (amsl). It showed that groundwater was flowing towards the river from the tannery clusters. The computed vs. observed heads with statistical analysis of the steady state calibration results are shown in Figure 6. The model was considered as sufficiently calibrated as 82% of the targets satisfied the ±2.5 m criterion. Four targets showed high residual errors with a maximum value of 5.26 m (see Figure 6) but these can be justified. Target OA-15 was towards the northern boundary for which there was no data available. Targets OA-22 and 41 were being completely controlled by the abstractions. The GHB was controlling the target 83029. Finally, the numerical error in the model was minimal, since the mass balance error was 0.0108%, as shown in Table 2. The table shows that groundwater abstraction was about 80.43% of annual groundwater recharge, but 10.25% of the annual recharge was being used up by evapotranspiration. There was a combination of inflow and outflow for river leakage. But the outflow of river leakage was comparatively higher than inflow, which was about 3.17% of the annual groundwater recharge. In this steady state, the minimum and maximum groundwater velocities were calibrated in order of 0.01 and 0.83 m/d with a mean of 0.29 m/d and it varied from 0.21 to 0.41 m/d in the tannery cluster.

3.4.7.4. Transient Calibration

The transient model was simulated for the period between April 2001 and February 2009, using initial heads computed in April 2001 under the steady state model. Each year in the transient model was divided into 4 stress periods: one dry period, which lasts 270 days and has 9 stress periods–one for each simulated month; and three distinct 30 days wet periods, one for each recharge month (i.e., October, November & December). This was done with the objective of better simulating the difference in rainfall and consequently in recharge between the three wet months. Each stress period was divided into 10-time steps in order to make it easier for the model to converge. In total there are 95-stress periods with 950 time steps in the transient model. The results of sensitivity analysis, summarized in Figure 7, shown that the model was insensitive to the recharge in the zone-5 and specific yield in the zone-3 due to the presence of Kodaganar river, as there was almost no change in the RMS error for a quite large change in the parameter multipliers. Nevertheless the model was sensitive to all the other calibration parameters in the different zones (recharge, hydraulic conductivity, specific yield and slightly evapotranspiration). Sensitivity analysis showed that even lower values in the evpotranpiration rate minimized the error. The recharge and hydraulic conductivity values used, on the other hand, gave the smallest RMS errors. Although the curves in Figure 7 indicated the use of lower evpotranpiration rate, the latter was not changed due to the fact that calibrated and observed heads matched sufficiently closely. In the transient model of dry stress period the evapotranspiration rates reduced to 73.4 m^3/d in 2008 from 250.1 m^3/d in 2001 due to decline of groundwater level in the study area.

Computed versus observed heads for each calibration target, during all time steps considering a total of 243 observed data points in the entire simulation show that maximum and minimum residuals were –18.93 m and 0.027 m observed at well OA-40 in April 2001 and PWD well 83029, respectively, with RMS = 6.756 m in the transient calibration statistics. The numerical error is order of ±0.0032% in the water mass balance during the model run. Two representative well hydrographs (in which only historical data were available) reproduced by the transient model in the calibration targets. It could be observed from the well hydrographs that the model effectively simulated groundwater flow in the central and northern parts. In the southern part however, the water table hydrograph was continually noticeably lower than observed values (i.e., hydrographs at PWD wells 83029 and 83029A), but their trends were similar to observed hydrographs, which was controlled by GHB.

The groundwater head distribution for February 2009 (see Figure 8) was reproduced using the transient model (at the stress period 95). It shows that groundwater was moving towards NW from the tannery cluster and NE in the northern part of the area. In this period computed and observed heads had a RMS error of 3.29 m for 20 targets with a correlation coefficient of 0.979. In February 2009, the minimum and maximum groundwater velocities were 0.02 and 0.91 m/d with a mean of 0.34 m/d (see Figure 9). The calibrated groundwater heads varied from 239.14 to 286.02 m (amsl), whereas observed heads from 240.94 to 289.28 m (amsl) in the study area. The velocity varies from 0.32 to 0.52 m/d in the tannery cluster.

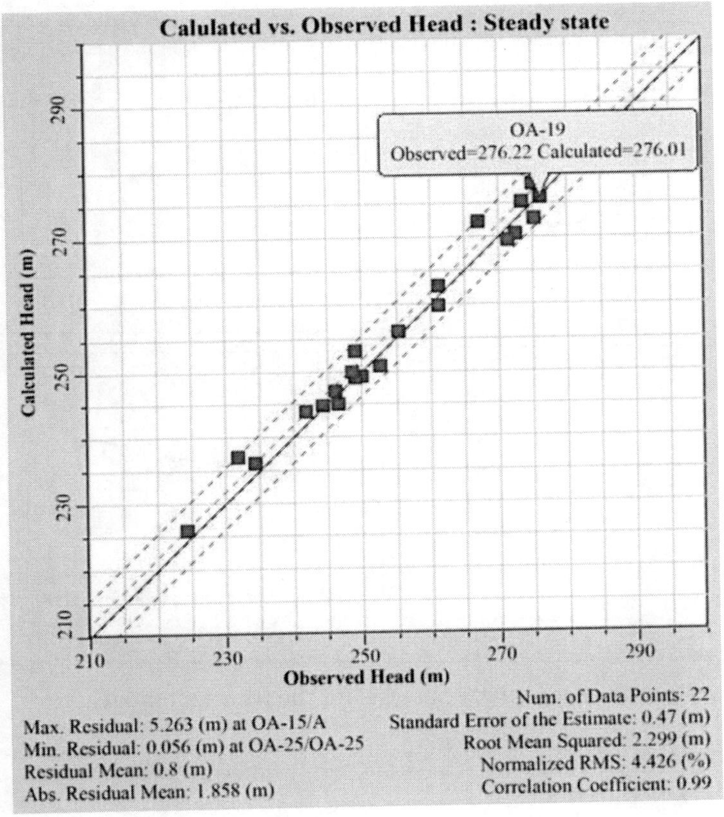

Figure 6. Computed vs. observed heads in the steady state calibration (April 2001).

Table 2. Steady state calibration mass balance results (m^3/day).

Parameters	Inflow (m3/day)	Outflow (m3/day)	Outflow as a percentage of recharge (%)
Recharge	29467.6035	0	0
GHB	515.0643	– 2332.5027	6.17
River leakage	2409.947	– 3342.7209	3.17
Evapotranspiration	0	– 3020.9141	10.25
Well abstractions	0	– 23700.00	80.43
Total	32392.6148	– 32396.1377	---
Error (m^3/day)	– 3.5229		
Error (%)	– 0.0108		

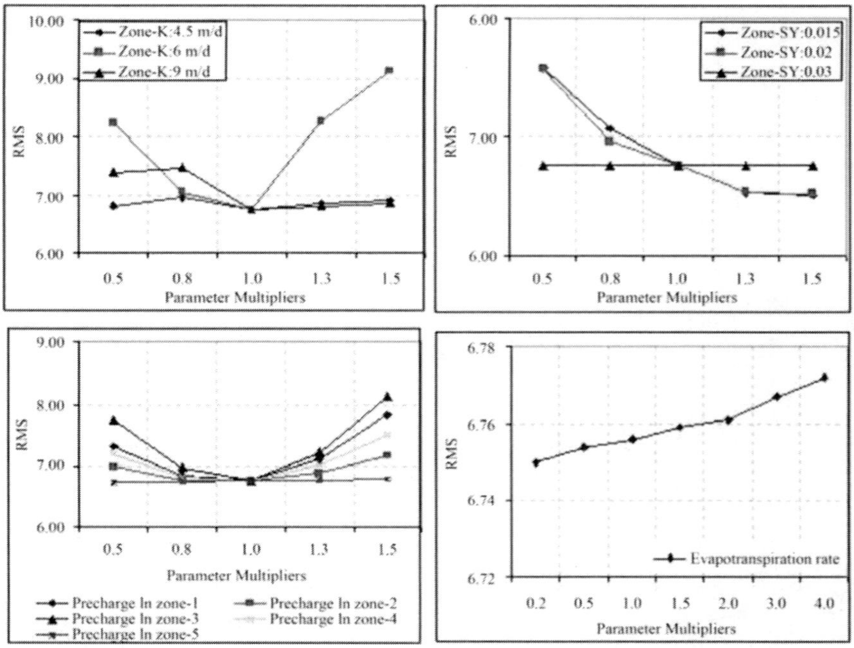

Figure 7. Results of sensitivity analysis for the transient model.

3.4.7.5. Predictive Model

The transient model was used to perform predictive simulations in order to assess the sustainability of groundwater system. Although a predictive simulation could last up to twice the calibrated period [29], it was decided in this case to simulate for 6 years in the future, instead of 16, because of the high uncertainty in predicting future hydraulic conditions and stresses to the system. Hence predictive simulations were performed for the period between March 2009 and December 2015.

Extrapolation of past and present stresses to future

Similar to the transient model, each year was divided into a dry stress period (January-September) and three wet periods (i.e., October, November and December). The extrapolation of rainfall and abstractions to the future was being discussed.

Rainfall and recharge: The generation of future rainfall values would require extensive statistical analysis. Since the objective of this study was not to predict exactly what will happen in the future but to examine the sustainability of the system it was decided to generate future rainfall values by simply replicating the past and present rainfall pattern. Since rainfall for the year 2001-2007 was known, the same value was repeated once more for the wet months until year 2015. Then recharge from rainfall was calculated using the same percentages as in the transient model.

Abstractions: Prediction of groundwater demands in future would require a socio-economic study, which is beyond the scope of this study. Since the main objective was to examine the response of the aquifer under different abstraction stresses, predictive simulations were performed under four different schemes, i.e., in Scheme I: abstraction was completely stopped; Scheme II: abstraction was kept at a current rate of year 2009; Scheme III: abstraction was increased by about 10%; and Scheme IV: 20% of the abstraction in 2009 was for domestic/Industrial and irrigation purposes.

Prediction Model

It was presumed that the aquifer would receive the same recharge just like the previous 6 years, while the abstraction for domestic/industrial usages was kept zero meaning there is no abstractions throughout the area. The calibrated well hydrographs are indicated a progressive increase of water level. The increased water table varied from 0.09 to 14.29 m from March 2009 to 2015. Maximum changes were observed on the eastern side and minimum towards the river. In Scheme II, the abstraction was kept constant in year 2009. The computed well hydrographs indicated a slow decline of the water table. It varied from –0.001 to –0.89 m with an average decline of 0.23 m throughout the area, but the changes of water level were almost zero along the river side. When the abstraction was increased by 10% of abstraction in 2009, then well hydrographs declined from –0.009 to –2.28 m (see Figure 10(a)). But in the case of increased abstraction of 20% in year 2009, the water level declined comparatively more. The changes of water level varied from –0.02 to –3.74 m (see Figure 10(b)). This decline would take place at an alarming rate with a significant decline of 1.49 m within 6 years.

In order to explore the possibility of increasing the groundwater level in this area through various means such as artificial recharge, the net recharge in the model was increased, keeping in view the groundwater potential recharge areas. The abstraction rate was kept same as in Schemes III & IV. It was found that even at a moderate rate of increase in recharge, the groundwater level in many areas would rise in both the schemes.

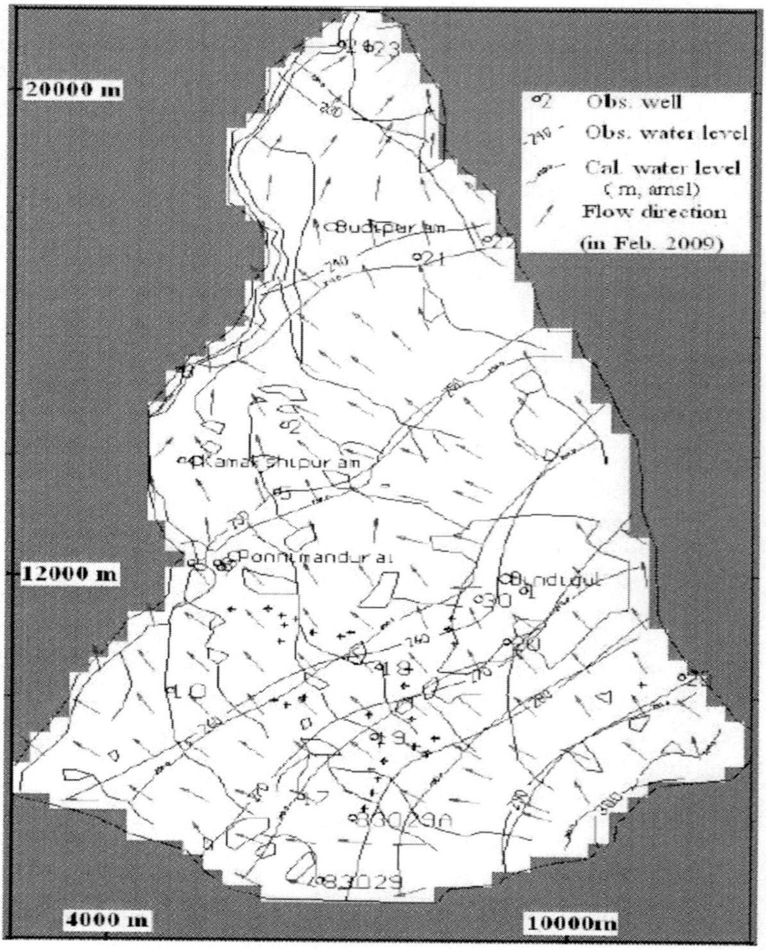

Figure 8. Groundwater head distribution (produced by the transient model) in Feb. 2009.

4. CONCLUSIONS

Groundwater flow model was simulated for a tannery belt in Southern India by assigning input and output stresses. The simulated results show the following:

- Groundwater flows towards northwest in the southern part, and north and northeast in the northern part of the study area. In the tannery cluster the groundwater velocity varies from 0.21 to 0.41 m/d in April

2001 and 0.32 to 0.52 m/d in February 2009, which is important for mass transport modeling.

- The estimated average groundwater recharge is about 80-250 mm/year, which is equivalent to 12-37% of annual rainfall. The total groundwater abstraction is about 80.43% of annual groundwater recharge, but 10.25% is being used up by evapotranspiration.
- Groundwater velocity is more sensitive to recharge from rainfall, hydraulic conductivity, and specific yield.
- The shallow aquifer, without further decline in water level, can sustain a pumping rate of 24892 m^3/day under a certain constraint. Any additional increase of withdrawal from this aquifer would result in a progressive decline of water level.

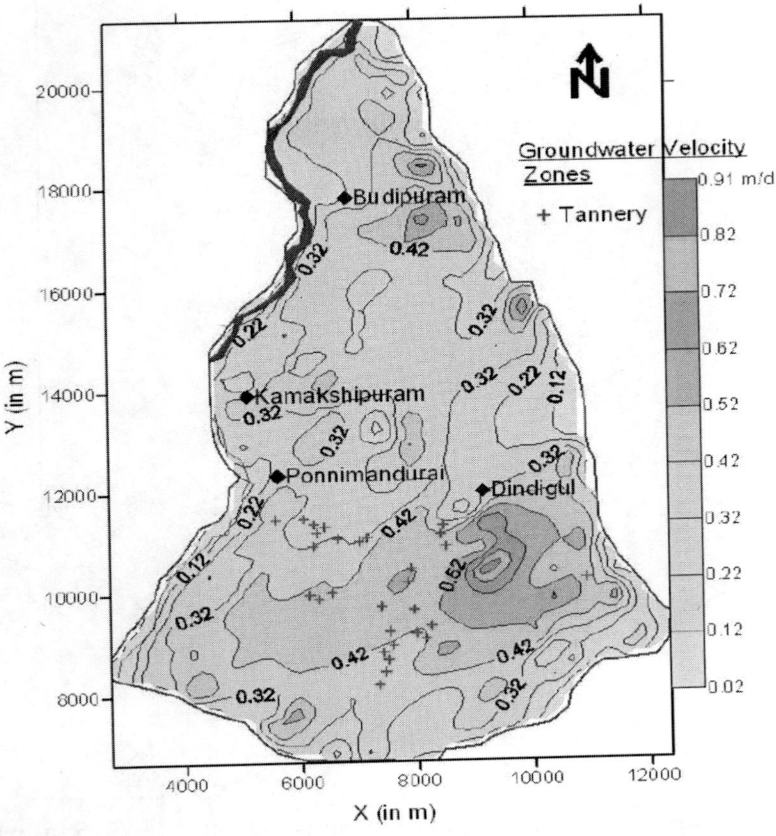

Figure 9. Groundwater velocity distribution in February 2009.

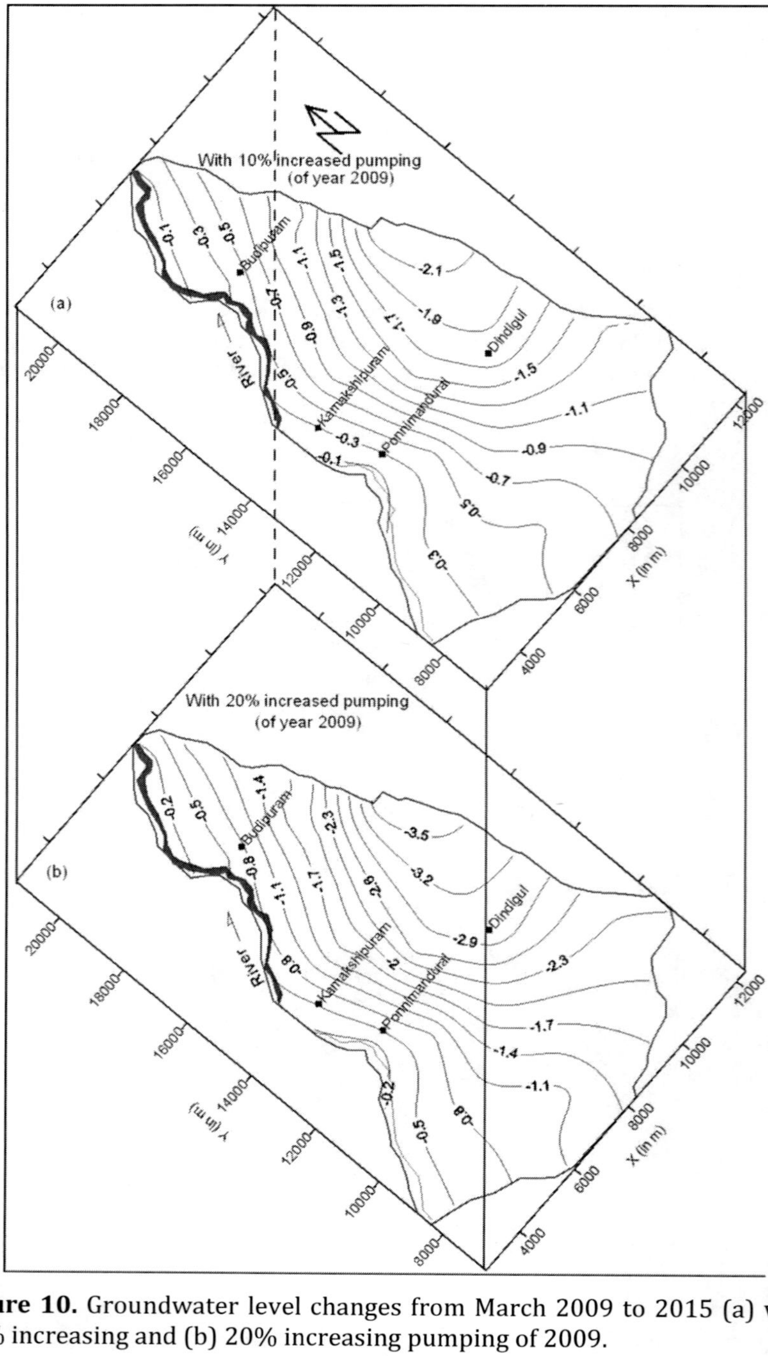

Figure 10. Groundwater level changes from March 2009 to 2015 (a) with 10% increasing and (b) 20% increasing pumping of 2009.

It is, therefore, suggested that groundwater resources be augmented through artificial recharge methods. As a first step, removal of silt in the existing irrigation tanks will considerably improve vertical infiltration, and construction of check dams in the upland area will enhance recharge. The present model is only a preliminary one and the obtained information represents a base for future groundwater work that will help in the planning, protection and decision-making regarding the groundwater management in this area. But it should be updated with additional field data. Then optimal utilization schemes can be evolved.

ACKNOWLEDGEMENTS

This work was performed, in part, under the BOYSCAST Fellowship of the first author (Dr. N. C. Mondal) funded by Department of Science & Technology (Government of India), New Delhi (Ref. No. SR/BY/A-05/ 2008, Date: 16-19th January 2009). The officers of PWD, Chennai provided suitable data and the members of Groundwater Division, NGRI helped during the field work. The authors are thankful to them.

REFERENCES

1. Public Works Department, "Groundwater Perspectives: A Profile of Dindigul District, Tamilnadu," PWD (Governmentt of India) Reptort, Chennai, 2000.
2. V. S. Singh, N. C. Mondal, R. Barker, M. Thangarajan, T. V. Rao and K. Subramaniyam, "Assessment of Groundwater Regime in Kodaganar River Basin (Dindigul district), Tamilnadu," Technical Report, 2003.
3. N. C. Mondal, V. K. Saxena and V. S. Singh, "Assessment of Groundwater Pollution Due to Tanneries in and around Dindigul, Tamilnadu, India," Environmental Geology, Vol. 48, 2005, pp. 149-157. doi:10.1007/s00254-005-1244-z
4. N. C. Mondal and V. P. Singh, "Need of Groundwater Management in Tannery Belt: A Scenario about Dindigul Town, Tamil Nadu," Journal of the Geological Society of India, Vol. 76, 2010, pp. 303-309.
5. M. G. McDonald and A. W. Harbaugh, "A Modular Three-Dimensional Finite Difference Ground-Water Flow Model," U.S.Geological Survey Techniques of Water-Resources Investigations, Book 6, Chapter A1,1988, p.586.
6. N. C. Mondal and V. S. Singh, "A New Approach to Delineate the Groundwater Recharge Zone in Hard Rock Terrain," Current Science, Vol. 87, 2004, pp. 658-662.
7. R. Chakrapani and P. M. Manickyan, "Groundwater Resources and Developmental Potential of Anna District, Tamil Nadu State," CGWB Report, Southern Region, Hyderabad, India, 1988,p.49.
8. M. S. Krishnan, "Geology of India and Burma," CBS Publishers and Distributions, India, 1982.
9. V. S. Singh, "Well Storage Effect during Pumping Test in an Aquifer of Low Permeability," Hydrological Science Journal, Vol. 45, 2000, pp. 589-594. doi:10.1080/ 0262666 0009492359

10. E. Orellana and H. M. Mooney, "Master Tables and Curves for Vertical Electrical Sounding over Layered Structures," Interciencia, Madrid, Spain, 1966.
11. B. Vander Velpen, "RESIST: a Computer Processing Package for DC Resistivity Interpretation for the IBM PC and Compatibles," M. Sc. Thesis, ITC-Delft, The Netherlands, 1988.
12. N. C. Mondal and V. P. Singh, "Entropy-Based Approach for Estimation of Natural Recharge in Kodaganar River Basin, Tamil Nadu, India," Current Science, Vol. 99, No. 11, 2010, pp. 1560-1569.
13. K. R. Rushton and S. C. Redshaw, "Seepage and Groundwater Flow," Wiley, Chichester [Eng.], New York, 1979.
14. Compagnie General de Geophysique, "Master Curves for Electrical Sounding," European Association of Exploration Geophysicists (EAEG), The Hague, 1963, p. 49.
15. P. K. Bhattacharya and H. P. Patra, "Direct Current Geoelectric Sounding-Principles and Interpretation," Amsterdam, Elsevier, 1968.
16. C. W. Fetter, "Applied Hydrogeology," Prentice Hall, Inc. New York, 2001.
17. J. C. Nonner, "Introduction to Hydrogeology," IHE Delft Lecture Notes Series, Taylor and Francis, London, 2006, p. 258.
18. U. Zimmermann, K. O. Munnich and W. Roether, "Downward Movement of Soil Moisture Traced by Means of Hydrogen Isotopes," American Geophysical Union, Geophysical Monograph Series, Vol. 11, 1967, pp. 28-36.
19. V. M. Chowdary, D. Ramakrishnan, Y. K. Srivastava, V. Chandran and A. Jeyaram, "Integrated Water Resource Development Plan for Sustainable Management of Mayurakshi Watershed, India Using Remote Sensing and GIS," Water Resource Management, Vol. 23, 2009, pp. 1581-1602.doi:10.1007/s11269-008-9342-9
20. R. B. Salama, I. Tapley, T. Ishii and G. Hawkes, "Identification of Areas of Recharge and Discharge Using Landsat-TM Satellite Imagery and Aerial-Photography Mapping Techniques," Journal of Hydrology, Vol. 162, No. 1-2, 1994, pp. 119-141. doi:10.1016/0022-1694(94)90007-8
21. P. Raj, "Trend Analysis of Groundwater Fluctuations in a Typical Groundwater Year in Weathered and Fractured Rock Aquifers in Parts of Andhra Pradesh," Journal of the Geological Society of India, Vol. 58, 2001, pp. 5-13.
22. D. Muralidharan and G. B. K. Shanker, "Various Methodologies of Artificial Recharge for Sustainable Groundwater in Quantity and Quality for Developing Water Supply Schemes," In: Proceedings of the All Indian Seminar on Water Vision for the 21st Century, IAH, Jadavpur University, Kolkata, 2000, p.208-229.
23. D. B. Bredenkamp and J. C. Vogel, "Study of a Dolomitic Aquifer with Carbon-14 and Tritium," In: Isotope Hydrology, Proceedings Symposium of International Atomic Energy Agency and UNESCO, Vienna, March 1970. Vienna, International Atomic Energy Agency STI/PUB/255, Paper No SM-129/21, 1970, pp. 349-372.
24. T. Sibanda, J. C. Nonner and S. Uhlenbrook, "Comparison of Groundwater Recharge Estimation Methods for the Semi-Arid Nyamandhlovu Area, Zimbabwe," Hydrogeology Journal, Vol. 17, 2009, pp. 1427-1441. doi:10.1007/s10040-009-0445-z
25. R. Chand, G. K. Hodlur, R. Ravi Prakash, N. C. Mondal and V. S. Singh, "Reliable Natural Recharge Estimates in Granite Terrain," Current Science, Vol. 88, 2005, pp. 821-824.
26. R. Rangarajan, N. C. Mondal, V. S. Singh and S. V. Singh, "Estimation of Natural Recharge and Its Relation With Aquifer Parameters in and around Tuticorin Town, Tamil Nadu, India," Current Science, Vol. 97, 2009, pp. 217-226.
27. V. P. Singh, "Entropy Based Parameters Estimation in Hydrology," Kluwer Academic Publishers, Boston, 1998.
28. R. Rangarajan and R. N. Athavale, "Annual Replenishable Groundwater Potential of India - An Estimate Based on Injected Tritium Studies," Journal of Hydrology, Vol. 234, 2000, pp. 38-53. doi:10.1016/S0022-1694(00)00239-0
29. M. P. Anderson and W. Woessner, "Applied Groundwater Modeling-Simulation of Flow and Advective Transport," Academic Press, San Diego, USA, 1992.

30. K. R. Rushton and J. Weller, "Response to pumping of a weathered-fractured granite aquifer," Journal of Hydrology, Vol. 80, 1985, pp. 299-309. doi:10.1016/0022-1694(85)90123-4
31. Groundwater Resource Estimation Committee (GREC), "A Report on Ground Water Resource Estimation Methodology-1996," Ministry of water resources (Government of India), 1996.
32. L. F. Konikow and J. D. Bredehoeft, "Computer Model of Two-Dimensional Solute Transport and Dispersion in Groundwater," Techniques of Water-Resources Investigations of the USGS; Chapter C2, Book 7,1978, p.90.

CHAPTER 2

Prediction of Stream Flow and Sediment Yield of Lolab Watershed Using SWAT Model

Sarvat Gull, Ahangar MA and Ayaz Mohmood Dar*

Department of Civil Engineering, National Institute of Technology, Srinagar, Jammu and Kashmir, India

ABSTRACT

The SWAT model was used to estimate the runoff and sediment yield of Lolab watershed. The model was calibrated, validated, and assessed for evaluation to model ambiguity using Nash–Sutcliffe coefficient (N_{SE}) and coefficient of determination (R^2). Ten highly sensitive parameters were recognized for stream flow simulation of which CN2 (Initial SCS CN II value) factor was the most sensitive one and four highly sensitive parameters were recognized for sediment yield simulation of which SPCON (Linear parameters for sediment re-entrainment) was most sensitive one. The model was calibrated for a time period between 1993 to 2000 and validated from 2001 to 2004 for flow and sediment yield. The predicted and observed stream flow and sediment yields generally matched well. The results of the model calibration and validation showed reliable estimates of monthly stream flow (R^2=0.74 and E_{NS}=0.68) and yearly stream flow (R^2=0.90 and E_{NS}=0.68) during the calibration period and monthly stream-flow (R^2=0.85 and E_{NS}=0.83) and yearly stream-flow (R^2=0.99 and E_{NS}=0.91) during the validation period. For sediment yield, this study shows antremendous model efficiency of monthly sediment yield (R^2=0.80 and E_{NS}=0.79) and yearly sediment yield (R^2=0.86 and E_{NS}=0.78) during the calibration period and monthly sediment yield (R^2=0.88 and E_{NS}=0.86) and yearly sediment yield (R^2=0.83 and E_{NS}=0.58) during the validation period. This study showed that the SWAT model is competent of predicting sediment yields and hence can be used as a tool for water resources planning and management in the study watershed.

Keywords: SWAT model; Model evaluation; Stream flow; Sediment yield

1. INTRODUCTION

A Watershed is an area of land that drains all the streams and rainfall to a common outlet. Essentially a watershed is all the land and water area which contributes runoff to an outlet in the main flow channel [1]. Watersheds are significant because the stream flow and the water quality of a river are affected, human-induced or natural, happening in the land area "above" the river-outflow point. The environment worsening of a watershed is a common occurrence in most parts of the world. Amongst several causes the major one are improper and unwise utilization of watershed resources observed in developing countries [2]. Watershed management implies rational utilization of land, soil, and water resources for optimum and sustained production with minimum hazards to natural resources and environment [3]. Soil erosion is a serious global issue because of its rigorous adverse economic and environmental impacts. Economic impacts on productivity may be due to direct effects on crops/plants both on-site and off-site, and environmental consequences are primarily off-site related to the damage to civil structure, siltation of water ways and reservoirs, and additional costs involved in water treatment. Even though the adverse influences of soil erosion on soil degradation have long been recognized as a key problem for human sustainability, estimation of soil erosion is often complicated due to the complex interplay of many factors such as climate, land cover, soil, topography, lithology, and human activities. This study employs the SWAT (Soil and Water Assessment Tool) model to take advantage of its integration with GIS (Geographic Information System) and locally available data and data from similar areas that can be used to calibrate and validate the model, so that runoff and sediment yield from the watershed can be predicted and most problematic sub-basins can be identified.

1.1 Objectives of the Study

1) To analyze the Land use/Land cover of Lolab watershed.
2) To estimate the catchment runoff and sediment yield of Lolab watershed of Pohru catchment using SWAT model.
3) To evaluate the performance of Swat model by comparing its predicted flow and sediment yield with corresponding observed values at the study watershed.
4) To identify the most problematic sub-basins with respect to sediment contribution.

2. METHODS AND MATERIALS

Lolab watershed is one of the watersheds of Pohru catchment with an area of about 45,250 hectares. The study area lies between 34° 41' to 34° 24' N Latitude and 74° 09' to 74° 23' E Longitude. The watershed can be divided into three distinct physiographic units i.e., the Mountains, the Karewas and the Flood plains. Elevation in the watershed varies from approximately 1,500 to 3,900 meters above mean sea level.

2.1 Model description

Arc SWAT: ArcSWAT is a public domain graphical user interface program. It is designed to link the hydrologic model SWAT (Soil and Water Assessment Tool) and the GIS package ARCINFO. The SWAT model is limited in that it does not explicitly allow for the inclusion of spatial data as model inputs. Data must be processed into a form that the model can use. Processing these data, even with the use of a GIS is tiresome and time consuming due to the large number of model parameters required for executing SWAT. The development of ArcSWAT aims at an effective use of spatial data to enhance hydrological modeling. The interface performs the following tasks

- To streamline GIS processes tailored toward SWAT modeling needs.
- To automate data communication between Arc/Info and SWAT.
- To provide a user-friendly data entry and editing environment for SWAT.

The SWAT model: The Soil and Water Assessment Tool (SWAT) is a river basin scale, continuous time and spatially distributed physically based model developed to predict the impact of land management practices on water, sediment, and agricultural chemical yields in complex catchments with varying soils, land use and management conditions over long periods of time [4,5]. In this study, the Arc SWAT 2005 version of the SWAT model is applied to predict stream flow and sediment yield. SWAT uses Hydrologic Response Units (HRUs) to describe spatial heterogeneity in terms of land cover, soil type and slope within a catchment. The SWAT model uses two steps for the simulation of hydrology: the land phase and routing phase. The land phase controls the amount of sediment, nutrient and pesticides loading to the main channel in each sub-basin. SWAT offers two methods for estimating surface runoff: The Soil Conservation Service (SCS) Curve Number (CN) procedure and the Green and Ampt infiltration method. Using daily or sub-daily rainfall amounts, SWAT simulates surface runoff volumes and peak runoff rates for each HRU. SCS curve number method is less data intensive than the Green-Ampt method [6]. In this study, the SCS curve number method was used to estimate surface runoff volumes because of the unavailability of sub-daily data for the Green and Ampt method. Sediment yield is estimated using a Modified Universal Soil Loss Equation (MUSLE).

2.2 Preparation of model inputs

The basic spatial input datasets used by the model include the Digital Elevation Model (DEM), stream network, land use/cover data, soil data and climatic data. DEM was derived from the NASA 15-meter Shutter Radar Topography Mission (SRTM) dataset. The DEM was used to delineate the catchment and analyze the drainage patterns of the land surface as well as derive slope parameter. The stream network data set was digitized from topographic map of scale (1:50000 and 1:250000) of the study area. The stream network dataset was superimposed onto the DEM to define the location of the stream network. Burning-in stream network operation is most important in situations where the DEM does not provide enough detail to allow the interface to accurately predict the location of the stream network. The land use is one of the most important factors that affect runoff, evapotranspiration, and surface erosion in a watershed. The land use map was obtained from the Department of Agricultural Engineering, SKUAST-K. Agriculture was the dominant land use category in the Lolab watershed followed by the sparse forest.

Since SWAT has pre-defined land use types which identified by four-letter codes, the Land use/Land cover map was reclassified in order to correspond with the parameters in the SWAT database.

The SWAT model requires different soil textural and physicalchemical properties such as soil texture, available water content, hydraulic conductivity, bulk density, and organic carbon content for different layers of soil. In this particular study, Soil Conservation Department, Kashmir global available soil data (1:50,00,000 scales) in vector format had been obtained from Earth Science Department, Kashmir University. It is converted into a grid format for SWAT model input parameters. A total of three soil texture classes were delineated out of which the fine loamy soil was the major class which accounted for 79.58 percent of the total area of the watershed and was followed by fine silty soil with 16.53 percent of the total area.

Meteorological data is needed by the SWAT model to simulate the hydrological conditions of the basin. The meteorological data required for this study were obtained from Indian Meteorological Department. The meteorological data collected were precipitation, maximum and minimum temperature, wind speed and sunshine hours. The records between 1993 and 2004 were obtained. The problem in the weather data was inconsistency in the data record, in some periods there is a record for precipitation but temperature data are missing, and vice versa. The hydrological data was required for performing sensitivity analysis, calibration, and validation of the model. The hydrological data was collected from Irrigation and Flood Control Department, Kashmir. Sediment data was obtained from Soil and Water Conservation Department, Jammu and Kashmir.

2.3 Arc SWAT model setup

The ArcSWAT interface was used for the setup and parameterization of the model. Digital Elevation Model was imported into the SWAT model to automatically delineate the watershed into several hydrologically connected sub-watersheds. After the DEM grid was loaded and the stream networks superimposed, the DEM map grid was processed to remove the non-draining zones. The initial stream network and sub-basin outlets were defined based on drainage area threshold approach. Besides those sub-basin outlets created by the interface, outlets were also manually added at the gauging stations where sensitivity analysis, calibration and validation tasks were later performed. Then watershed delineation activity was finalized by calculating the geomorphic sub-basin parameter. A total of 43 sub basins were created.

The land use and the soil data in a projected Grid file format were loaded into the Arc SWAT interface to determine the area and hydrologic parameters of each land-soil category simulated within each sub-watershed. The DEM data used during the watershed delineation was also used for slope classification. The multiple slope discretization operation was preferred over the single slope discretization as the sub-basins have a wide range of slopes between them. Based on the suggested minimum, maximum, mean and median slope statistics of the watershed, three slope classes (0-10, 10-20, and >20) were applied and slope grids reclassified. Then land use, soil and slope grids were overlaid. In multiple HRU definition, a threshold level was used to eliminate minor land uses, soils, or slope classes in each sub-basin. Land uses, soils or slope classes which cover less than the threshold level was eliminated and the area of the remaining land use, soil, or slope class was reapportioned so that 100% of the land area in the subbasin was modeled. The threshold levels set is a function of the project goal and amount of detail required. In the SWAT user manual, it is suggested that it is better to use a larger number of sub-basins than larger number of HRUs in a sub-basin; a maximum of 10 HRUs in a sub-basin is recommended. Hence, taking the recommendations into consideration, 5%, 5%, and 5% threshold levels for the land use, soil and slope classes were applied, respectively so as to encompass most of spatial details. A total of 244 HRU's were created. The climatic variables required by SWAT daily precipitation, maximum and minimum temperature, solar radiation, wind speed and relative humidity were prepared in the appropriate dbase format and imported into the model.

SWAT is a complex model with many parameters that makes manual calibration difficult. Hence, sensitivity analysis was performed to limit the number of optimized parameters to obtain a good fit between the simulated and measured data. Sensitivity analysis helps to determine the relative ranking of which parameters most affect the output variance due to input variability [7] which reduces uncertainty and provides parameter estimation guidance for the calibration step of the model. SWAT model has an embedded tool to perform sensitivity analysis and provides recommended ranges of parameter changes. SWAT2005 uses a combination of Latine Hypercube Sampling and One-At-a-Time sensitivity analysis methods (LH-OAT method) [8]. The concept of the Latin-Hypercube Simulation is based on the Monte Carlo Simulation to allow a robust analysis but uses a stratified sampling approach that allows efficient estimation of the output statistics while the One-Factor-At-a-Time is an integration of a local to a global sensitivity method [8]. In local methods, each run has only one parameter changed per simulation which aides in the clarity of a change in outputs related directly to the change in the parameter altered [8].

2.4 Model calibration and validation

There are three calibration approaches widely used by the scientific community. These are the manual calibration, automatic calibration, and a combination of the two. Manual calibration is the most widely used approach. However, it is tedious, time consuming, and success of it depends on the experience of the modeler and knowledge of the watershed being modeled [9]. Automatic calibration involves the use of a search algorithm to determine best-fit parameters. It is desirable as it is less subjective and due to extensive search of parameter possibilities can give results better than if done manually.

The manual calibration approach helps to compare the measured and simulated values, and then to use the expert judgment to determine which variable to adjust, how much to adjust them, and ultimately assess when reasonable results have been obtained. The auto- calibration technique is used to obtain an optimal fit of process parameters which is based on a multi-objective calibration and incorporates the Shuffled Complex Evolution Method algorithms [8]. The model was calibrated in order to make the simulation result more realistic for independent calibration period (1993 to 2000). In this study, both the manual and auto-calibration techniques were employed to get the best model parameters.

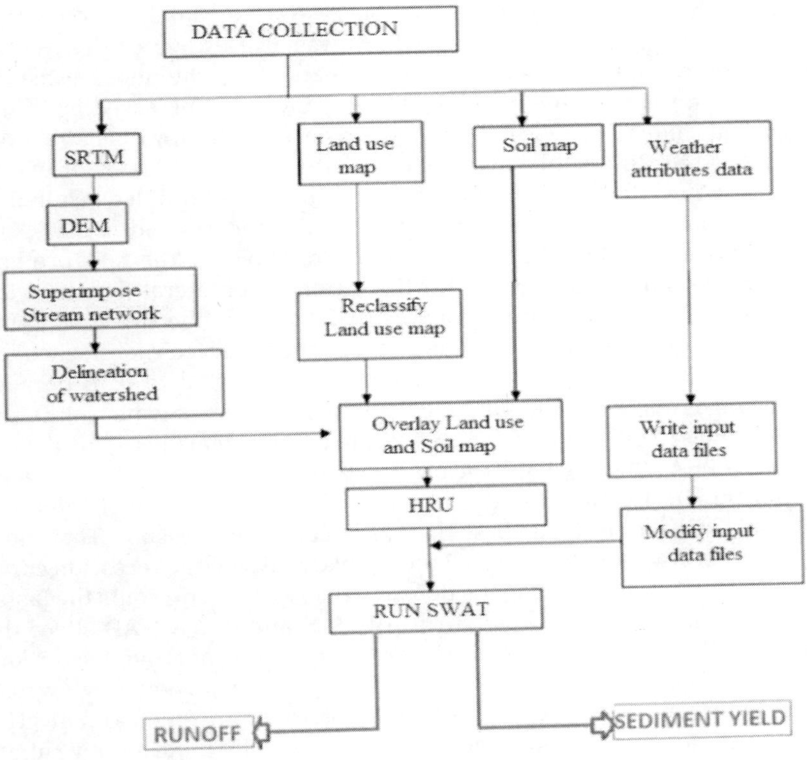

Figure 1: SWAT methodology for simulation of runoff and sediment yield.

In order to utilize the calibrated model for estimating the effectiveness of future potential management practices, the model was tested against an independent set of measured data. This testing of a model on an independent set of data set was used. As the model, predictive capability was demonstrated as being reasonable in both the calibration and validation phases, the model can be used for future predictions under different management scenarios. In this study, the model was validated with independent validation period (2001-2004). The simulation flowchart is shown in Figure 1.

3. RESULTS AND DISCUSSION

3.1 Sensitivity analysis and parameters calibration

The first simulation using default parameters was not able to correctly reproduce the runoff in the watershed because the actual discharge peaks were underestimated and the base flow was overestimated. Therefore, parameters calibration was need after identifying the most sensitive parameters for runoff shown in Table 1. The CN2 was increased by 7% of the original value to increase the runoff and decrease infiltration. SOL_AWC was reduced by 20% in all soils to increase the movement of water within the soil profiles. The default value of SOL_K for the soil profiles resulted in an overestimation of the lateral flow in the watershed; therefore, the SOL_K was decreased by 13% to reduce the lateral flow in the soil profiles. SLSUBBSN was also used to control the estimates of lateral flow, and the calibrated values were varied from 25 to 228 m for the individual sub-basins. The simulated runoff become closer to the observed runoff when the ESCO was adjusted to 0.15 and the EPCO was adjusted to 0.38 because these changes allow us to modify the depth distribution of water in the soil layers to meet soil evaporation and plant uptake demands. The default value of ALPHA_BF leaded to large base flow; however, the Lolab watershed has a characteristically rapid response of runoff to rainfall, and the watershed has zero or a low base discharge during dry seasons. Therefore, adjusting ALPHA_BF to 0.70 caused the simulated discharge recession curve to be steeper than when using the default value, which represents the faster drainage behavior of the watershed. GWQMN and GW_REVAP affect the amount of groundwater flow and control the upwelling of groundwater into the unsaturated soil zone. The effects of these parameters on base flow also affect runoff, and low values of GWQMN correspond to high runoff [10]; thus, this parameter was adjusted to 37 to realistically predict daily runoff. In addition, SOL_Z and CH_K2 were determined to be 2,400 mm and 31 mm h-1 to obtain better model accuracy.

Table 1: Sensitive parameters for stream flow simulation and calibrated values.

The most sensitive parameters	Parameter definition	Range	Ranks of the parameters	Calibrated value
CN2	Initial SCS CN II value	± 25%	1	7%[a]
SOL_AWC	Soil available water capacity (%)	± 25%	2	-20%[a]
SOL_K	Saturated hydraulic conductivity(mm/h)	± 15%	3	-13%[a]
SOL_Z	Depth from soil surface to the bottom of the layer(mm)	0 to 3000	4	2400
ESCO	Soil evaporation compensation factor	0.1 to 1	5	0.15
ALPHA BF	Base flow alpha factor(days)	0 to 1	6	0.7
GWQMN	Threshold depth of water in a shallow aquifer for return flow (mm)	0 to 1000	7	32
SLSUBBSN	Average slope length (m)	10 to 300	8	25-228 (varies by subbasins)
EPCO	Plant evaporation compensation factor	0.01 to 1	9	0.38
CH_K2	Channel effective hydraulic conductivity(mm/h)	0 to 500	10	31

Table 2: Sensitive parameters for sediment load simulation and calibrated values.

The most sensitive parameters	Parameter definition	Range	Ranks of the parameters	Calibrated value
SPCON	Linear parameters for sediment re-entrainment	0.0001-0.01	1	0.008
CH_EROD	Channel erodibility factor	0-1	2	0.55
SPEXP	Exponent parameter for sediment re-entrainment	01-Feb	3	1.45
CH_COV	Channel cover factor	0-1	4	0.3

Table 2 presents an overview of the parameter changes applied during sediment calibration. The parameters were modified based on appropriate ranges as defined in the SWAT model documentation. The values of SPCON and SPEXP were adjusted to increase channel sediment transport capacity and lower the amount of sediment deposition. Two other sensitive parameters affecting sediment transport at the watershed level, CH_EROD and CH_COV, were adjusted to 0.55 and 0.30 respectively (Table 2).

3.2 Evaluation of stream flow simulation

The observed and simulated monthly and yearly stream flow for the calibration period from 1993-2000 are shown in Figures 2 and 3. The statistical results for the model performance displayed satisfactory efficiency with R^2=0.74 and N_{SE}=0.68 for monthly stream flow and R^2=0.90 and N_{SE}=0.68 for annual flows between the simulated and observed data shown in Table 3. However, the simulated flow was generally lower than the corresponding observed values during periods with concentrated rainfall. The calibrated model was then run from 2001- 2004 to validate the model. The observed and simulated monthly and yearly stream flow for the validation period from 2001-2004 are shown in Figures 4 and 5. The statistical analysis results also demonstrated reasonable agreement between the observed and simulated stream flow with monthly R^2 value of 0.85 and N_{SE} value of 0.83 and yearly R^2 value of 0.99 and N_{SE} value of 0.91. Although the statistical evaluation showed the satisfactory stream flow simulation for both calibration and validation periods, SWAT tended to underestimate the runoff during high-flow periods. This could be partly because the present curve number technique is unable to generate accurate runoff prediction for a day that experience several storms.

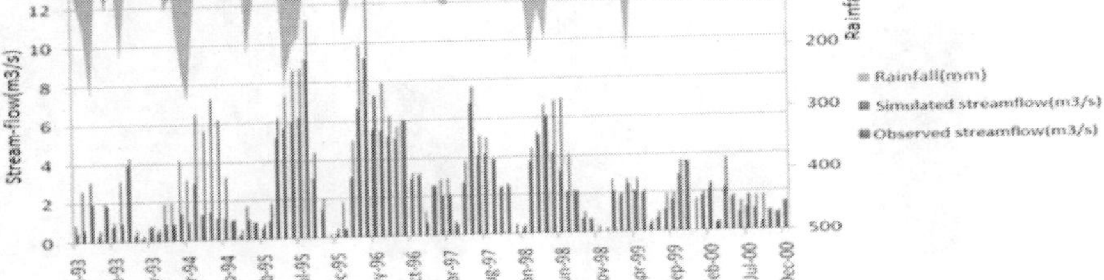

Figure 2: Bar chart showing monthly observed and simulated stream flow of Lolab watershed during calibration Period.

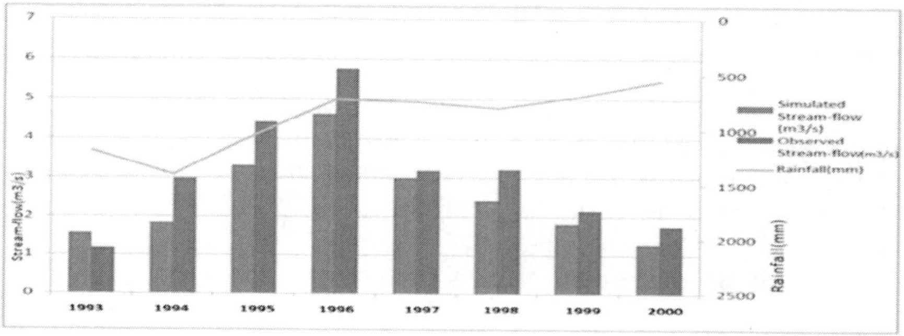

Figure 3: Bar chart showing yearly observed and simulated stream flow of Lolab watershed during calibration Period.

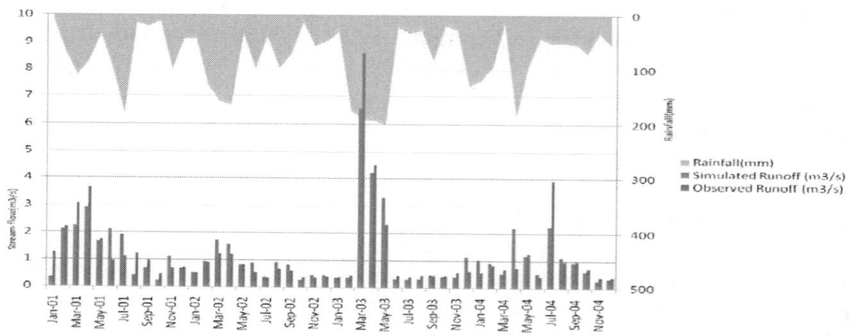

Figure 4: Bar chart showing monthly observed and simulated stream flow of Lolab watershed during validation Period.

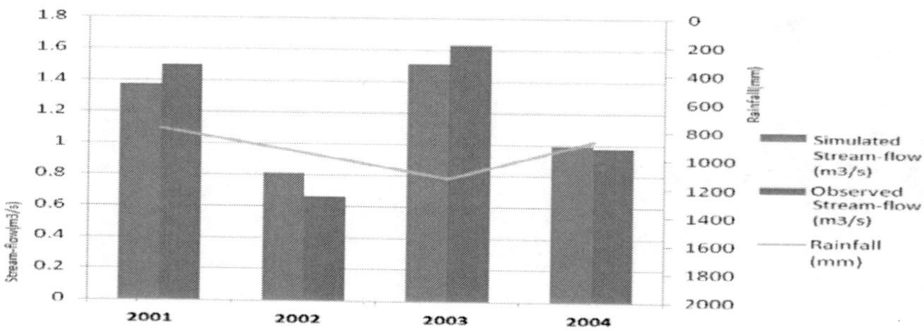

Figure 5: Bar chart showing yearly observed and simulated stream flow of Lolab watershed during validation Period.

When several storms occur during a single day, the soil moisture level and the corresponding runoff curve number vary from storm to storm [11]. However, SCS-CN methods define a rainfall event as the sum of all rainfall that occurs during one day, and this might lead to underestimation of runoff [12].

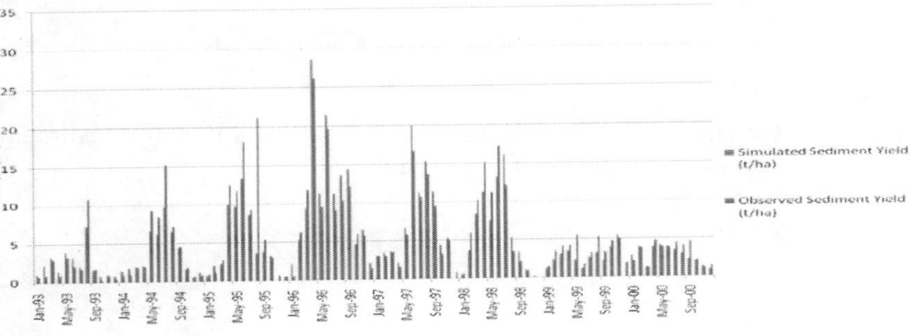

Table 3: Model efficiencies for stream flows during calibration and validation periods.

	Nash-Sutcliffe model efficiency (N_{SE})		Coefficient of Determination (R_2)	
	Monthly	Yearly	Monthly	Yearly
CAL	0.68	0.68	0.74	0.9
VAL	0.83	0.91	0.85	0.99

3.3 Evaluation of sediment load simulation

The results obtained during calibration periods of sediment load simulation are shown in Figures 6 and 7. The majority of sediment transport occurs during high-flow events; therefore, capturing these events accurately is vital to model calibration [13]. The monthly and yearly R^2 and the N_{SE} demonstrated that the correlation and agreement between them was acceptable, shown in Table 4. For validation period, the observed and simulated sediment loads during 2001-2004 are shown in Figures 8 and 9. Although the validation period showed a better match than the calibration period, the SWAT tended to underestimate many of sediment loads in both calibration as well as validation periods.

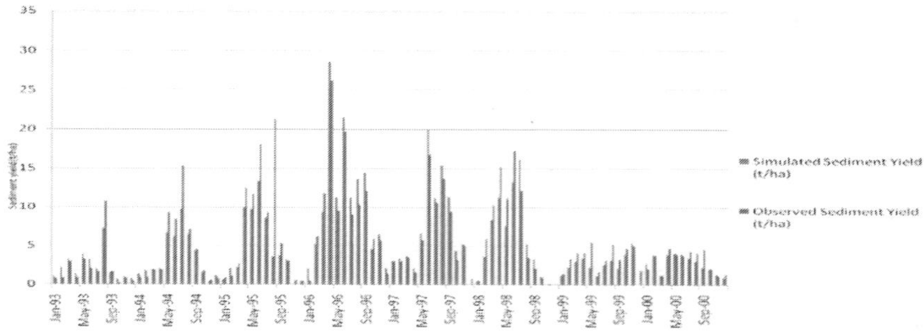

Figure 6: Bar chart showing monthly observed and simulated sediment yield of Lolab watershed during calibration Period.

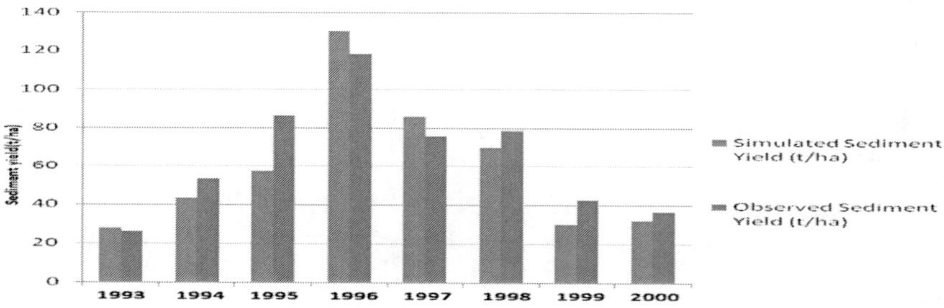

Figure 7: Bar chart showing yearly observed and simulated sediment yield of Lolab watershed during calibration Period.

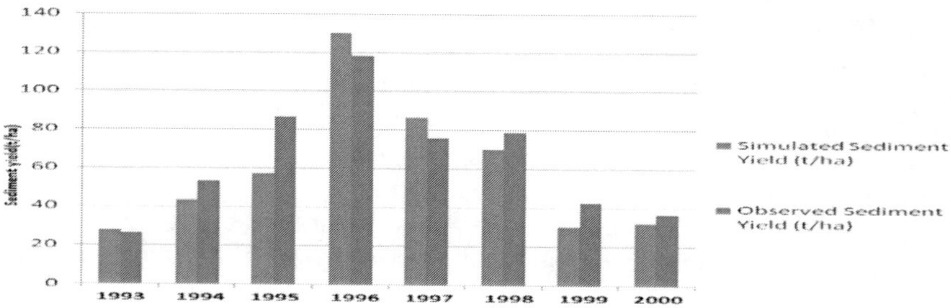

Figure 8: Bar chart showing monthly observed and simulated sediment yield of Lolab watershed during validation Period.

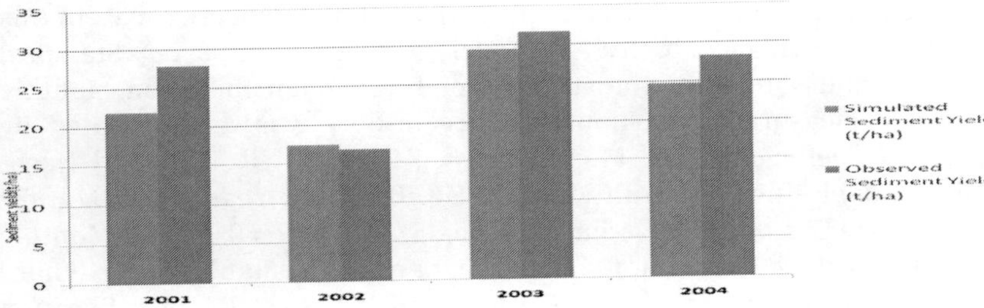

Figure 9: Bar chart showing yearly observed and simulated sediment yield of Lolab watershed during validation Period.

Table 4: Model efficiencies for Sediment yield during calibration and validation periods.

	Nash-Sutcliffe model efficiency (N_{SE})		Coefficient of Determination (R_2)	
	Monthly	Yearly	Monthly	Yearly
CAL	0.79	0.78	0.80	0.86
VAL	0.86	0.58	0.88	0.83

3.4 Hydrology-current-research-sediment-yield-watershed Sediment distribution

The spatial variability of sedimentation rates was identified and based on which the potential area of intervention was identified. The average annual yield of sedimentation for each sub-basin was used to generate sediment source map shown in Figure 10. The study showed that Sub-basins 24, 25, 28 and 38 of Lolab watershed at the existing condition generates maximum annual average sediment yields of 139.54 t/ha, 132.98 t/ha, 138.06 t/ha, 129.23 t/ha respectively. This was attributed due to the topographic slope of these sub-basins. It was a land with more than 42.72% of which has a slope greater than 23 degrees.

4. CONCLUSION

Although the problem of soil erosion is recognized from gross erosion estimates and field observations, quantitative information and data are required at micro watershed level to develop alternative watershed management plans and for decision making. In this study, attempts were made to characterize the Lolab watershed in terms of stream flow and sediment yield and identification of potential sediment source areas. In this study the performance of SWAT model was evaluated using Standard calibration and validation statistics. A good agreement between monthly and yearly measured and simulated stream-flow was demonstrated by correlation coefficients Viz. Monthly R^2=0.74 Yearly R^2=0.90 and Nash-Sutcliffe model efficiency with Monthly E_{NS}=0.68 and Yearly E_{NS}=0.68 for calibration period while as Monthly R^2=0.85, Yearly R^2=0.99 and Monthly E_{NS}=0.83, Yearly E_{NS}=0.91 for validation period. In simulating sediment yield, the monthly and yearly correlation coefficients i.e., R^2 for calibration period were 0.80 and 0.86 respectively while as the monthly and yearly Nash-Sutcliffe model efficiency coefficients i.e., E_{NS} were 0.79 and 0.78 respectively. The values of monthly and yearly correlation coefficients i.e., R^2 for validation period were 0.88 and 0.83 respectively while as the monthly and yearly Nash-Sutcliffe model efficiency coefficients i.e., E_{NS} were 0.86 and 0.58 respectively. A good performance of the model in the Validation period indicates that the fitted parameters during calibration period can be taken as a representative set of parameters for Lolab watershed and further simulation and evaluation of alternative scenario analysis can be carried out for other periods using the SWAT model. The SWAT model prediction verified that the Lolab watershed is highly erosion potential area contributing high sediment yield exceeding the tolerance limit (soil formation rate) in the study area.

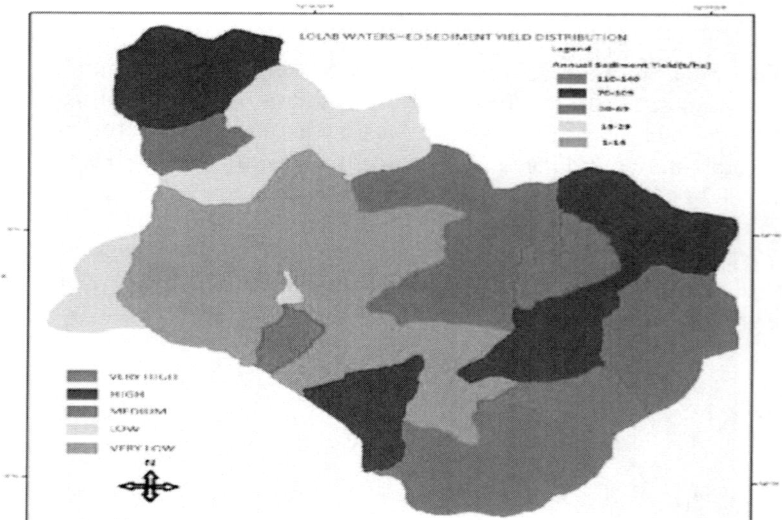

Figure 10: Shows the spatial variability of sedimentation rate.

In general, the methodology presented in this paper can be used in other watersheds, and the results are instructive for further use of SWAT in evaluating different management scenarios in hilly-gullied regions of Kashmir. Sub-basin no's 24, 25, 28, 38 of Lolab watershed at the existing condition generates a maximum annual average sediment yield, this can be reduced by using sediment yield intervention strategies such as land slope stabilization, construction bench terraces, changing the land use of steep area and afforestation.

ACKNOWLEDGEMENTS

We are thankful to the Department Earth Sciences, Kashmir University and Department of Irrigation and Flood Control, Jammu and Kashmir for providing the moral support as well as necessary data. We are also greatly thankful to the National Institute of Technology, Srinagar for providing us the lab facilities and our regards also go to the Meteorological Department, Kashmir as well as SKUAST-K for a moral support.

REFERENCES

1. Rajora R (1998) Integrated watershed management. Chapter 24. Rawat Publications, Jaipur, India.
2. Food and Agriculture Organization (FAO) (1985) Guidelines for economic appraisal of watershed management projects. In: Gregersen HM, Brooks KN, Dixon JA, Hamilton LS. F'AO Conservation Guide No 16, Rome, Italy.
3. Pulle AA (1988) African Rice (Oryza glaberimma Stued.): Lost crop of the enslaved Africans Discovered in Suriname. An article published by Tinde Van Andel, Mar 5, 2010.
4. Arnold JG, Srinivasan R, Muttiah RS, Williams JR (1998) Large area hydrologic modeling and assessment - Part 1: Model development. Journal of the American Water Resources Association 34: 73-89.
5. Setegn SG, Srinivasan R, Dargahi B (2008) Hydrological Modeling in the Lake Tana Basin, Ethiopia Using SWAT Model. The Open Hydrology Journal 2: 49-62.
6. Fontaine TA, Cruickshank TS, Arnold JG, Hotchkiss RH (2002) Development of a snowfall-snowmelt routine for mountainous terrain for the soil water assessment tool (SWAT). Journal of Hydrology 262: 209-223.
7. Van Griensven A (2005) Sensitivity, auto-calibration, uncertainty and model evaluation in SWAT2005.
8. Van Griensven A, Green CH (2007) Auto-calibration in Hydrological Modeling: Using SWAT2005 in small-scale watersheds. Environmental Modeling & Software 23: 422-434.
9. Eckhardt K, Arnold JG (2001) Automatic calibration of a distributed catchment model. Journal of Hydrology 251: 103-109.
10. Kannan N, White SM, Worrall F, Whelan MJ (2007) Sensitivity analysis and identification of the best evapotranspiration and runoff options for hydrological modelling in SWAT-2000. Journal of Hydrology 332: 456-466.
11. Kim NW, Lee J (2008) Temporally weighted average curve number method for daily runoff simulation. Hydrological Processes 22: 4936-4948.

12. Chow VT, Maidment DR, Mays LW (1988) Applied Hydrology. McGraw Hill, New York, USA, pp: 47-61.
13. Benaman J, Shoemaker CA (2005) An analysis of high-flow sediment event data for evaluating model performance. Hydrological Processes 19: 605-620.

CHAPTER 3

Understanding the Effects of Parameter Uncertainty on Temporal Dynamics of Groundwater-Surface Water Interaction

Gopal Chandra Saha [1,2,*], Jianbing Li [2] and Ronald W. Thring[2]

[1]*Institute for Environmental Sustainability, Mount Royal University, Calgary, AB. T3E 6k6, Canada*

[2]*Environmental Engineering Program, University of Northern British Columbia, Prince George, B.C. V2N 4Z9, Canada*

AUTHOR CONTRIBUTIONS

Gopal Chandra Saha and Jianbing Li designed the idea; Gopal Chandra Saha performed the simulation, and analyzed the results; Gopal Chandra Saha, Jianbing Li, and Ronald W. Thring wrote the paper.

ABSTRACT

This study presents the understanding of temporal dynamics of groundwater-surface water (GW-SW) interaction due to parameter uncertainty by using a physically-based and distributed gridded surface subsurface hydrologic analysis (GSSHA) model combined with a Monte Carlo simulation. A study area along the main stem of the Kiskatinaw River of the Kiskatinaw River watershed, Northeast British Columbia, Canada, was used as a case study. Two different greenhouse gas (GHG) emission scenarios (i.e., A2: heterogeneous world with self-reliance and preservation of local identities, and B1: a more integrated and environmental-friendly world) of the Special Report on Emissions Scenarios (SRES) from the Fourth Assessment Report of the Intergovernmental Panel on Climate Change (IPCC) for 2013 were used as case scenarios. Before conducting uncertainty analysis, a sensitivity analysis was performed to find the most sensitive parameters to the model output (i.e., mean monthly groundwater contribution to stream flow). Then, a Monte Carlo simulation was used to conduct the uncertainty analysis. The uncertainty analysis results under both case scenarios revealed that the pattern of the cumulative relative frequency distribution of the mean monthly and annual groundwater contributions to stream flow varied monthly and annually, respectively, due to the uncertainties of the sensitive model parameters. In addition, the pattern of the cumulative relative frequency distribution of a particular month's groundwater contribution to the stream flow differed significantly between both scenarios. These results indicated the complexities and uncertainties in the GW-SW interaction system. Therefore, it is of necessity to use such uncertainty analysis results rather than the point estimates for better water resources management decision-making.

Keywords: parameter uncertainty; groundwater-surface water interaction; Monte Carlo simulation

1. INTRODUCTION

Groundwater-surface water (GW-SW) interaction plays a vital role in the functioning of riparian ecosystems [1]. During wet periods, surface water can recharge groundwater, but during dry periods groundwater can act as an important source to feed the surface water flow. As a result, groundwater and surface water are closely-linked components of the hydrologic system due to their interdependency to each other. The development and exploitation of any one component can affect the other. Therefore, for developing sustainable water resources management, it is crucial to understand and quantify the exchange processes between these two components [2]. During the last decade, many researchers have used different hydrologic models to quantify these exchange processes. For instance, Scibek and Allen [3] used the coupled Hydrologic Evaluation of Landfill Performance (HELP) and visual MODular three-dimensional finite-difference groundwater FLOW (MODFLOW) model; Van Roosmalen et al. [4] used the DK model (The National Water Resource model for Denmark); Goderniaux et al. [5] used the HydroGeoSphere model; Stoll et al. [6] used the MIKE Système Hydrologique Européen (MIKE-SHE) model; Jackson et al. [7] used the coupled Zoom Object-Oriented Distributed Recharge Model (ZOODRM) and Zoom Object-Oriented Quasi-3-Dimensional Model (ZOOMQ3D); Vansteenkiste et al. [8] used the MIKE-SHE, and Water and Energy Transfer between Soil, Plants and Atmosphere (WetSpa) models; El Hassan et al. [9] used the Gridded Surface Subsurface Hydrologic Analysis (GSSHA) model; Wu et al. [10] used Groundwater and Surface-water FLOW (GSFLOW) model; Faramarzi et al. [11] used the Soil and Water Assessment Tool (SWAT) model. Most of the parameters (e.g., soil properties, surface roughness) in hydrologic models used for GW-SW interaction simulations require intensive field measurements [12], and they are always associated with uncertainty. Such uncertainty could also lead to uncertainty in modeling outputs [13], which could jeopardize the decision-making of water resources management. The uncertainty analysis could provide a range of outputs instead of one output, and enable the watershed manager to take proper action with respect to water withdrawal from the river, and allocation to the stakeholders for future water supply depending on the month and season.

Due to the importance of parameter uncertainty in hydrologic models, many researchers used different uncertainty analysis methods in different hydrologic models to conduct uncertainty analysis of modeling outputs during the last several decades. For example, Beven and Binley [14] used the Generalized Likelihood Uncertainty Estimation (GLUE) method in the Institute of Hydrology Distributed Model (IHDM) to investigate how the stream flow hydrograph varies under the parameter uncertainty during a number of storms in the Gwy catchment, Wales. Kuczera and Mroczkowski [15] assessed the use of multinormal approximation to parameter uncertainty for the exploration of multiresponse data (i.e., stream flow, stream chloride concentration, groundwater level) in the CATPRO hydrosalinity model in the Wights catchment in Western Australia. Vrugt et al. [16] used the Markov Chain Monte Carlo (MCMC) method in the HYdrological MODel (HYMOD) to find out the variation of stream flow under the parameter uncertainty in the Leaf River watershed, Mississippi, USA. Benke et al. [12] used the Monte Carlo simulation (MCS) method as an uncertainty analysis method in the 2C hydrological model to investigate the impacts of parameter uncertainty on the prediction of stream flow in Eastern Australia. Mishra [17] used the first-order second-moment (FOSM) and MCS methods in the Natural Systems Regional Simulation Model (NSRSM) to compare the impacts of parameter uncertainty on stream flow prediction in Southern Florida, USA. Shen et al. [18,19] used GLUE and MCS methods, respectively, in the SWAT model to quantify the effects of parameter uncertainty on stream flow and sediment in the Daning River watershed of the Three Gorges Reservoir Region, China. Wu et al. [10] used the Probabilistic Collocation Method (PCM) in the GSFLOW model for integrated modeling of GW-SW systems (i.e., annual stream flow, annual GW-SW exchange, and annual groundwater level) in the Heihe River Basin, China. Fan et al. [20] used PCM and MCS methods in the HYMOD to compare the uncertainty of stream flow predictions due to the parameter uncertainty in the Xiangxi River basin of the Three Gorges Reservoir Region, China. Fan et al. [21] used the Hybrid Sequential Data Assimilation and Probabilistic Collocation (HSDAPC) method in the HYMOD for analyzing uncertainty of stream flow predictions due to the parameter uncertainty in the Xiangxi River basin of the Three Gorges Reservoir Region, China. Wu et al. [22] used CoupModel combined with the GLUE method to investigate the water and energy balance in seasonally-frozen soils due to parameter uncertainty in Northern China. Faramarzi et al. [11] used multivariate uniform distribution in the SWAT model to assess uncertainty of freshwater scarcity in semi-arid watersheds of Alberta, Canada. However, there is little knowledge regarding the uncertainty analysis of mean monthly (i.e., intra-annual) and annual groundwater contributions to stream flow due to parameter uncertainty in a watershed. The objective of the study was to understand and find out the variation of the mean monthly and annual groundwater contributions to stream flow due to parameter uncertainty. These contributions information will be useful in determining the temporal status of groundwater resources and site conditions for groundwater-dependent terrestrial ecosystems [23]. They will also determine the temporal variations of stream flow dependency on groundwater, and these will provide useful information for both short and long-term water withdrawal from the river and water supply decision-making. This study also presents a new technique to understand temporal dynamics of GW-SW interaction systems by using temporal mean groundwater contributions to the stream flow. A GSSHA hydrologic model was developed for this study in a study area along the main stem of the Kiskatinaw River of the Kiskatinaw River watershed (KRW) in Northeastern British Columbia, Canada. The A2 and B1 greenhouse gas (GHG) emission scenarios of SRES (Special Report on Emissions Scenarios) of the Intergovernmental Panel on Climate Change (IPCC) of 2013 were used as case scenarios. Before conducting uncertainty analysis, a sensitivity analysis was conducted to find out the most sensitive parameters to the model outputs. Then Monte Carlo realizations of the most sensitive modeling parameters were generated for the GSSHA model to conduct the uncertainty analysis of the temporal dynamics of the GW-SW interaction.

2. STUDY AREA

The Kiskatinaw River watershed (KRW) is a multi-use watershed located in Northeast British Columbia, Canada, as shown in Figure 1. The KRW provides water for drinking water supply and household activities, timber harvesting, oil and gas, agriculture, wildlife, recreation, and mineral resources. The City of Dawson Creek has been using water from Kiskatinaw River for drinking purpose since the mid-1940s because of the presence of high total hardness in groundwater of this region [24]. The water demand in the KRW increased significantly after 2005 because of shale gas exploration/production activities. The large scale of shale gas exploration/production, timber harvesting, and agricultural activities in recent years have resulted in water conflicts among various water users.

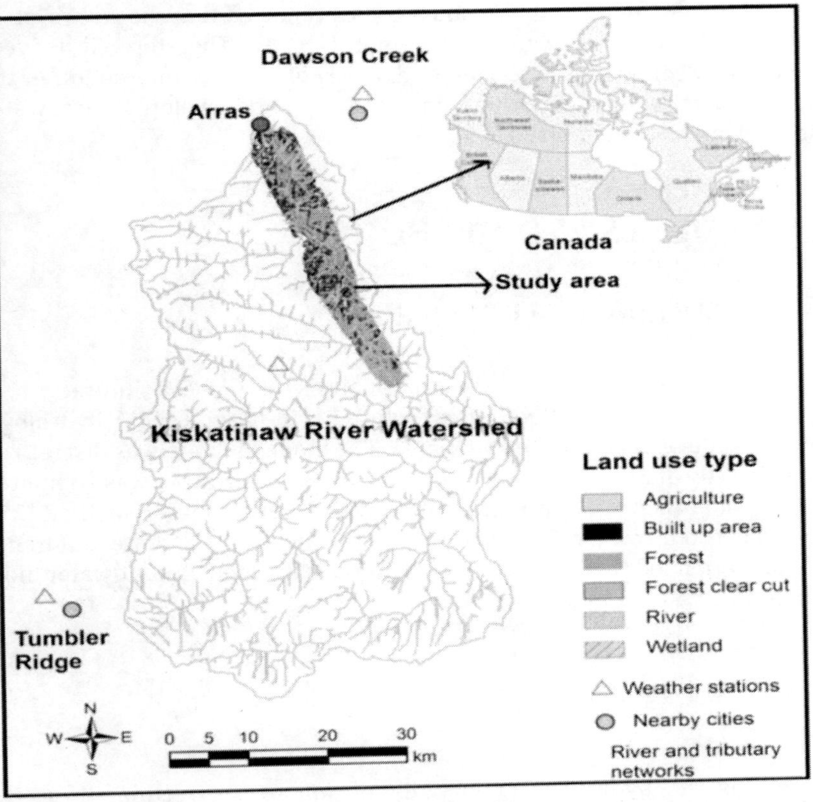

Figure 1. Land use map of the study area and its location in the Kiskatinaw River Watershed, as well as in Canada.

The study area (213.82 km^2) is situated along the main stem of the Kiskatinaw River in the KRW (Figure 1). The City of Dawson Creek's drinking water intake of the water supply system is located at Arras in the study area. The elevation of the study area ranges from 687 m to 950 m, and the mean slope is 7.8% [25]. The dominant soil texture in the study area is mainly clay loam (91%), although silt loam and sandy loam occupy 6% and 3% of the study area, respectively. The study area is mainly a forested area (68%) (Figure 1). Other land use types are forest clear cut (18.7%), agriculture (8%), wetland (2%), water (1.8%), and built up area (1.5%).

The hydrologic system in the KRW is flashy, with peak flows occurring from late June to early July, but in January the flow drops to 0.052 m^3/s [24]. The mean annual precipitation and temperature in the KRW is 499 mm, and 2.7 °C, respectively (mean over the period of 2000–2011). The major portion of the precipitation is rainfall (320 mm), therefore, the hydrologic system in the KRW is rain dominated. Previous studies reported that the river system in the KRW is groundwater dependent [25,26]. Therefore, it is very important to conduct an uncertainty analysis of GW-SW interaction in the study area to understand those interactions and develop future water resources management plans.

3. MATERIALS AND METHODS

3.1. GW-SW Interaction Modeling

In order to conduct an uncertainty analysis of GW-SW interaction, a physically-based, distributed and structured grid-based GSSHA hydrologic model was developed for the study area. The model domain was discretized into grid cells of 30 m by 30 m. In GSSHA model, infiltration was computed by using the Green and Ampt infiltration with a redistribution method [27]. Saturated groundwater flow was computed by using a finite difference representation of the following 2-D lateral saturated groundwater flow equation:

$$\frac{\partial}{\partial x}\left(K_x b\frac{\partial h}{\partial x}\right) + \frac{\partial}{\partial y}\left(K_y b\frac{\partial h}{\partial y}\right) = S\frac{\partial h}{\partial t} + W(x, y, t), \tag{1}$$

where $K_x$is the hydraulic conductivity of soil in x direction, K_y is the hydraulic conductivity of soil in y direction, h is the hydraulic head, S is the storage term, b is the depth of the saturated media, and W is the flux term for sources and sinks. Water flux between the stream and the saturated groundwater was estimated by using Darcy's law and the following equation:

$$f = \frac{K_{sb}}{M_{sb}}(E_{gw} - E_{sw}), \quad (2)$$

where f is the flux between the stream and the saturated groundwater, $K_{sb}$is the hydraulic conductivity of the streambed material, $M_{sb}$is the depth of the streambed material, $E_{gw}$is the elevation of groundwater surface, and $E_{sb}$is the elevation of stream water surface. Groundwater discharge was then calculated by multiplying this f by the top width and length of the channel segment. Overland flow was simulated by using the Alternating Direction Explicit (ADE) finite difference method, while channel flow was simulated by using an explicit solution of the diffusive wave equation. The details of GSSHA can be found from Downer [28].

The GSSHA model requires a number of inputs, such as watershed specific data (i.e., elevation, channel geometry, soil type, and land use/land cover), hydrological (i.e., stream flow, and groundwater level) data, and climate (i.e., precipitation and temperature) and meteorological (i.e., wind speed, relative humidity, barometric pressure, direct and global radiations) data. The details of these data for this study were listed in Table 1. In this study, width and length of the channel segment were assumed constant. Observed hourly climate and meteorological data of three nearby weather stations were averaged by an arithmetic method to obtain daily distribution of those parameters for 2000–2011. In addition, in this study the soils of the study area were assumed as isotropic because the stratification of the soil layers of the study area were not available. Since GSSHA is a grid-based model, three index maps were prepared for assigning parameters at the grid level. One is a land use index map, and others are a soil type index map and a combined land use and soil type index map. In addition, initial groundwater table and aquifer (unconfined) bottom maps were prepared by using the inverse distance weighted (IDW) interpolation method. The distance between ground elevation and aquifer bottom elevation was used to make the vertical layer of the modeling domain. A similar approach was used by Woldeamlak et al. [32] in their developed model for investigating GW-SW interaction. The barometric pressure correction was estimated as per the technical guidelines of Solinst [33] to correct the groundwater table data because groundwater table fluctuates by atmospheric pressure with altitude change [34]. Around the perimeter of the study area, no-flow boundary condition was chosen for the developed model based on previous studies results [35]. In addition, the flux river boundary condition was chosen for the stream because of the significant exchange between groundwater and the stream network. Then the model was calibrated using Shuffled Complex Evolution method (automated calibration). Due to the limited observed stream flow data, the model calibration was conducted for the time period from 15 October 2010 to 31 December 2011. A simulation time step of 1 min was used for calibration and validation based on the temporal convergence study of observed and simulated stream flows at the outlet of the study area. There was flood in the KRW in 2011, and the model calibration was conducted to evaluate the performance of the model during flooding year. During calibration (Figure 2a), R^2 (coefficient of determination) = 0.67, and NSE (Nash-Sutcliffe efficiency) = 0.62 were obtained. Santhi et al. [36] and Van Liew et al. [37] suggested an acceptable model evaluation when R^2 value of greater than 0.5 obtained. These evaluation statistics criteria indicated satisfactory GSSHA model calibration. The calibrated parameters were tabulated in Table 2. Model validation was performed for the time period from 15 October 2006 to 15 October 2010, and this time period was normal annual precipitation years. The modeling validation results (Figure 2b) presented R^2 = 0.63, and NSE = 0.58. Therefore, satisfactory model validation was also achieved. During calibration and validation periods, the flow direction was from groundwater to surface water, which indicates gaining stream or groundwater discharging to the stream.

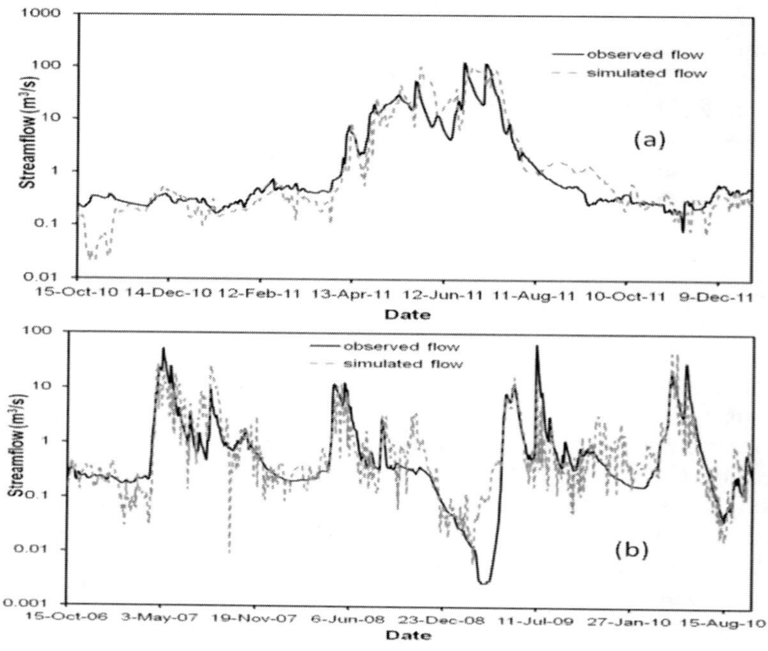

Figure 2. Observed and simulated stream flows by the GSSHA model at the outlet of the study area during (**a**) calibration and (**b**) validation periods.

Table 1. Details of input data for GSSHA model development.

Data Type	Data and Format	Source
Watershed	• Canadian Digital Elevation Data of 13.74 m by 23.81 m grid • Channel geometry • Land use/land cover of 30 m by 30 m grid for year 2010 • Soil	• Geo Base of Natural Resources Canada [29] • Field survey and Google maps • Paul [30] • Soil survey report of Land Resource Research Institute [31], and Saha [25]
Hydrological	• Daily stream flow from 2006 to 2011 • Groundwater level	• Water Survey of Canada station • Groundwater monitoring network in the KRW
Climate and Meteorological	• Observed hourly precipitation and temperature and other meteorological data (i.e., relative humidity, wind speed, barometric pressure, direct and global radiations) from 2000 to 2011	• Three weather stations from British Columbia Ministry of Forests, Lands and Natural Resources Operations

Table 2. Calibrated parameters' values used in the GSSHA model. Here GW means groundwater.

Process	Process Parameter	Unit	Value
Infiltration and GW	Saturated hydraulic conductivity (Ks) (clay loam-forest)	cm/hr	0.15
Infiltration and GW	Ks (clay loam-built-up area)	cm/hr	0.03
Infiltration and GW	Ks (clay loam-forest clear cut area)	cm/hr	0.08
Infiltration and GW	Ks (clay loam-agriculture)	cm/hr	0.10
Infiltration and GW	Ks (clay loam-wetland)	cm/hr	0.05
Infiltration and GW	Ks (sandy loam-forest)	cm/hr	0.93
Infiltration and GW	Ks (sandy loam-forest clear cut area)	cm/hr	0.34
Infiltration and GW	Ks (sandy loam-agriculture)	cm/hr	0.42
Infiltration and GW	Ks (silt loam-forest clear cut area)	cm/hr	0.7
Infiltration and GW	Ks (silt loam-forest)	cm/hr	0.81
Infiltration and GW	Ks (silt loam-built-up area)	cm/hr	0.09
Infiltration	Initial moisture (clay loam)	-	0.21
Infiltration	Initial moisture (silt loam)	-	0.15
Infiltration	Initial moisture (sandy loam)	-	0.11
Overland flow	Manning's n (built-up area)	-	0.011
Overland flow	Manning's n (agriculture)	-	0.035
Overland flow	Manning's n (forest)	-	0.1
Overland flow	Manning's n (forest clear cut area)	-	0.03
Infiltration and GW	Porosity (silt loam)	-	0.501
Infiltration and GW	Porosity (clay loam)	-	0.464
Infiltration and GW	Porosity (sandy loam)	-	0.453
Channel flow	Manning's n (river)	-	0.025
Soil moisture	Soil moisture depth	m	0.5
GW-stream	Ks (stream bed material)	cm/hr	1.1
GW-stream	Stream bed material's thickness	cm	15
Retention	Retention depth (agriculture)	mm	0.1
Retention	Retention depth (forest)	mm	0.12
Retention	Retention depth (forest clear cut area)	mm	0.1

3.2. Generation of Case Scenarios

Two case scenarios were generated to understand and find out the variation of the mean monthly and annual groundwater contributions to stream flow due to parameter uncertainty. Those scenarios were generated by using two climate change scenarios for 2013, which were obtained from CRCM 4.2 (Canadian Regional Climate Model) modeling outputs of the CCCma (Canadian Centre for Climate Modeling and Analysis). Those scenarios were A2 and B1 GHG emission scenarios of SRES of the IPCC [38]. The A2 scenario was used because it corresponds to regional economic development, and the large-scale shale gas exploration/production activities began in the KRW during 2005, and enhanced regional economy [39]. The B1 scenario was selected because it represents a more integrated and environmental friendly world [38]. These CRCM outputs are monthly means for two GHG emission scenarios [40]. For each scenario, three CRCM outputs were averaged as a mean. A similar approach was applied in various hydrological studies [3,8]. Then those outputs were distributed as daily values in the particular month by using the delta change method. This method is a commonly used simple downscaling method to deal with biases when using climate model outputs in hydrological studies at the catchment or sub-watershed scale [41,42]. This method has been used in a number of hydrological impact studies [4,40,43,44,45]. The details of delta change method can be found in Graham et al. [44], van Roosmalen et al. [4], and van Roosmalen et al. [45].

The temperature outputs from the delta change method showed that the trend of the mean monthly temperatures in 2013 under the A2 and B1 scenarios was similar to those of 2000–2011; with the highest and lowest mean monthly temperatures occurring in July and January, respectively, (Figure 3a). It was also found that the mean monthly temperatures under the A2 scenario were associated with higher values than under the B1 scenario for most of the months, except in July, August, October, and December. The mean annual temperature in 2013 under the A2 and B1 scenarios was 3.29 °C and 3.05 °C, respectively, which was above by 0.59 °C and 0.35 °C as compared to those of 2000–2011, respectively. This variability occurred because of the anthropogenic increases in the atmospheric concentrations of greenhouse gases [38,46].

The precipitation outputs from the delta change method showed that the monthly precipitations of 2013 under the A2 and B1 scenarios had variable patterns (Figure 3b) due to the anthropogenic increases in the atmospheric concentrations of greenhouse gases [38,46]. The peak monthly precipitation under the B1 scenario followed the similar trend of the peak mean monthly precipitations of 2000–2011; on the other hand, it followed an opposite trend under the A2 scenario. It was also found that the monthly precipitations were higher in most of the months under the B1 scenario than under the A2 scenario, except in March, April, July, September, and November. The annual precipitation of 2013 under the A2 and B1 scenarios was 509 mm and 524 mm, respectively, and these numbers were greater than the mean annual precipitation of 2000–2011 by 10 mm (2%) and 25 mm (5%), respectively.

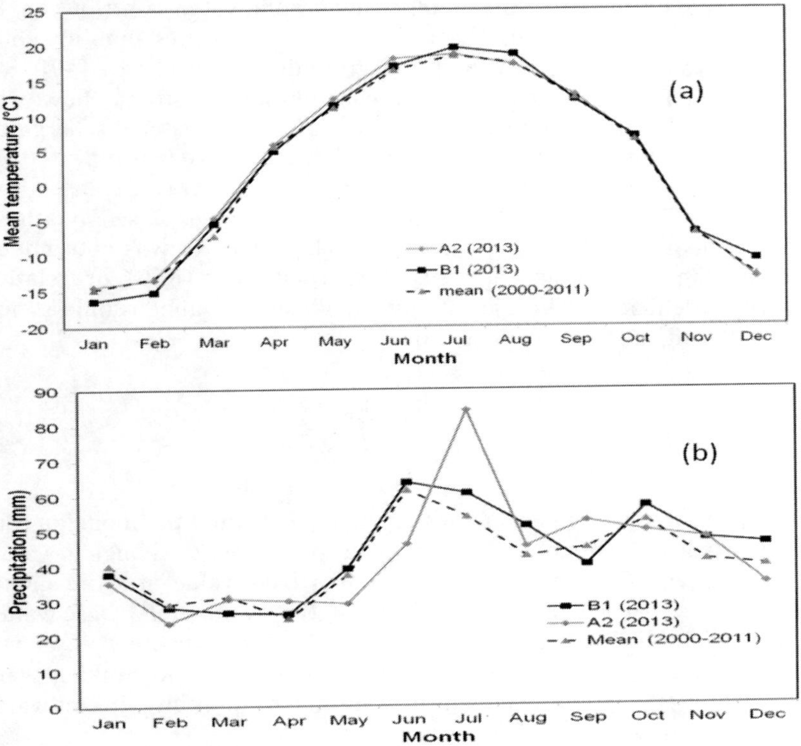

Figure 3. Projected (**a**) mean monthly temperatures and (**b**) monthly precipitations under the A2 and B1 scenarios in 2013.

3.3. Sensitivity Analysis of GW-SW Interaction

Before conducting the uncertainty analysis of any hydrological model, it is important to conduct a sensitivity analysis of the modeling input parameters because sensitivity analysis indicates the assessment of uncertainty importance [17]. In this study, sensitivity analysis was conducted using the OAT (One-factor-At-a-Time) method [47]. The OAT method was chosen because it is the simplest method for conducting a sensitivity analysis [48], and there were 28 calibrated parameters in the developed GSSHA model which required a large number of numerical runs for conducting the sensitivity analysis using other methods (e.g., factorial design). Using the OAT method, each calibrated parameter was changed by a small amount at a time from a reference (i.e., base) value while keeping the remaining parameters constant, and then the corresponding change in the modeling output (i.e., mean monthly groundwater contribution to stream flow) was computed. The GSSHA model estimates monthly total volume of stream discharge (flow) and groundwater discharge [49], and then the mean monthly groundwater contribution to stream flow was calculated by dividing monthly total volume of groundwater discharge by monthly total volume of stream discharge. This procedure was repeated for three times, and every time the calibrated parameter was increased and decreased by a factor of 20% of the reference value. Based on these computed changes, relative sensitivity of each parameter was determined by calculating the normalized sensitivity coefficient (NSC) or relative sensitivity coefficient. This sensitivity coefficient is dimensionless and calculated using the following equation:

$$NSC = \frac{(R_a - R_n)/R_n}{(P_a - P_n)/P_n}, \tag{3}$$

where P_a and R_a are the parameter's changing value and the model output after a particular model run using the parameter's changing value, respectively, and P_n and R_n are parameter's base value and the model output after a particular model run using the parameter's base value, respectively. Table 3 lists all of the parameters' relative sensitivities, as well as their sensitivity rankings. Here, relative sensitivity and ranking were evaluated based on the change in the mean monthly groundwater contributions to the stream flow.

Table 3 showed that 12 parameters out of 28 calibrated parameters of the developed model had small to large impacts on the modeling outputs. Eleven of these twelve parameters control infiltration and soil moisture, which, in turn control groundwater flow, whereas the other parameter controls stream (i.e., channel) flow routing. Manning's n (river), which controls stream flow routing, had the highest relative sensitivity. Rushton [50] also indicated that, in many regional groundwater model studies, aquifer (i.e., groundwater) to river flows are not very sensitive to changes in the river coefficient (in the GSSHA model, hydraulic conductivity of the streambed material and streambed material's thickness), but only groundwater heads beneath the river show greater influence on the aquifer to river flows. Therefore, the output of sensitivity analysis of this study also supported the findings of Rushton [50]. Two criteria (mean and standard deviation) of the normalized sensitivity coefficients were selected to identify the most sensitive parameters, which influence the modeling outputs [51]. Mean and standard deviation were calculated based on the estimated NSC values of the four runs in the OAT method. Soil moisture depth, initial soil moisture (clay loam), Ks (clay loam-forest), porosity (clay loam), and Ks (clay loam-forest clear cut area) ranked second, third, fourth, fifth, and sixth, respectively, based on their relative sensitivities' values. The parameters (i.e., Ks (clay loam-agriculture), Ks (sandy loam-forest), Ks (clay loam-built up area), Ks (clay loam-wetland), porosity (silt loam), and Ks (silt loam-forest)), ranked from seventh to twelfth, and their changes had very little impact on the change in the mean monthly groundwater contributions to the stream flow, as well as the relative sensitivity because clay loam-agriculture, sandy loam-forest, clay loam-built up area, clay loam-wetland, silt loam, and silt loam-forest occupy 8%, 2%, 1%, 2%, 6%, and 6% of the study area, respectively. As a result, these six parameters (i.e., ranked seventh to twelfth) were not considered for uncertainty analysis in this study. Benke et al. [12] also noted that when parameters have little impact on the modeling output value, they can be easily ignored for simplification of the model structure.

Table 3. Calibrated parameters' relative sensitivities and their sensitivity rankings.

Parameter	Unit	Relative Sensitivity	Rank
Manning's n (river)	-	0.39	1
Soil moisture depth	m	0.32	2
Initial soil moisture (clay loam)	-	0.29	3
Ks (clay loam-forest)	cm/hr	0.24	4
Porosity (clay loam)	-	0.12	5
Ks (clay loam-forest clear cut area)	cm/hr	0.08	6
Ks (clay loam-agriculture)	cm/hr	0.006	7
Ks (sandy loam-forest)	cm/hr	0.0052	8
Ks (clay loam-built up area)	cm/hr	0.005	9
Ks (clay loam-wetland)	cm/hr	0.003	10
Porosity (silt loam)	-	0.002	11
Ks (silt loam-forest)	cm/hr	0.0001	12

3.4. Uncertainty Analysis of GW-SW Interaction

Uncertainty analysis of GW-SW interaction was conducted using the six most sensitive parameters. These input parameters' values were obtained from text books, journals, and other field-collected results conducted by scientific institutes under similar conditions [52,53,54,55,56,57,58,59,60]. The values of these parameters were assumed to be normally distributed. This distribution was used for these parameters in other studies (e.g., Chen et al. [61] and Cheng et al. [62] for hydraulic conductivity; Le et al. [63] for porosity; Bora and Rajput [58] and Morikawa et al. [64] for Manning's n; Choi and Jacobs [65] for initial soil moisture; and Adams et al. [66] for soil moisture depth). In this study, a Monte Carlo simulation was used as the uncertainty analysis method because it is the most popular reliability-analysis-based stochastic method for evaluating uncertainties in hydrology studies [67]. Table 4 presented all these parameters' mean and standard deviation values which were used for generating Monte Carlo realizations of these parameters based on the assumed probabilistic distributions. Using these realizations, uncertainty analysis of GW-SW interaction was conducted for the A2 and B1 GHG emission scenarios of 2013.

Table 4. Mean and standard deviation values of the most sensitive parameters used for uncertainty analysis.

Parameter	Unit	Mean	Standard Deviation
Manning's n (river)	-	0.032	0.01
Soil moisture depth	m	0.60	0.125
Initial soil moisture (clay loam)	-	0.18	0.04
Ks (clay loam-forest)	cm/hr	0.20	0.08
Porosity (clay loam)	-	0.45	0.02
Ks (clay loam-forest clear cut area)	cm/hr	0.12	0.05

4. RESULTS AND DISCUSSION

4.1. Uncertainty Analysis of GW-SW Interaction under the A2 Scenario

The uncertainty analysis results of GW-SW interaction under the A2 GHG emission scenario were analyzed by using the cumulative relative frequency distribution of temporal (i.e., monthly, and annual) groundwater contributions to the stream flow. The results showed that the pattern of the cumulative relative frequency distribution of the mean monthly groundwater contributions to the stream flow varied monthly due to the sensitive modeling parameter uncertainty (Figure 4). It also demonstrated the complexities and uncertainties in the GW-SW interaction system. Table 5 shows the detailed uncertainty analysis results (i.e., range, mean, and standard deviation) of the mean monthly groundwater contributions to stream flow in different months of 2013 under the A2 scenario against the simulated value for the corresponding month using the calibrated parameters' values that were used in model calibration and validation. These results indicated the necessity of using uncertainty analysis in modeling parameters rather than point estimates. The results also illustrated that the output, generated by the calibrated model using the calibrated parameters' values, fell within the range of modeling outputs from uncertainty analysis. It was also found that the calculated highest range (i.e., interval) of the mean monthly groundwater contribution to the stream flow occurred during low flow months in winter (December–February), and early spring (March). During winter and early spring, precipitation occurred mainly as snow in the study area. This snow accumulated, and resulted in low surface runoff before complete snow melt. Therefore, the range of the mean groundwater contribution to the stream flow was higher during winter months. On the other hand, the lowest range of the mean groundwater contribution to the stream flow occurred during summer months (June and July) because higher rainfall during summer resulted in a greater increase of surface runoff [4,68,69] compared to that of groundwater discharge.

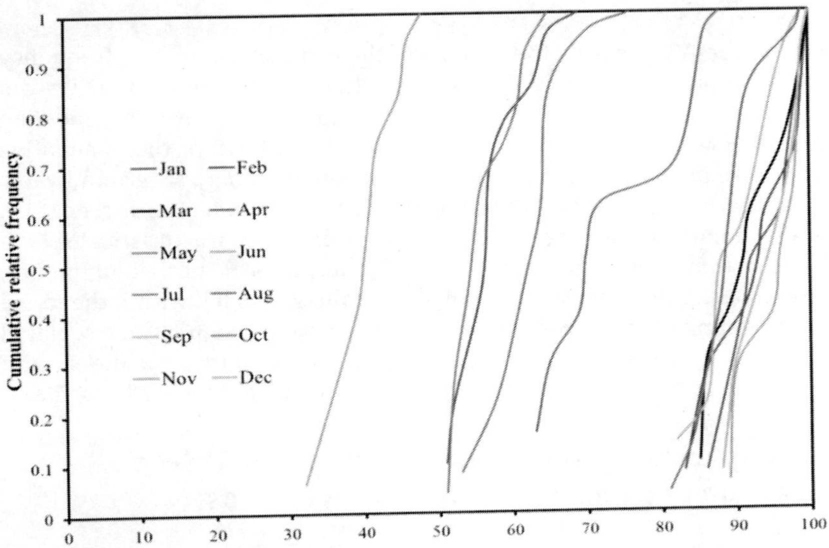

Figure 4. Cumulative relative frequency distribution of mean groundwater contributions to stream flow in different months of 2013 under the A2 GHG emission scenario.

Table 5. Uncertainty analysis results of mean monthly groundwater contributions to stream flow under the A2 GHG emission scenario in 2013 against the simulated value for corresponding month using the calibrated parameters' values.

Month	Range of Mean Groundwater Contribution to Stream Flow (%)	Mean (%)	Standard Deviation (%)	Output Using the Calibrated Parameters' Values (%)
January	86–100	94.19	5.74	93.16
February	83–100	93.22	6.32	90.79
March	85–100	92.44	5.63	93.53
April	51–69	58.19	6.08	54.26
May	51–65	56.78	4.71	57.23
June	32–48	41.10	4.45	44.04
July	53–76	64.40	5.55	59.20
August	63–88	76.76	10.01	71.89
September	82–99	90.41	6.33	88.94
October	81–99	90.06	5.53	93.81
November	89–99	94.73	4.35	95.90
December	88–100	95.23	4.55	97.50

4.2. Uncertainty Analysis of GW-SW Interaction under the B1 Scenario

Similar to the A2 scenario, the pattern of the cumulative relative frequency distribution of the mean monthly groundwater contributions to the stream flow under the B1 scenario varied monthly due to the sensitive modeling parameter uncertainty (Figure 5). However, the pattern of the cumulative relative frequency distribution of particular month's groundwater contribution to stream flow differed significantly between both scenarios. This also demonstrated the complex and uncertain characteristics of GW-SW interaction system in watershed. The output of a particular month, generated using the calibrated parameters' values, also fell within the range of the modeling outputs from uncertainty analysis (Table 6). The calculated highest range of mean monthly groundwater contribution to the stream flow also occurred during low flow months in the winter and early spring.

4.3. Comparison of Uncertainty Analysis of GW-SW Interaction between the A2 and B1 Scenarios

Figure 6 illustrates the cumulative relative frequency distribution of the mean annual groundwater contribution to stream flow in 2013 under the A2 and B1 scenarios. Similar to the pattern of the cumulative relative frequency distribution of mean monthly groundwater contributions to stream flow, the pattern of the cumulative relative frequency distribution of mean annual groundwater contributions to stream flow under both scenarios varied due to the modeling parameter uncertainty. Under the A2 scenario, the mean annual groundwater contribution to stream flow ranged from 72% to 87%, with a mean of 79% and a standard deviation of 4%. On the other hand, the mean annual groundwater contribution to stream flow in 2013 was 78% using the calibrated parameters' values. Under the B1 scenario, the mean annual groundwater contribution to stream flow ranged from 70% to 82%, with a mean of 75% and a standard deviation of 3.5%. On the other hand, the mean annual groundwater contribution to stream flow in 2013 was close to 77% using the calibrated parameters' values. Compared to the A2 scenario, the range of the mean annual groundwater contributions to stream flow under the B1 scenario was lower. This occurred because more precipitation was predicted under the B1 scenario than under the A2 scenario in 2013. This increased precipitation resulted in surface water and groundwater levels increasing. However, the major increase occurred in surface water levels due to increased surface runoff because of the steep topography of the study area (average slope of 7.8%). Therefore, the gradient between the groundwater and surface water was decreased, and resulted in lower range of mean annual groundwater contribution to stream flow under the B1 scenario.

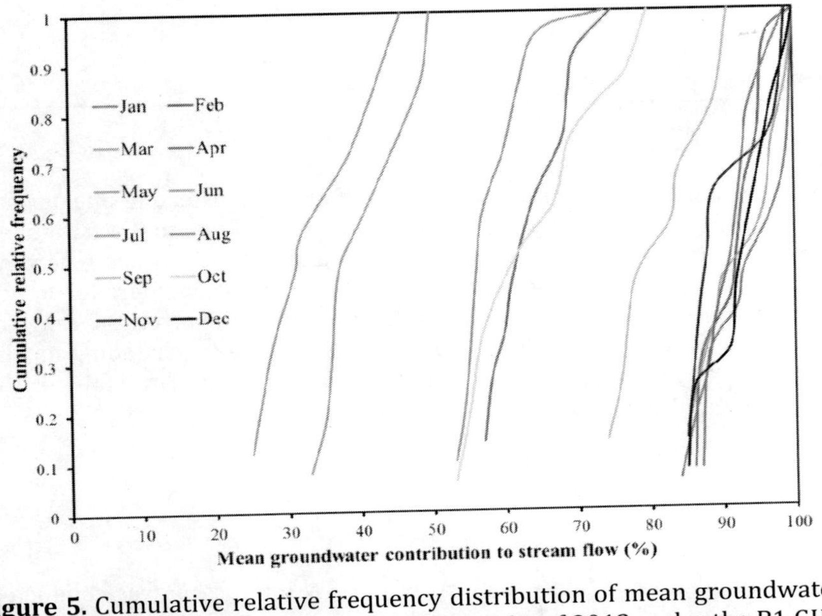

Figure 5. Cumulative relative frequency distribution of mean groundwater contributions to stream flow in different months of 2013 under the B1 GHG emission scenario.

It also demonstrated that uncertainty in the mean annual groundwater contribution to the stream flow under the A2 scenario was higher than that in the B1 scenario because of the higher precipitation in the B1 scenario. Therefore, lower uncertainties are associated in GW-SW interaction system during wet years compared to dry years. This reflects the complexities in GW-SW interaction system.

Table 6. Uncertainty analysis results of mean monthly groundwater contributions to stream flow under the B1 GHG emission scenario in 2013 against simulated value for corresponding month by using the calibrated parameters' values.

Month	Range of Mean Groundwater Contribution to Stream Flow (%)	Mean (%)	Standard Deviation (%)	Output Using the Calibrated Parameters' Values (%)
January	87–100	94.29	5.56	99.79
February	86–100	93.53	4.94	98.32
March	85–100	93.27	5.82	98.46
April	57–75	64.15	6.35	63.16
May	25–46	35.67	7.97	28.30
June	33–50	41.35	6.36	47.21
July	53–74	60.53	6.43	61.30
August	84–99	91.64	5.19	96.20
September	74–91	81.99	6.91	87.70
October	53–80	66.02	10.10	56.48
November	85–99	91.15	5.49	87.82
December	85–100	92.65	5.64	97.89

The uncertainty analysis provides a range of outputs instead of one output, and this range of outputs would provide more information to the watershed manager to take particular action with a certain degree of risks regarding water withdrawal from the river and allocation to stakeholders for future water supplies depending on the month and season. However, the range of outputs and prediction would differ from year to year depending on the amount and pattern (i.e., only rainfall or both snow and rainfall) of the annual precipitation.

5. CONCLUSIONS

In this study an uncertainty analysis of GW-SW interaction was conducted to understand and determine the variation of temporal (i.e., monthly, and annual) mean groundwater contributions to stream flow due to parameter uncertainty by using a new technique, which was based on temporal mean groundwater contributions to streamflow.

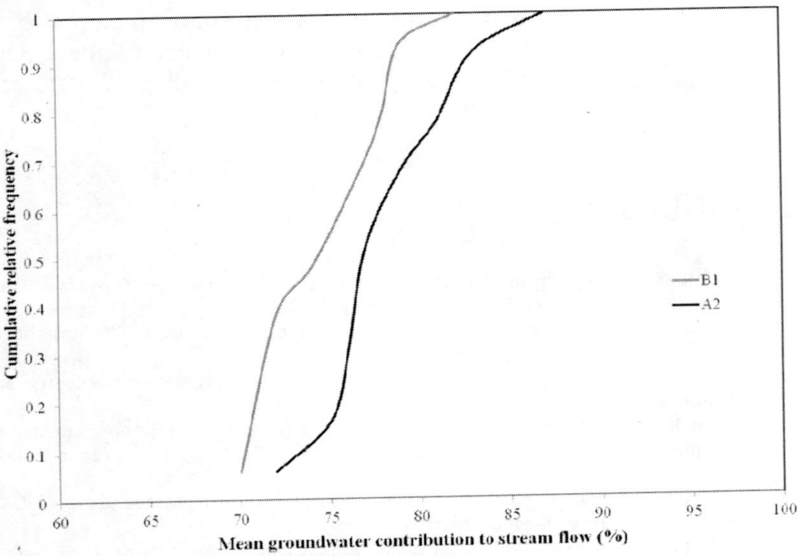

Figure 6. Cumulative relative frequency distribution of mean annual groundwater contributions to the stream flow in 2013 under the A2 and B1 GHG emission scenarios.

The uncertainty analysis results showed that the pattern of the cumulative relative frequency distribution of mean monthly and annual groundwater contributions to stream flow under both case scenarios varied monthly and annually, respectively, because of the uncertainties of the sensitive model parameters. The results also demonstrated that uncertainties in the sensitive parameters had significant effects on the predicted mean monthly and annual groundwater contributions to stream flow. The comparison of two case scenarios also indicated that lower uncertainties are associated in the GW-SW interaction system during wet years compared to dry years. In general, these uncertainty analysis results represent a new way to indicate the complexities and uncertainties in GW-SW interaction system. Therefore, it is very essential to use the uncertainty analysis results for better water resources management decision-making.

ACKNOWLEDGMENTS

The authors would like to thank two anonymous reviewers for providing their valuable comments and suggestions, which contributed much to improve the paper. The authors would also like to thank Peter Caputa, Faye Hirshfield, Siddhartho Shekhar Paul, Malyssa Maurer, Reg Whiten, and Chelton van Geloven for their help and support.

REFERENCES

1. Kalbus, E.; Reinstorf, F.; Schirmer, M. Measuring methods for groundwater-surface water interactions: A review. *Hydrol. Earth Syst. Sci.***2006**,*10*, 873–887.
2. Sophocleous, M. Interactions between groundwater and surface water: The state of the science. *Hydrogeol. J.***2002**,*10*, 52–67.
3. Scibek, J.; Allen, D.M. Modeled impacts of predicted climate change on recharge and groundwater levels. *Water Resour. Res.* **2006**, *42*, W11405.
4. Van Roosmalen, L.; Christensen, B.S.B.; Sonnenborg, T.O. Regional differences in climate change impacts on groundwater and stream discharge in Denmark. *Vadose Zone J.* **2007**, *6*, 554–571.
5. Goderniaux, P. Impact of climate change on groundwater reserves. Ph.D. Thesis, University of Liege, Liège, Belgium, 2010.
6. Stoll, S.; Hendricks Franssen, H.J.H.; Butts, M.; Kinzelbach, W. Analysis of the impact of climate change on groundwater related hydrological fluxes: A multi-model approach including different downscaling methods. *Hydrol. Earth Syst. Sci.***2011**,*15*, 21–38.
7. Jackson, C.R.; Meister, R.; Prudhomme, C. Modeling the effects of climate change and its uncertainty on UK Chalk groundwater resources from an ensemble of global climate model projections. *J. Hydrol.* **2011**, *399*, 12–28.
8. Vansteenkiste, T.; Tavakoli, M.; Ntegeka, V.; Willems, P.; De Smedt, F; Batelaan, O. Climate change impact on river flows and catchment hydrology: A comparison of two spatially distributed models.*Hydrol. Process.***2012**.
9. El Hassan, A.A.; Sharif, H.O.; Jackson, T.; Chintalapudi, S. Performance of a conceptual and physically based model in simulating the response of a semi-urbanized watershed in San Antonio, Texas. *Hydrol. Process.* **2013**, *27*, 3394–3408.
10. Wu, B.; Zheng, Y.; Tian, Y.; Wu, X.; Yao, Y.; Han, F.; Liu, J.; Zheng, C. Systematic assessment of the uncertainty in integrated surface water-groundwater modeling based on the probabilistic collocation method. *Water Resour. Res.***2014**, *50*, 5848–5865.
11. Faramarzi, M.; Abbaspour, K.C.; Adamowicz, W.L.; Lu, W.; Fennell, J.; Zehnder, A.J.B.; Goss, G.G. Uncertainty based assessment of dynamic freshwater scarcity in semi-arid watersheds of Alberta, Canada. *J. Hydrol. Reg. Stud.***2017**,*9*, 48–68.
12. Benke, K.K.; Lowell, K.E.; Hamilton, A.J. Parameter uncertainty, sensitivity analysis and prediction error in a water-balance hydrological model. *Math. Comput. Model.* **2008**, *47*, 1134–1149.
13. Muleta, M.K.; Nicklow, J.W. Sensitivity and uncertainty analysis coupled with automatic calibration for a distributed watershed model.*J. Hydrol.***2004**,*306*, 127–145.
14. Beven, K.J.; Binley, A.M. The future of distributed models: Model calibration and uncertainty prediction. *Hydrol. Process.***1992**,*6*, 279–298.
15. Kuczera, G.; Mroczkowski, M. Assessment of hydrological parameter uncertainty and the worth of multiresponse data. *Water Resour. Res.* **1998**, *34*, 1481–1489.

16. Vrugt, J.A.; Gupta, H.V.; Bouten, W.; Sorooshian, S. A Shuffled Complex Evolution Metropolis algorithm for optimization and uncertainty assessment of hydrologic model parameters. *Water Resour. Res.* **2003**, *39*, 1201.
17. Mishra, S. Uncertainty and sensitivity analysis techniques for hydrologic modeling. *J. Hydroinform.* **2009**, *11*, 282–296.
18. Shen, Z.Y.; Chen, L.; Chen, T. Analysis of parameter uncertainty in hydrological and sediment modeling using GLUE method: A case study of SWAT model applied to Three Gorges Reservoir Region, China. *Hydrol. Earth Syst. Sci.* **2012**, *16*, 121–132.
19. Shen, Z.Y.; Chen, L.; Chen, T. The influence of parameter distribution uncertainty on hydrological and sediment modeling: A case study of SWAT model applied to the Daning watershed of the Three Gorges Reservoir Region, China. *Stoch. Env. Res. Risk* **2013**, *27*, 235–251.
20. Fan, Y.R.; Huang, W.; Huang, G.H.; Huang, K.; Zhou, X. A PCM-based stochastic hydrological model for uncertainty quantification in watershed systems. *Stoch. Env. Res. Risk* **2015**, *29*, 915–927.
21. Fan, Y.R.; Huang, G.H.; Baetz, B.W.; Li, Y.P.; Huang, K.; Li, Z.; Chen, X.; Xiong, L.H. Parameter uncertainty and temporal dynamics of sensitivity for hydrologic models: A hybrid sequential data assimilation and probabilistic collocation method. *Environ. Modell. Softw.* **2016**, *86*, 30–49.
22. Wu, M.; Jansoon, P.; Tan, X.; Wu, J.; Huang, J. Constraining Parameter Uncertainty in Simulations of Water and Heat Dynamics in Seasonally Frozen Soil Using Limited Observed Data. *Water* **2016**, *8*, 64.
23. Naumburg, E.; Mata-Gonzalez, R.; Hunter, R.; Mclendon, T.; Martin, D. Phreatophytic vegetation and groundwater fluctuations: A review of current research and application of ecosystem response modeling with an emphasis on Great Basin vegetation. *Environ. Manag.* **2005**, *35*, 726–740.
24. Dobson Engineering Ltd.; Urban Systems Ltd. Kiskatinaw River Watershed Management Plan. 2003. File 0714.0046.01. Available online: http://www.dawsoncreek.ca/cityhall/departments/water/watershed/background-watershed-management-plans/ (accessed on 12 November 2016).
25. Saha, G.C. Groundwater-surface water interaction under the effects of climate and land use changes. Ph.D. Thesis, University of Northern British Columbia, Prince George, BC, Canada, 2014.
26. Saha, G.C.; Paul, S.S.; Li, J.; Hirshfield, F.; Sui, J. Investigation of land-use change and groundwater-surface water interaction in the Kiskatinaw River Watershed, British Columbia (parts of NTS 093P/01, /02, /07-/10) Report 2013-1. In *Geoscience BC Summary of Activities 2012*; Geoscience BC: Vancouver, BC, Canada, 2013; pp. 139–148.
27. Ogden, F.L.; Saghafian, B. Green and ampt infiltration with redistribution. *J. Irrig. Drain. Eng.* **1997**, *123*, 386–393.
28. Downer, C.W. Identification and modeling of important stream flow producing processes in watersheds. Ph.D. Thesis, University of Connecticut, Storrs, USA, 2002.
29. GeoBase. Available online: www.geobase.ca (accessed on 15 January 2016).
30. Paul, S.S. Analysis of land use and land cover change in Kiskatinaw River watershed: A remote sensing, GIS & modeling approach. M.Sc. Thesis, University of Northern British Columbia, Prince George, BC, Canada, 2013.
31. Land Resource Research Institute. Soils of Fort St. John-Dawson Creek, British Columbia. Soil Survey report No. 42; Agriculture Canada: Vancouver, BC, USA, 1985.
32. Woldeamlak, S.T.; Batelaan, O.; De Smedt, F. Effects of climate change on the groundwater system in the Grote-Nete catchment, Belgium. *Hydrogeol. J.* **2007**, *15*, 891–901.
33. Solinst. Available online: www.solinst.com (accessed on 20 May 2014).
34. McWhorter, D.B.; Sunada, D.K. *Ground-water Hydrology and Hydraulics*; Water Resources Publications: Fort Collins, Colorado, USA, 1977.
35. Kala Groundwater Consulting Ltd. Groundwater potential evaluation Dawson Creek. British Columbia. Reference. 2001, R01332-0622. Available online: http://www.dawsoncreek.ca/wordpress/wpcontent/uploads/watershed/2001DCk_Kala_GWPotential.pdf (accessed on 18 July 2016).

36. Santhi, C.; Arnold, J.G.; Williams, J.R.; Dugas, W.A.; Srinivasan, R.; Hauck, L.M. Validation of the SWAT model on a large river basin with point and nonpoint sources. *J. Am. Water Resour. Assoc.* **2001**, *37*, 1169–1188.
37. Van Liew, M.W.; Arnold, J.G.; Garbrecht, J.D. Hydrologic simulation on agricultural watersheds: Choosing between two models. *Trans. ASAE* **2003**, *46*, 1539–1551.
38. IPCC. *IPCC Special Report. Emission Scenarios*; Cambridge University Press: Cambridge, UK, 2000; p. 570.
39. British Columbia Ministry of Energy and Mines. The status of exploration and development activities in the Montney Play region of northern BC. 2012. Available online: http://www.offshore oilandgas.gov.bc.ca/OG/oilandgas/petroleumgeology/ UnconventionalGas /Documents/C%20Adams.pdf (accessed on 19 December 2015).
40. Bergström, S.; Carlsson, B.; Gardelin, M.; Lindström, G.; Pettersson, A.; Rummukainen, M. Climate change impacts on runoff in Sweden – assessments by global climate models, dynamical downscaling and hydrological modeling. *Clim. Res.***2001**,*16*, 101–112.
41. Hay, L.E.; Wilby, R.L.; Leavesley, G.H. A comparison of delta change and downscaled GCM scenarios for three mountainous basins in the United States. *J. Am. Water Resour. Assoc.* **2000**, *36*, 387–398.
42. Xu, C.Y.; Widen, E.; Halldin, S. Modelling hydrological consequences of climate change - Progress and challenges. *Adv. Atmos. Sci.***2005**,*22*, 789–797.
43. Andreasson, J.; Bergström, S.; Carlsson, B.; Graham, L.P.; Lindström, G. Hydrological Change - Climate change impact simulation for Sweden. *J. Hum. Environ.* **2004**, *33*, 228–234.
44. Graham, L.P.; Andréasson, J.; Carlsson, B. Assessing climate change impacts on hydrology from an ensemble of regional climate models, model scales and linking methods – a case study on the Lule River basin. *Clim. Change* **2007**, *81*, 293–307.
45. Van Roosmalen, L.; Christensen, J.H.; Butts, M.; Jensen, K.H.; Refsgaard, J.C. An intercomparison of regional climate model data for hydrological impact studies in Denmark. *J. Hydrol.* **2010**, *380*, 406–419.
46. PCC. *Climate Change 2007: The Physical Science Basis, Contribution of working Group I to the 4th assessment report (AR4) of the Intergovernmental Panel on Climate Change*; Cambridge University Press: Cambridge, UK, 2007; p. 996.
47. Saltelli, A. What is Sensitivity Analysis? In*Sensitivity Analysis*; Saltelli, A., Chan, K., Scott, E.M., Eds.; Wiley: New York, NY, USA, 2000.
48. Hamby, D.M. A review of techniques for parameter sensitivity analysis of environmental models. *Environ. Monit. Assess.***1994**, *32*, 135–154.
49. Downer, C.W.; Ogden, F.L. *Gridded Surface Subsurface Hydrologic Analysis (GSSHA) User's Manual*; Version 1.43 for Watershed Modeling System 6.1; U.S. Army Engineer Research and Development Center: Vicksburg, MS, USA, 2006; ERDC/CHL SR-06-1.
50. Rushton, K. Representation in regional models of saturated river-aquifer interaction for gaining/losing rivers. *J. Hydrol.***2007**,*334*, 262–281.
51. Nejadhashemi, A.P.; Wardynski, B.J.; Munoz, J.D. Evaluating the impacts of land use changes on hydrologic responses in the agricultural regions of Michigan and Wisconsin. *Hydrol. Earth Syst. Sci. Discuss.* **2011**, *8*, 3421–3468.
52. Clapp, R.B.; Hornberger, G.M. Empirical equations for some soil hydraulic properties. *Water Resour. Res.***1978**,*14*, 601–604.
53. Rawls, W.J.; Brakensiek, R.L.; Saxton, K.E. Estimation of soil water properties. *Trans. ASAE***1982**,*25*, 1316–1320.
54. Rawls, W.J.; Brakensiek, D.L.; Miller, N. Green–Ampt infiltration parameters from soil data. *J. Hydraul. Eng.* **1983**, *109*, 62–70.
55. Reynolds, W.D.; Drury, C.F.; Yang, X.M.; Fox, C.A.; Tan, C.S.; Zhang, T.Q. Land management effects on the near-surface physical quality of a clay loam soil. *Soil Tillage Res.***2007**,*96*, 316–330.
56. Minhas, P.S.; Sharma, D.R. Hydraulic conductivity and clay dispersion as affected by application sequence of saline and simulated rain water. *Irrig. Sci.***1986**,*7*, 159–167.
57. Miller, L.L.; Hinkel, K.M.; Nelson, F.E.; Paetzold, R.F.; Outcalt, S.I. Spatial and temporal patterns of soil moisture and thaw depth at barrow, Alaska USA. In Proceedings of the

Seventh International Conference on Permafrost, Yellowknife, NT, Canada, 23-27 June 1998. Collection Nordicana No. 55.

58. Bora, P.K.; Rajput, T.B.S. Spatial and temporal variability of manning's n in irrigation furrows. *J. Agric. Eng.* **2003**, *40*, 35–42.
59. Hatch, C.E.; Fisher, A.T.; Ruehl, C.R.; Stemler, G. Spatial and temporal variations in streambed hydraulic conductivity quantified with time-series thermal methods. *J. Hydrol.* **2010**, *389*, 276–288.
60. Saskatchewan Ministry of Agriculture. Irrigation scheduling manual. November 2008. Available online: http://www.agriculture.gov.sk.ca/Default.aspx?DN=288f45ad-5bab-41b1-9878-80acf555e3bf (accessed on 19 May 2013).
61. Chen, X.; Mi, H.; He, H.; Liu, R.; Gao, M.; Huo, A.; Cheng, D. Hydraulic conductivity variation within and between layers of a high floodplain profile. *J. Hydrol.* **2014**, *515*, 147–155.
62. Cheng, C.; Song, J.; Chen, X.; Wang, D. Statistical Distribution of Streambed Vertical Hydraulic Conductivity along the Platte River, Nebraska. *Water Resour. Manag.* **2011**, *25*, 265–285.
63. Le, T.M.H.; Eiksund, G.R.; Strom, P.J. Statistical characterisation of soil porosity. In *Safety, Reliability, Risk and Life-Cycle Performance of Structures and Infrastructures*; Deodatis, G., Ellingwood, B.R., Frangopol, D.M., Eds.; Taylor & Francis Group: London, UK, 2013; pp. 5395–5402.
64. Morikawa, H.; Iiyama, K.; Ohori, M. A study of the stochastic properties of auto-correlation coefficients for micrometer data simultaneously observed at two sites. In *Applications of Statistics and Probability in Civil Engineering*; Faber, M.H., Kohler, J., Nishijima, K., Eds.; Taylor & Francis Group: London, UK, 2011; pp. 723–727.
65. Choi, M.; Jacobs, J.M. Soil moisture variability of root zone profiles within SMEX02 remote sensing footprints. *Adv. Water Res.* **2007**, *30*, 883–896.
66. Adams, J.R.; Berg, A.A.; McNairn, H. Field leve; soil moisture variability at 6- and 3-cm sampling depths: Implications for microwave sensor validation. *Vadose Zone J.* **2013**, *12*.
67. Lahkim, M.B.; Garcia, L.A. Stochastic modeling of exposure and risk in a contaminated heterogeneous aquifer, 1: Monte Carlo uncertainty analysis. *Environ. Eng. Sci.* **1999**, *16*, 315–328.
68. Clow, D.W.; Schrott, L.; Webb, R.; Campbell, D.H.; Torizzo, A.; Dornblaser, M. Groundwater occurrence and contributions to stream flow in an Alpine catchment, Colorado Front Range. *Groundw.– Watershed Issue* **2003**, *41*, 937–950.
69. Koeniger, P.; Leibundgut, C.; Stichler, W. Spatial and temporal characterization of stable isotopes in river water as indicators of groundwater contribution and confirmation of modelling results; a study of the Weser river, Germany. *Isot. Environ. Health Stud.* **2009**, *45*, 289–302.

CHAPTER 4

Highlighting the Role of Groundwater in Lake–Aquifer Interaction to Reduce Vulnerability and Enhance Resilience to Climate Change

Yohannes Yihdego [1,2,*], John A Webb [2] and Babak Vaheddoost [3]

[1] *Snowy Mountains Engineering Corporation (SMEC), Sydney, New South Wales 2060, Australia*

[2] *Environmental Geoscience, La Trobe University, Melbourne, Victoria 3086, Australia*

[3] *Hydraulic Lab., Istanbul Technical University, Istanbul 34467, Turkey*

ABSTRACT

A method is presented to analyze the interaction between groundwater and Lake Linlithgow (Australia) as a case study. A simplistic approach based on a "node" representing the groundwater component is employed in a spreadsheet of water balance modeling to analyze and highlight the effect of groundwater on the lake level over time. A comparison is made between the simulated and observed lake levels over a period of time by switching the groundwater "node "on and off. A bucket model is assumed to represent the lake behavior. Although this study demonstrates the understanding of Lake Linlithgow's groundwater system, the current model reflects the contemporary understanding of the local groundwater system, illustrates how to go about modeling in data-scarce environments, and provides a means to assess focal areas for future data collection and model improvements. Results show that this approach is convenient for getting first-hand information on the effect of groundwater on wetland or lake levels through lake water budget computation via a node representing the groundwater component. The method can be used anywhere and the applicability of such a method is useful to put in place relevant adaptation mechanisms for future water resources management, reducing vulnerability and enhancing resilience to climate change within the lake basin.

Keywords: lake–groundwater interaction; water balance; wetland; ecosystem; hydrology; climate change; adaptation

1. INTRODUCTION

When water demand exceeds water availability, water scarcity is inevitable. Climate change, population growth, and economic development add to water scarcity risks mainly in arid regions [1,2,3,4,5,6]. Concerted data collection efforts are usually lacking in developing countries, especially when it comes to groundwater systems. Many lake studies encountered difficulties in estimating groundwater or defining a plausible, appropriate conceptual model of the aquifer system at hand [7,8,9,10,11]. Prudent understanding of the (ground) water system forms a major inhibiting factor for effective water management. Water management based on such models may have unintended or even detrimental consequences.

Wetland's importance has been recognized by the Ramsar convention due to the possible impacts it may have on people in the next decades and how its conservation can ameliorate poverty conditions. Recent research has been done to assess the relationship between groundwater and surface water, due to the dependability of ecosystems on groundwater contributions [12,13,14,15,16]. The pressure on groundwater resources by these activities has alerted the interest of environmental authorities, who need to assess the hydrodynamic between wetlands, adjacent groundwater, and surface water [17, 18, 19,20,21,22,23,24,25,26,27,28,29,30,31,32,33,34,35,36,37,38,39,40,41,42 ,43,44,45,46,47,48,49,50,51,52,53,54,55,56,57,58,59,60,61,62,63,64].

Many activities have been carried out to validate the hydrogeological conceptual model in the wetland vicinity. A group of observation wells installed around the wetland has enabled the assessment through nonlinear Darcy's expression to approximate volumes of recharge and discharge from the aquifer to the wetland.

Groundwater is important for understanding lake systems due to its influence on a lake's water budget, nutrient budget, and acid buffering capacity [21,22]. As a result, groundwater flows to and from lakes have often been estimated using simple flow grids and one-dimensional Darcian calculations. Groundwater interaction with lakes can be spatially and temporally variable, however. Other water balance approaches have also been used that employed a representative groundwater head underneath the lake for calculating the flux over the entire lake area. The most sophisticated way of investigating lake–groundwater interactions is by explicitly including lakes in groundwater flow models [10,28,29,30]. Water resources managers rely on tools that assist with streamlining supply and demand [25,26,27,31,49].

Modeling lake–aquifer interaction is an essential milestone in limnological studies [32]. Groundwater is one of the most important hydrological variables of the water budget in lakes; however, this variable, due to its nature, cannot be addressed without uncertainties [32,33]. Many scientists, however, have tried to model the interaction between groundwater and surface water using various statistical, conceptual, or empirical models (e.g., Lohman [34]; Edelman [2]; Lewis et al. [35]; Post et al. [36]; Jakovovic et al. [37]; Yihdego et al. [38]).

The role of groundwater in wetland water budget is of great concern to ecologists, water managers, and environmental scientists [39]. Groundwater is critical for understanding most lake systems because it influences a lake's water budget and nutrient budget [40]. Several studies have reported on the use of a mass balance approach to simulate lake levels from hydrological and meteorological data [41,42]. There have been many different empirical, analytical, and numerical approaches for simulating lake–groundwater interactions (e.g., fixed lake stages, High-K nodes, and LAK3 package through numerical modeling). The advantages and disadvantages of these approaches have been documented by many researchers. Many highly sophisticated models like LAK Package are not widely used due to data and resource constraints [17,35,43]. The limited use of LAK Package is attributable to the lack of standardization and associated graphical user interfaces, complex three-dimensional discretization and data needs, and spatial and temporal complexity inherent in including surface water features in a groundwater model. While sophisticated approaches offer more detail, their advantages may be offset by associated complexity and even instability of the solution procedure [44]. Consequently, a full-featured LAK package is not an automatic choice for practitioners, regulators, and the wider community;

rather, the chosen method should depend on both the hydrogeological conditions and the modeling objectives [39,45,46,47,48].

Within the Glenelg Hopkins Catchment Management Authority area, Lake Linlithgow and the nearby shallow lakes are considered to be of environmental importance due to the drainage and/or degradation of many other wetlands throughout this region [32,52,53]. However, in recent years Lake Linlithgow has been affected by algal blooms and high lake salinities, and has become dry over the summer during most years since 2000. The normal seasonal fluctuation for Lake Linlithgow is 10,000 μS/cm in winter and spring, rising to 16,000 μS/cm in late autumn [60]. However, during 1999, the lowest salinity levels recorded were 27,000 μS/cm and they peaked at 58,000 μS/cm (seawater) in the autumn, before sharply increasing to 63,000 μS/cm prior to drying out in February 2000. Extensive research was carried out to characterize future changes in groundwater salinization within the basalt aquifers in the Hamilton area, including Lake Linlithgow, using hydrogeological, chemical, and isotopic techniques [14,32,50]. The aim of this study is to improve our understanding of lake–aquifer interaction through analyzing the groundwater component of Lake Linlithgow as a case study.

2. STUDY AREA

Lake Linlithgow covers an area of 9.65 km^2 and is the largest in a series of highly to moderately saline wetlands, located approximately 16 km east of Hamilton, western Victoria (Figure 1). The lake is fed by Boonawah Creek and there is no surface outlet. The catchment area for Lake Linlithgow is 85 km^2. The volcanic plains surrounding Lake Linlithgow are topographically subdued, comprising a flat to undulating plain with average elevation of ~200 m Australian Height Datum (AHD), gently increasing in elevation from south to north (Figure 1). The plain is dotted with several prominent eruption points and a number of smaller low-relief volcanic cones. The volcanic plain is deeply dissected (up to 80 m) by streams on its western and southwestern margins.

The pre-European vegetation comprised River Red Gum, Swamp Gum, Manna Gum, Blackwood, and Lightwood along moderately incised drainage lines, and Tea-tree, Silver Banksia, and Lightwood on the black self-mulching clays associated with the margins of swamps and lakes (i.e., Buckley Swamp) [55]. The poorly developed drainage lines associated with the headwaters of Grange Burn, Violet Creek, and Muddy Creek were often dominated by grasslands. Lake Linlithgow and Lake Kennedy (Figure 2) were presumably vegetated with salt-tolerant species such as plantain, Australian salt grass, and streaked arrow grass, and are relatively unaltered since European settlement.

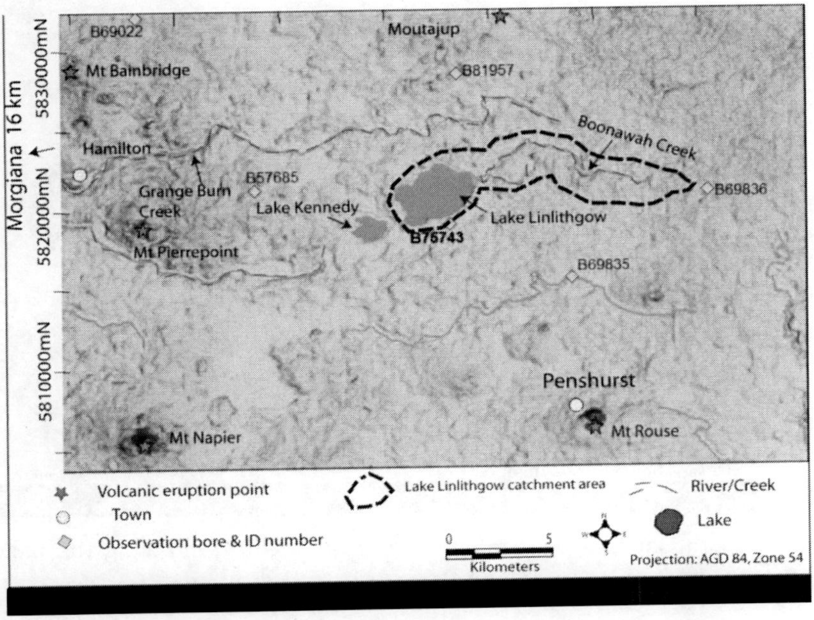

Figure 1. Digital elevation model showing the relative elevation of the terrain surrounding Lake Linlithgow.

The most significant land use change occurred between 1900 and 1920, with the conversion to introduced pasture [55].

2.1. Hydrology

The average rainfall over the catchment is about 689 mm, recorded at Hamilton Research Centre. Rainfall records over the last 45 years clearly show that rainfall is highly variable, but over the last decade there has been a substantial drop, with annual rainfall below the long-term average. Maximum rainfall is received over winter (the wettest months are July and August) and exceeds or equals evaporation for May to September, when groundwater recharge is most likely to occur. June to September rainfall contributes 45% to the mean annual precipitation in the catchment; pan evaporation is highest from October to April and totals 1053 mm.

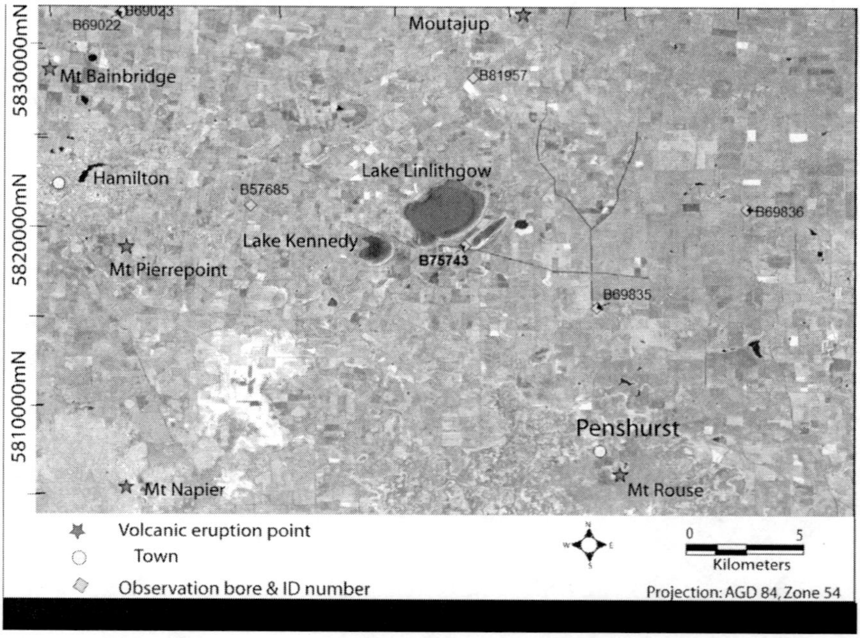

Figure 2. Landsat imagery (taken in February 2004) showing the Lake Linlithgow area.

2.2. Geology

The basement geology of the area consists of Cambrian volcanics, Early Paleozoic turbidites, and Silurian sandstone, intruded by Devonian granites (Figure 3). These basement rocks do not outcrop around Lake Linlithgow.

The disruption of the drainage system by the basalt flows formed the lakes in the center of the catchment. Lake Linlithgow sits on first phase (~4 Ma) basalts, and is encircled by second phase (~2 Ma) basalt flows, forming a relatively flat lakebed with steep banks. Lunettes occur on the eastern and northeastern margin of the lake, ~4 m above the current maximum lake level [50]. There is a local groundwater flow system around Lake Linlithgow [32,55], and a regional to intermediate groundwater flow system in the surrounding volcanic plain (Figure 4).

Potentiometric surface contours for the Newer Volcanics aquifer indicate that Lake Linlithgow is a groundwater through-flow lake, with groundwater flow entering the lake from the east and leaving towards the west (Figure 4). Groundwater enters through a series of springs and seeps along the eastern lake margin, mostly from the second phase basalt aquifer, which terminates here [50,54,55,56]. Hydrographs from bores in the basalt aquifer surrounding Lake Linlithgow (Figure 5) show a strong correlation with the lake level. The declining trend of 20 cm/year from 1997 to 2001 is clearly a response to below-average rainfall [14]. The palaeosols lying between successive basalt phases hinder outflow from the lake and act as barriers to groundwater flow (Figure 6).

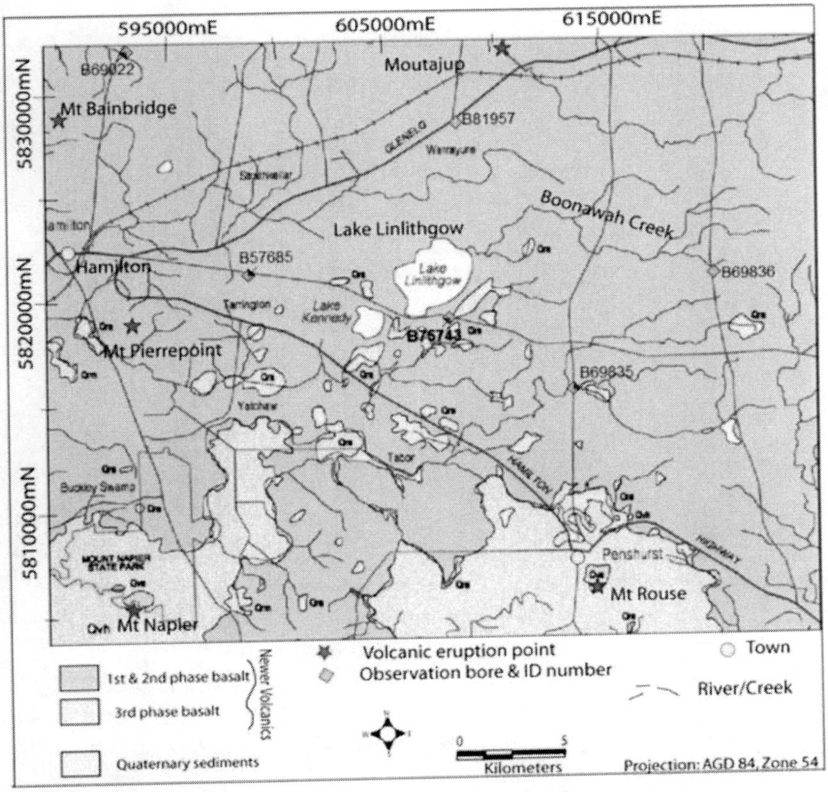

Figure 3. Geological map of the area surrounding Lake Linlithgow.

2.3. Lake Hydrology

Monthly lake level and salinity data are available for Lake Linlithgow from 1964 to 2007, when the lake dried out, although the salinity measurements are available only sporadically from 1964 to 1974 (Figure 7). Lake Linlithgow has a median depth of 1.45 m, but typically varies seasonally by up to 1 m (Figure 7). A graph of cumulative deviation of rainfall from the mean for Hamilton Research Station shows a strong correlation with lake levels; peaks in the lake level generally correspond to periods of above-average rainfall (Figure 7). In years of below-average rainfall, the lake dries out during summer, as occurred in January–April of 1983, 2000, and 2001, but after a succession of high rainfall years, lake levels rise substantially (Figure 7).

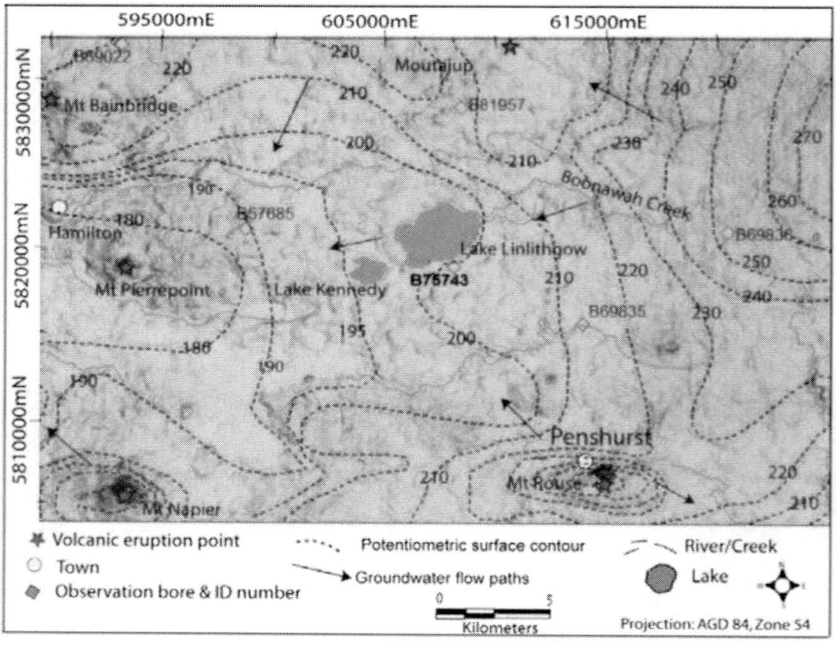

Figure 4. Lake Linlithgow and associated wetlands (gray) and basalt aquifer potentiometric surface contours (m AHD).

Lake Linlithgow is typically saline (median 12400 μS/cm). The large seasonal variation in lake level is correlated with a substantial range in lake salinity (3300–98,900 μS/cm; Figure 8). There is a general tendency for Lake Linlithgow to become more saline until the lake dries out (Figure 8).

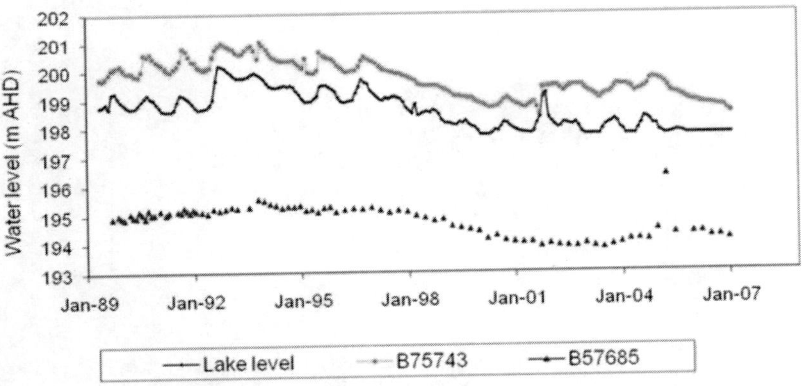

Figure 5. Comparison of bore hydrographs with the lake level. The positions of the bores are shown in Figure 4; screened depth of B57685 is from 82.5 to 88.5 m.

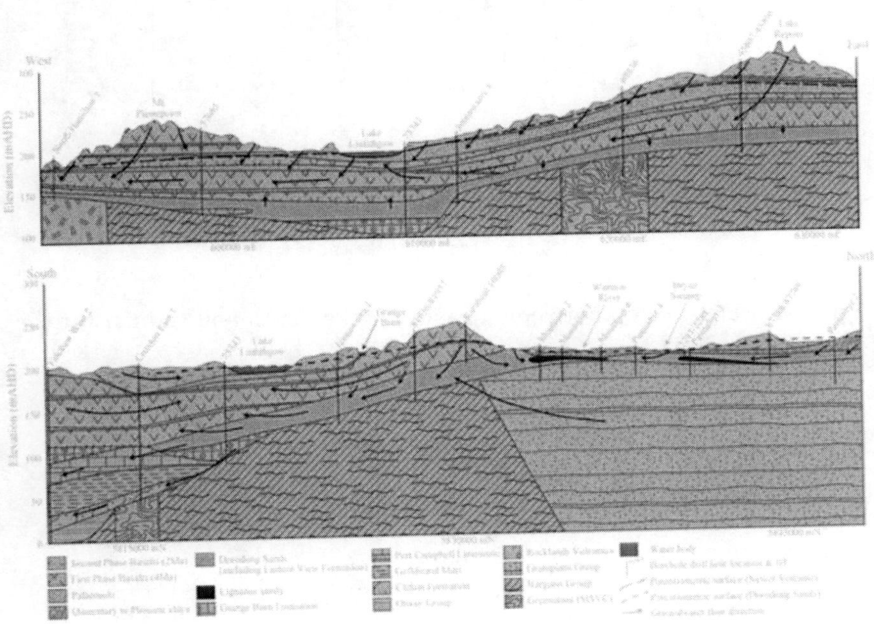

Figure 6. West–East and South–North hydrogeological cross sections through Lake Linlithgow [50,55].

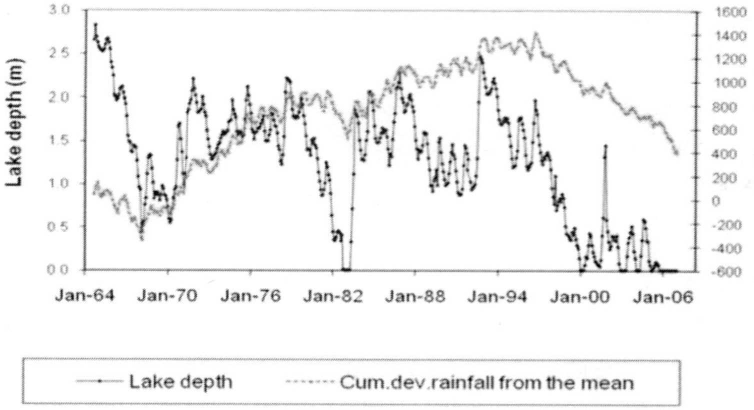

Figure 7. Lake level with variation in rainfall.

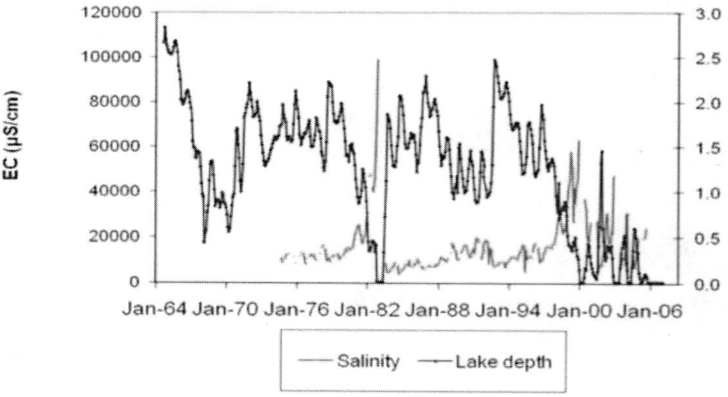

Figure 8. Relationship between measured lake depth and salinity for the period 1964–2007.

3. METHODOLOGY

A spreadsheet model was employed to analyze for the current study [32]. An explanation of the model is given below.

3.1. Water Balance Model

Monthly time-step modeling of the lake water level was carried out using Excel spreadsheets, and the resulting water budget was used as input for the salt budget. The lake water balance is calculated by estimating all the lake's water gains and losses, and the corresponding change in volume is expressed as:

$$\text{Volume change} = \text{Surface water inflow} + \text{Rainfall} + Q_{in} - \text{Evaporation} - \text{Surface water outflow} - Q_{out}, \quad (1)$$

where Q_{in} is the groundwater inflow and Q_{out} the groundwater outflow.

The net flux of the groundwater flow ($Q_{in} - Q_{out}$) can be calculated as:

$$Q = C\,(H_{lake} - H_{aquifer}) \text{ in } m^3 \cdot month^{-1} \quad (2)$$

where C is the conductance of the lakebed sediments ($m^2 \cdot month^{-1}$) and H is the water level in the lake and surrounding aquifer (m). Hydraulic conductivity can be expressed as

$$C = K \times A/L \text{ in } m^2 \cdot month^{-1} \quad (3)$$

where K is the hydraulic conductivity of the lakebed sediments ($m \cdot month^{-1}$) and A and L are the lake area (m^2) and lakebed sediment thickness (m), respectively. The formulation is similar to that used in the river (RIV) and lake packages, which specifies the flux through the riverbed or lakebed as a function of stage, potentiometric head in the connected cells, and the riverbed or lakebed conductance in which the lakebed conductance, COND, at each cell is either specified by the user in the lake package input file, or calculated from the lakebed geometry and hydraulic conductivity. As with the river package, flow from the lake to the groundwater in the LAK package is limited when the head in a cell falls below the lakebed bottom. Also, if the stage of the lake is below the top of the lakebed, the lake cell is dry and seepage into the groundwater is cut off for that cell. In this spreadsheet model, a similar procedure was applied to formulate the boundary conditions that control the solution of potentiometric head.

The temporal area is estimated from the lake stage–area–volume relationship built in [42], while the lakebed sediment is estimated using the soil erosion model of the catchment. The water level in the surrounding aquifer is updated ($H_{aquifer\text{-}new}$) using the inflow and outflow calculated for the previous month ($H_{aquifer\text{-}pre}$) is

$$H_{aquifer\text{-}pre} = Q/A \times S_y \ (m) \quad (4)$$

$$H_{aquifer\text{-}new} = H_{in\text{-}old} + H_{in} \ (m), \quad (5)$$

where A is the surface area of the interacting aquifer and S_y is the specific yield of the aquifer. The model requires known hydrometeorological data (inflow from the rivers, rainfall on the lake surface, aquifer area, and evaporation from the lake) and estimates the unknown net groundwater flux due to interaction of the lake with the surrounding aquifer by comparing the simulated and recorded lake levels and calculating a residual. The model was calibrated using solver and iteration. The net groundwater component is represented as a node in the equation to switch on and off and assess the significance of the groundwater in the overall lake water budget, as explained through the fluctuation and lake level trend over time.

3.2. Lake Water Budget and Model Parameters

3.2.1. Lake Storage

Bathymetry data are not available for the lake; however, the area–depth relationship was estimated. The lake area was measured from 12 Landsat images taken between 1972 and 2004 and correlated with the measured lake depth in the month when the images were taken (Figure 9). The lake area was estimated using ENVI software (with threshold method and grow button); this is more accurate than the GIS (vector) method used by [50] because it finds similar pixels selected for the water body and thereby better identifies the natural boundary. The line of best fit (with $r^2 = 0.99$) to the area/depth data gives the relationship for Lake Linlithgow as:

$$A(t) = 1.545(D(t))^5 - 23.09(D(t))^4 + 135.9(D(t))^3 - 393.7(D(t)^2 + 561.1(D(t) - 305.8. \quad (6)$$

The polynomial relationship between lake area (A(t)) and lake depth (D(t)) is due to the lake's relatively flat lakebed and steep banks, and gives a better fit for the depth–area relationship than the logarithm function used by [50], which had a smaller r^2 value.

The lake volume at the beginning of a given month can therefore be calculated from the depth and area at the end of the preceding month. The lake depth must be first adjusted, because the minimum reading on the base of the lake level gauge is ~1.3 m. Thus, this value is deducted from the recorded lake level to get the true lake depth. For the modeling, the initial volume in September 1964 was calculated from the measured water depth in that month (2.79 m).

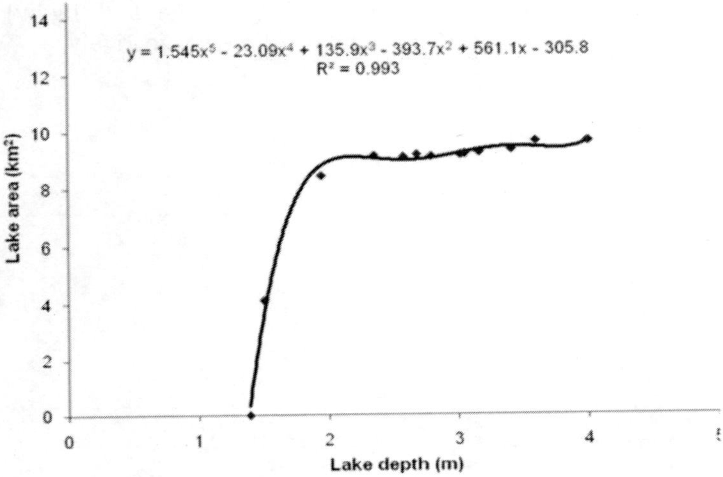

Figure 9. Relationship between lake area derived from Landsat imagery and measured lake depth.

3.2.2. Precipitation on Lake

Monthly precipitation is taken from the nearest rainfall station (Hamilton Research Centre), which lies approximately 13 km southwest of Lake Linlithgow, and is multiplied by the lake area to give the volume of direct rainfall into the lake. The precipitation data are only available up until 2001; values for the subsequent time period have been extrapolated from the rainfall at Carinya station (~65 km from Hamilton Research Centre) using the correlation between rainfall at Hamilton Research Centre and Carinya.

3.2.3. Surface Flow to the Lake

Surface inflow to Lake Linlithgow is received through Boonawah Creek, for which there are no gauging data available. Surface inflows can, however, be estimated by the *tanh* cumulative surplus rainfall approach [42] using the Grange Burn flow, which is gauged at Morgiana (gauge no. 238219; Figure 1), because the Grange Burn and Lake Linlithgow catchments are adjacent and have similar topography, soil type, vegetation, land use, and rainfall. The tanh cumulative surplus rainfall approach provides a tool with which runoff in Grange Burn can be predicted for any given monthly precipitation/evaporation (Figure 10). Using area scaling, this can be converted to a flow in Boonawah Creek;

the catchment areas for Boonawah Creek (at Lake Linlithgow) and Grange Burn at Morgiana are 85 km^2 and 997 km^2, respectively.

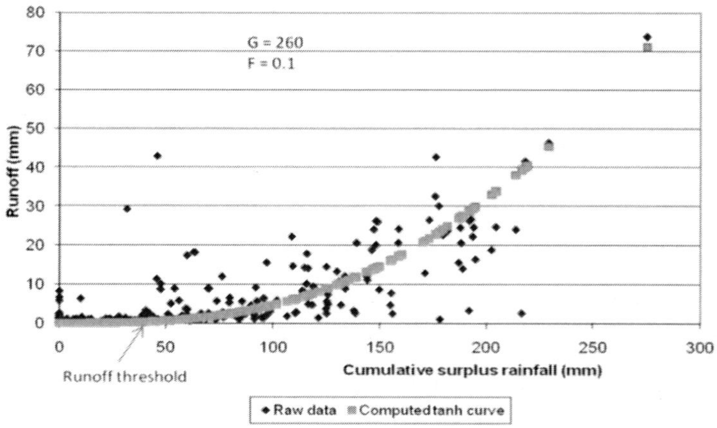

Figure 10. Stream flow modeling for the Grange Burn at Morgiana (gauge no. 238219). The location of the gauge station is shown in Figure 1.

3.2.4. Outflow from the Lake

The evaporation from Lake Linlithgow is estimated using monthly evaporation data at Hamilton Research Station, where pan evaporation was measured from 1968 to June 2000. The evaporation data from July 2000 has been extrapolated from White Swan Reservoir station by establishing a correlation between pan evaporation at Hamilton Research Centre and White Swan Reservoir. Evaporation data from 1964 to 1968 are lacking at White Swan Reservoir, so evaporation for this period was estimated from the correlation of evaporation at the Hamilton Research Centre with that at the Melbourne regional office. A local calibration coefficient was used to adjust the seasonal pan evaporation data from Hamilton Research Centre for the best fit of the model. The optimized local calibration coefficient is 1.18; this takes into account the spatial variation in position, elevation, and storage effect between Hamilton Research Centre and Lake Linlithgow.

3.2.5. Groundwater Inflow and Outflow Estimation

Groundwater inflow/outflow was initially estimated using Darcy's Law. The widths of the groundwater inflow and outflow zones along the lake perimeter are ~6.1 km and 3.6 km respectively. The cross-sectional area is calculated by multiplying the width of the groundwater inflow/outflow zone by the saturated thickness (8 m) of the first phase basalt aquifer in hydraulic contact with the lake. The average hydraulic conductivity of the basalt aquifer (0.09 m/day) is derived from groundwater flow rates calculated using groundwater radiocarbon ages by Bennetts [50]. The hydraulic gradient either side of the lake, derived directly from the potentiometric contours, is 3.7×10^{-3} (Figure 4). Similar to Lake Burrumbeet, an average value of 0.000864 m/d has been chosen for the permeability of the lake floor in order to estimate the groundwater outflow through the lakebed. Using these figures, the monthly groundwater outflow and inflow from/to the lake were estimated as ~0.29 ML and ~0.48 ML, respectively.

Model calibration, carried out by adjusting input parameters so that the simulated lake levels fit the observed lake level, gives the net groundwater flow estimation. In this case the optimized lake–aquifer conductance (C) value is $2.71 \times 10^3 m^2$/day, which is equivalent to a hydraulic conductivity value of lake shore sediments of about 7 m/day; this falls within the range of previous hydraulic conductivity estimates for these sand beaches. The optimized specific yield and aquifer area are 0.1 and 45 km^2, respectively [32].

The optimized specific yield compares reasonably with previous specific yield estimates of the basalt aquifer, which range up to 0.18 [51]. The aquifer area in hydraulic contact with the lake (45 km^2) is significant compared with the catchment area of Lake Linlithgow (85 km^2), indicating that lake level fluctuations directly impact about half of the basalt aquifer in the catchment area.

The best possible fit of the water budget model was attained at ~0.37×10^6 L and 3.1×10^6 L monthly average groundwater outflow and inflow, with the exception of dry periods. The value of optimized groundwater outflow compares closely with the initial estimate.

4. RESULTS AND DISCUSSION

4.1. Lake Water Levels

The predicted lake levels show good agreement with the measured lake level data (Figure 11), with an r^2 value of ~0.85 (Figure 12). The sum of the squared differences did not exceed 0.4 m, except for a few outliers (Figure 13).

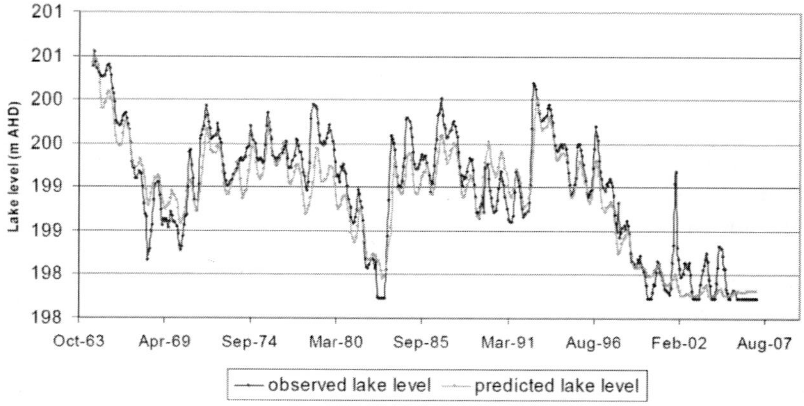

Figure 11. Lake Linlithgow measured water levels and a comparison to modeled results for the period 1964–2007.

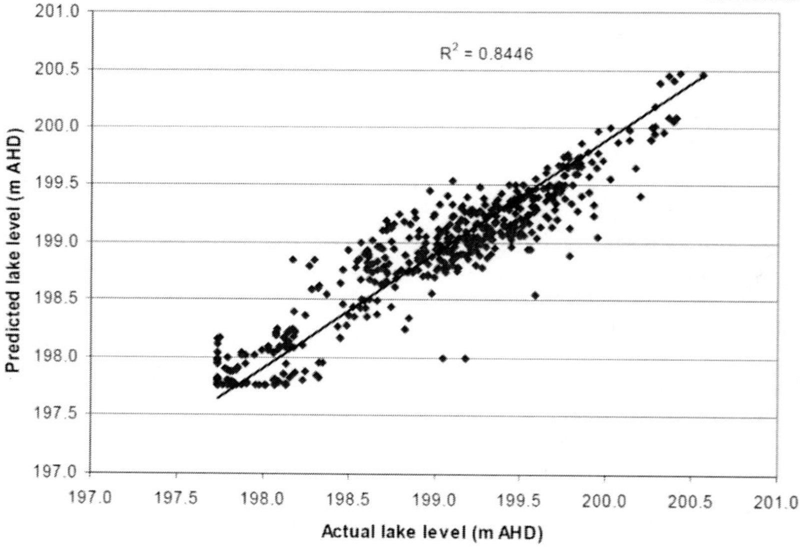

Figure 12. Correlation between observed and calculated lake elevation in AHD m.

The water balance shows that the major influence on lake levels is evaporation (Table 1), accounting for an average of 54% of the total water budget. It has the greatest influence in summer, reaching up to 99% of the total budget, but decreased during dry periods, because it is proportional to the lake area [32]. Groundwater inflow contributes about 1% of the lake water budget. Even though groundwater outflow is minor (0.1%), it dominates the lake water losses during early winter (June), a time when evaporation is very small (Figure 14).

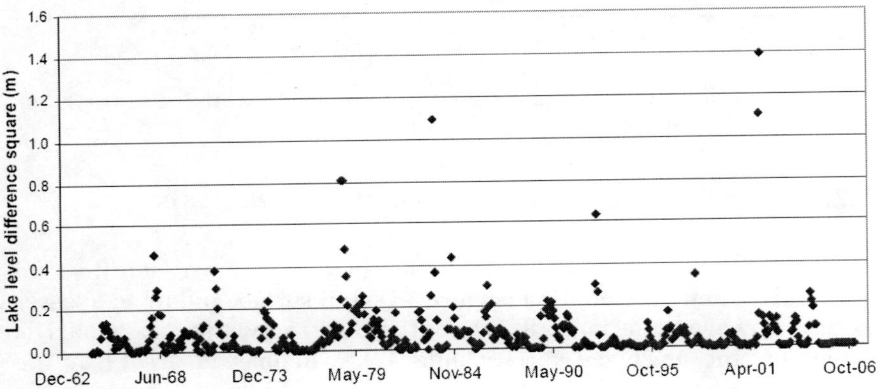

Figure 13. Temporal distribution of square difference between observed and calculated lake levels.

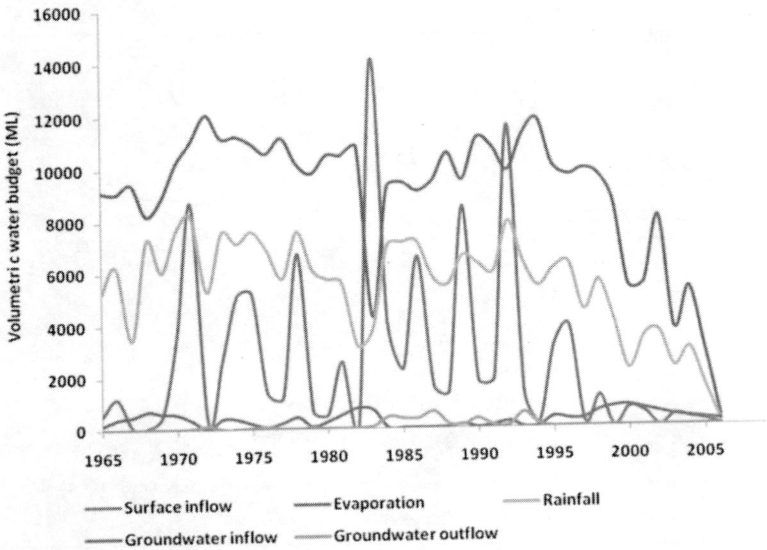

Figure 14. Annual water balance of Lake Linlithgow (1965–2006).

Table 1. Average long-term (1964–2007) monthly contribution (in percent) of each Lake Linlithgow water budget component to the overall lake water budget.

Evaporation (%)	Groundwater Outflow (%)	Precipitation (%)	Surface Inflow (%)	Groundwater Inflow (%)
54	0.1	37	8	1

Thus, Lake Linlithgow is a groundwater through-flow lake, and the lake–groundwater wetland.

4.2. Water Budget Errors and Sensitivity Analysis

A sensitivity analysis shows that the model is most sensitive to evaporation and precipitation (Figure 15), and surface inflow to a lesser extent. However, the most likely source of error within the model is estimating the ungauged surface inflow (i.e., Boonawah Creek) by the tanh relationship from Grange Burn; this could be in error if the threshold value at which cumulative surplus rainfall becomes runoff in Grange Burn at Morgiana is different to that of the smaller catchment of Boonawah Creek.

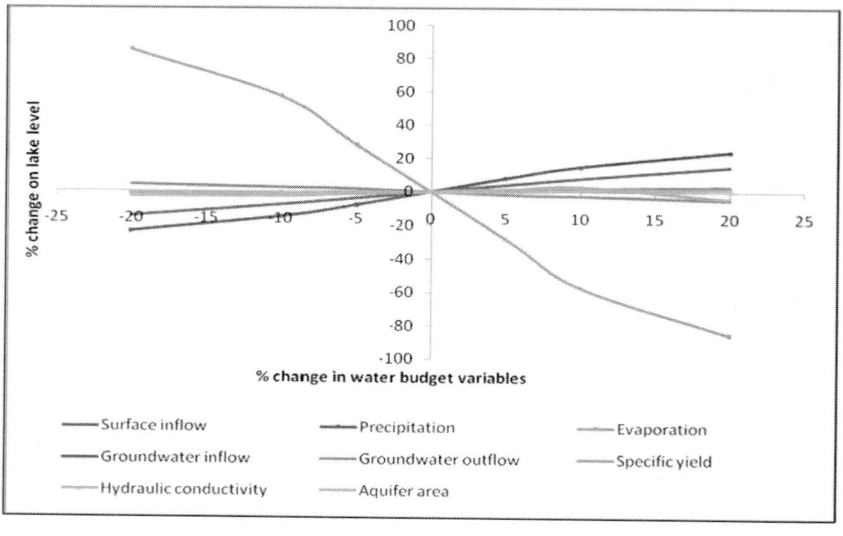

Figure 15. Relative sensitivity of Lake Linlithgow water balance model to changes in the water budget components.

4.3. Interpretation

To assess the effect of groundwater component, which is the main objective of this paper, further analysis was carried out through a second model run without a groundwater component. The calculated water levels followed the same trend as observed lake levels, but were on average lower than the observed values. The lower calculated lake level implies that the total observed lake storage is lower than would be expected if the lake did not have groundwater inflow (Figure 16).

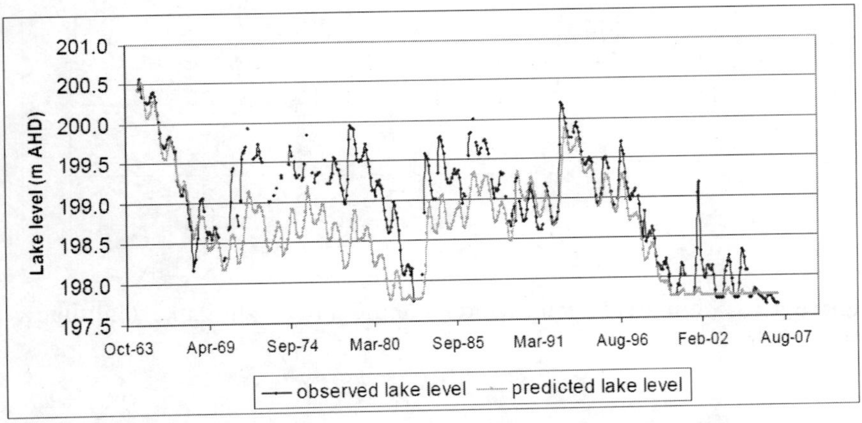

Figure 16. Observed and simulated lake level (deactivating the groundwater node from the model).

The effect of groundwater is evident from the graph. The simulated volumetric component of the groundwater of the lake water budget is shown in Figure 17. The base flow is estimated (G = 260) (Figure 10) using the tanh cumulative surplus rainfall approach. This simplistic model seems a reliable and modest tool to see the lake–groundwater relationship at a glance and will be useful if constrained by the mass balance and reliable estimate of parameters [9,32,51]. This approach could be adopted for much wetland management. This model will give insight into the groundwater–surface water relationship and guide our future data gathering from sensitive parameters. The parameters used are small and convenient for management practice.

The need for study of the interaction of lakes and groundwater stems from the fact that groundwater is commonly ignored or is the residual term in lake water. Because all variables in a lake water budget are rarely measured, it is impossible to adequately evaluate the errors or the residual.

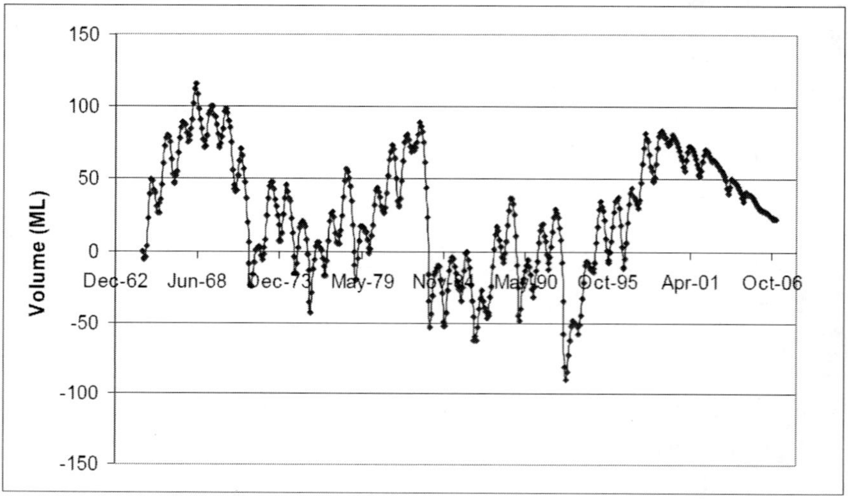

Figure 17. Volumetric groundwater component in the Lake Linlithgow water budget.

Even if groundwater is included as part of a lake water balance study, improper placement of wells can lead to a misunderstanding of the interaction between lakes and groundwater. No matter how many wells are used to define the water table in the settings, the maps shows a gradient towards the lake, and there would be no way to detect the out-seepage that occurs. If only one or a few wells are placed near a lake, the groundwater flow system would not be adequately defined and would be subject to misinterpretation [16,55,58].

Over a long period of falling lake levels from 1968 up to the drought period 1982/3, the simulated lake levels are more than one meter lower than the observed levels due to the lack of groundwater inflow. However, there is a good match between the simulated and actual lake levels during a period of rising lake levels, indicating that the groundwater outflow from Lake Linlithgow recharging the aquifer is very small. When the groundwater component is added to the model, it accurately follows the observed levels (Figure 11).

Groundwater is an important component of the water balance of Lake Linlithgow, and its flow is influenced by the lake level. If the lake level rises, groundwater outflow and therefore lake recharge into the surrounding aquifer will increase. If the lake level decreases, groundwater discharge from the aquifer into the lake will increase. This interaction causes inertia in the lake–groundwater system, delaying reactions to external (meteorological) stresses (Figure 17). This phenomenon can be shown using the lake water balance model. If the model is run with no groundwater components, it overshoots after periods of rise or recession (Figure 16).

5. CONCLUSIONS

This paper aims at an understanding of the groundwater system in wetlands, via a case study at Lake Linlithgow, Australia.

When the model was run based on surface water balance components (via switching the "node" off, i.e., in the absence of groundwater), there was a progressive separation between observed and calculated lake levels. The calculated levels imply that the lake should accumulate more storage than is actually observed. This separation could not be attributed to systematic errors in surface runoff, precipitation, and evaporation measurements. Rather, it is an indication of subterranean water fluxes (i.e., groundwater).

The results of such analysis lead to a better understanding of the groundwater system within a lake/wetland ecosystem.

Uncertainties associated with the groundwater component of the lake water budget increase unpredictability and unreliability and further complicate the future management of wetlands. No matter how many wells are used to define the water table in the settings, the groundwater flow system would not be adequately defined and would be subject to misinterpretation. The present method gives a better way of studying surface water–groundwater interaction, which is in great demand in the current strategy for better wetland and integrated water resources management. The result underscores the need to put in place relevant adaptation mechanisms to reduce vulnerability and enhance resilience to climate change within the lake basin.

ACKNOWLEDGMENTS

We would like to thank the anonymous reviewers. The manuscript has benefitted from the reviewers' and editors' comments.

AUTHOR CONTRIBUTIONS

Y.Y. and J.W. conceived and designed the research; Y.Y. performed the research; Y.Y. and J.W. analyzed the data; B.V. contributed analysis tools; Y.Y. wrote the paper.

CONFLICTS OF INTEREST

The authors declare no conflict of interest.

REFERENCES

1. Dausman, A.M.; Doherty, J.; Langevin, C.D.; Dixon, J. Hypothesis testing of buoyant plume migration using a highly parameterized variable-density groundwater model. *Hydrogeol. J.* **2010**, *18*, 147–160.
2. Edelman, J.H. *Groundwater Hydraulics of Extensive Aquifers (No. 13)*; International Livestock Research Institute, Wageningen University: Wageningen, The Netherlands, 1983.
3. Yihdego, Y.; Drury, L. Mine water supply assessment and evaluation of the system response to the designed demand in a desert region, central Saudi Arabia. *Environ. Monit. Assess. J.* **2016**, *188*, 619.
4. Yihdego, Y.; Drury, L. Mine dewatering and impact assessment: Case of Gulf region. *Environ. Monit. Assess. J.* **2016**, *188*, 634.
5. Yihdego, Y.; Paffard, A. Predicting open mine inflow and recovery depth in the Durvuljin soum, Zavkhan Province, Mongolia. In *Mine Water and the Environment*; Springer: Berlin/Heidelberg, Germany, 2016; pp. 1–10.
6. Yihdego, Y.; Paffard, A. Hydro-engineering solution for a sustainable groundwater management at a cross border region: Case of Lake Nyasa/Malawi basin, Tanzania. *Int. J. Geo-Eng.* **2016**, *7*, 23.
7. Christensen, S.; Doherty, J. Predictive error dependencies when using pilot points and singular value decomposition in groundwater model calibration. *Adv. Water Resour.* **2008**, *31*, 674–700.
8. Tonkin, M.; Doherty, J. Calibration-constrained Monte Carlo analysis of highly-parameterized models using subspace techniques. *Water Resour. Res.* **2008**, *45*, 12.
9. Yihdego, Y.; Webb, J.A. Hydrogeological constraints on the hydrology of Lake Burrumbeet, southwestern Victoria, Australia. In *21st VUEESC Conference, Victorian Universities Earth and Environmental Sciences Conference*; Hagerty, S.H., McKenzie, D.S., Yihdego, Y., Eds.; Abstracts No 88, 12; Geological Society of Australia: Sydney, Australia, 2007.
10. Yihdego, Y. Three Dimensional Groundwater Model of the Aquifer around Lake Naivasha Area, Kenya. Master's Thesis, Department of Water Resources, ITC, University of Twente, Enschede, The Netherlands, March 2005.
11. Yihdego, Y.; Webb, J.A. Characterizing groundwater dynamics in Western Victoria, Australia using Menyanthes software. In Proceedings of the 10th Australasian Environmental Isotope Conference and 3rd Australasian Hydrogeology Research Conference, Perth, Australia, 1–3 December 2009.

12. Rosenberry, D.O.; Lewandowski, J.; Meinikmann, K.; Nützmann, G. Groundwater-the disregarded component in lake water and nutrient budgets. Part 1: Effects of groundwater on hydrology. *Hydrol. Process.* **2015**, *29*, 2895–2921.
13. Scibek, J.; Allen, D.M.; Cannon, A.J.; Whitfield, P.H. Groundwater–surface water interaction under scenarios of climate change using a high-resolution transient groundwater model. *J. Hydrol.* **2007**, *333*, 165–181.
14. Yihdego, Y.; Webb, J.A. Modeling of bore hydrographs to determine the impact of climate and land use change in a temperate subhumid region of southeastern Australia. *Hydrogeol. J.* **2011**, *19*, 877–887.
15. Yihdego, Y.; Webb, J.A.; Leahy, P. Response to Parker: Rebuttal: ENGE-D-13-00994R2 "Modelling of lake level under climate change conditions: Lake Purrumbete in south eastern Australia". *Environ. Earth Sci. J.* **2016**, *75*, 1–4.
16. Yihdego, Y.; Webb, J.A. Assessment of wetland hydrological dynamics in a modified catchment basin: Case of Lake Buninjon, Victoria, Australia. *Water Environ. Res. J.* **2017**, *89*, 144–154.
17. Doherty, J.; Hunt, R.J. Response to comment on "Two statistics for evaluating parameter identifiability and error reduction". *J. Hydrol.* **2010**, *380*, 489–496.
18. Doherty, J.; Johnston, J.M. Methodologies for calibration and predictive analysis of a watershed model. *J. Am. Water Resour. Assoc.* **2003**, *39*, 251–265.
19. Yihdego, Y. Drought and Pest Management Initiatives (Book chapter 11). In *Handbook of Drought and Water Scarcity (HDWS): Vol. 3: Management of Drought and Water Scarcity*; Eslamian, S., Eslamian, F.A., Eds.; Francis and Taylor, CRC Group: Burlington, MA, USA, 2016.
20. Yihdego, Y. Drought and Groundwater Quality in Coastal Area (Book chapter 15). In *Handbook of Drought and Water Scarcity (HDWS): Vol. 2: Environmental Impacts and Analysis of Drought and Water Scarcity*; Eslamian, S., Eslamian, F.A., Eds.; Francis and Taylor, CRC Group: Burlington, MA, USA, 2016.
21. Lewandowski, J.; Meinikmann, K.; Nützmann, G.; Rosenberry, D.O. Groundwater–the disregarded component in lake water and nutrient budgets. Part 2: Effects of groundwater on nutrients. *Hydrol. Process.* **2015**, *29*, 2922–2955.
22. Shaw, G.D.; White, E.S.; Gammons, C.H. Characterizing groundwater–lake interactions and its impact on lake water quality. *J. Hydrol.* **2013**, *492*, 69–78.
23. Yihdego, Y.; Al-Weshah, R. Hydrocarbon assessment and prediction due to the Gulf War oil disaster, North Kuwait. *J. Water Environ. Res.* **2016**.
24. Yihdego, Y.; Al-Weshah, R. Gulf war contamination assessment for optimal monitoring and remediation cost –benefit analysis, Kuwait. *Environ. Earth Sci.* **2016**, *75*, 1–11.
25. Gallagher, M.R.; Doherty, J. Predictive error analysis for a water resource management model. *J. Hydrol.* **2007**, *34*, 513–533.
26. Hunt, R.J.; Doherty, J.; Tonkin, M.J. Are models too simple? Arguments for increased parameterization. *Ground Water***2007**, *45*, 254–262.
27. Yihdego, Y.; Eslamian, S. Drought Management Initiatives and Objectives (Book chapter 1). In *Handbook of Drought and Water Scarcity (HDWS): Vol. 3: Management of Drought and Water Scarcity*; Eslamian, S., Eslamian, F.A., Eds.; Francis and Taylor, CRC Group: Burlington, MA, USA, 2016; p. 41.
28. Shaw, G.D.; Mitchell, K.L.; Gammons, C.H. Estimating groundwater inflow and leakage outflow for an intermontane lake with a structurally complex geology: Georgetown Lake in Montana, USA. *Hydrogeol. J.* **2016**, 1–15.
29. Skahill, B.; Doherty, J. Efficient accommodation of local minima in watershed model calibration. *J. Hydrol.* **2006**, *329*, 122–139.
30. Tonkin, M.; Doherty, J. A hybrid regularised inversion methodology for highly parameterized models. *Water Resour. Res.* **2005**, *41*, W10412.
31. Yihdego, Y. Data visualization tool as a framework for groundwater flow and transport models. In *Proceedings of the International Conference on "MODFLOW and more 2013", Translating Science into Practice*; Integrated Groundwater Modelling Centre (IGWMC), Colorado School of Mines, University: Golden, CO, USA, 2013.

32. Yihdego, Y.; Webb, J.A. Modelling of seasonal and long-term trends in lake salinity in south-western Victoria, Australia. *J. Environ. Manag.* **2012**, *112*, 149–159.
33. Gallagher, M.R.; Doherty, J. Parameter estimation and uncertainty analysis for a watershed model. *Environ. Model. Softw.* **2006**, *22*, 1000–1020.
34. Lohman, S.W. *Ground-Water Hydraulics*; US Government Printing Office: Washington, DC, USA, 1972; p. 70.
35. Lewis, M.A.; Cheney, C.S.; ÓDochartaigh, B.E. *Guide to Permeability Indices*; CR/06/160N; British Geological Survey, Natural Environmental Council: London, UK, 2006.
36. Post, V.; Kooi, H.; Simmons, C. Using hydraulic head measurements in variable-density ground water flow analyses. *Ground Water* **2007**, *45*, 664–671.
37. Jakovovic, D.; Werner, A.D.; de Louw, P.G.; Post, V.E.; Morgan, L.K. Saltwater upcoming zone of influence. *Adv. Water Resour.* **2016**, *94*, 75–86.
38. Yihdego, Y.; Reta, G.; Becht, R. Hydrological analysis as a technical tool to support strategic and economic development: Case of Lake Navaisha, Kenya. *Water Environ. J.* **2016**, *30*, 40–48.
39. Yihdego, Y.; Reta, G.; Becht, R. Human impact assessment through a transient numerical modelling on The UNESCO World Heritage Site, Lake Navaisha, Kenya. *Environ. Earth Sci.* **2017**, *76*, 9.
40. Yihdego, Y.; Webb, J.A. Validation of a model with climatic and flow scenario analysis: Case of Lake Burrumbeet in Southeastern Australia. *Environ. Monit. Assess.* **2016**, *188*, 1–14.
41. Jones, R.N.; McMahon, T.A.; Bowler, J.M. Modelling historical lake levels and recent climate change at three closed lakes, Western Victoria, Australia (c.1840–1990). *J. Hydrol.* **2001**, *246*, 159–180.
42. Yihdego, Y.; Webb, J.A. An empirical water budget model as a tool to identify the impact of land-use change on stream flow in southeastern Australia. *Water Resour. Manag.* **2012**, *27*, 4941–4958.
43. James, S.C.; Doherty, J.; Eddebarh, A.A. Post-calibration uncertainty analysis: Yucca Mountain, Nevada, USA. *Ground Water* **2009**, *47*, 851–869.
44. Tonkin, M.; Doherty, J.; Moore, C. Efficient nonlinear predictive error variance analysis for highly parameterized models. *Water Resour. Res.* **2007**, *43*, 7.
45. Doherty, J.; Skahill, B. An advanced regularization methodology for use in watershed model calibration. *J. Hydrol.***2006**, *327*, 564–577.
46. Doherty, J. Groundwater model calibration using pilot points and regularization. *Ground Water* **2003**, *41*, 170–177.
47. Moore, C.; Doherty, J. The cost of uniqueness in groundwater model calibration. *Adv. Water Resour.* **2006**, *29*, 605–623.
48. Yihdego, Y.; Becht, R. Simulation of lake–aquifer interaction at Lake Navaisha, Kenya using a three-dimensional flow model with the high conductivity technique and a DEM with bathymetry. *J. Hydrol.* **2013**, *503*, 111–122.
49. Yihdego, Y.; Al-Weshah, R. Engineering and environmental remediation scenarios due to leakage from the Gulf War oil spill using 3-D numerical contaminant modellings. *J. Appl. Water Sci.* **2016**.
50. Bennetts, D. Hydrology, Hydrogeology and Hydro-Geochemistry of Groundwater Flow Systems within the Hamilton Basalt Plains, Western Victoria, and Their Role in Dry Land Salinisation. Ph.D. Thesis, Department of Earth Sciences, La Trobe University, Melbourne, Australia, September 2005.
51. Yihdego, Y.; Webb, J.A. Use of a conceptual hydrogeological model and a time variant water budget analysis to determine controls on salinity in Lake Burrumbeet in southeast Australia. *Environ. Earth Sci.* **2015**, *73*, 1587–1600.
52. Yihdego, Y.; Webb, J.A.; Leahy, P. Modelling of lake level under climate change conditions: Lake Purrumbete in southeastern Australia. *Environ. Earth Sci.* **2015**, *73*, 3855–3872.
53. Yihdego, Y.; Webb, J.A.; Leahy, P. Modelling water and salt balances in a deep, groundwater-throughflow lake—Lake Purrumbete, southeastern Australia. *Hydrol. Sci. J.* **2016**, *61*, 186–199.

54. Yihdego, Y. *Modelling of Lake Level and Salinity for Lake Purrumbete in Western Victoria, Australia*; A Co-Operative Research Project between La Trobe University and EPA Victoria; EPA Victoria: Melbourne, Australia, 2010.
55. Yihdego, Y. Modelling Bore and Stream Hydrograph and Lake Level in Relation to Climate and Land Use Change in Southwestern Victoria, Australia. Ph.D. Thesis, Faculty of Science, Technology and Engineering, Melbourne, La Trobe University, Melbourne, Australia, May 2010.
56. Yihdego, Y.; Webb, J.A. Characterizing groundwater dynamics using Transfer Function-Noise and auto-regressive modelling in Western Victoria, Australia. In Proceedings of the 5th IASME/WSEAS International Conference on Water Resources, Hydraulics and Hydrology (WHH '10), Cambridge, UK, 23–25 February 2010.
57. Yihdego, Y.; Al-Weshah, R. Assessment and prediction of saline sea water transport in groundwater using using 3-D numerical modelling. *Environ. Process. J.* **2016**.
58. Winter, T.C. Effects of water table configuration on seepage through lakebeds. *Limnol. Oceanogr.* **1981**, *26*, 925–934.
59. Yihdego, Y. Evaluation of Flow Reduction due to Hydraulic Barrier Engineering Structure: Case of Urban Area Flood, Contamination and Pollution Risk Assessment. *J. Geotech. Geol. Eng.* **2016**, *34*, 1643–1654.
60. Al-Weshah, R.; Yihdego, Y. Modelling of Strategically Vital Fresh Water Aquifers, Kuwait. *Environ. Earth Sci.* **2016**, *75*, 1315.
61. Yihdego, Y.; Danis, C.; Paffard, A. 3-D numerical groundwater flow simulation for geological discontinuities in the Unkheltseg Basin, Mongolia. *Environ. Earth Sci. J.* **2015**, *73*, 4119–4133.
62. Yihdego, Y.; Webb, J.A. Characterizing groundwater dynamics in Western Victoria, Australia using Menyanthes software. In Proceedings of the 10th Australasian Environmental Isotope Conference and 3rd Australasian Hydrogeology Research Conference, Perth, Australia, 1–3 December 2009.
63. Yihdego, Y.; Webb, J.A. Modelling of Seasonal and Long-term Trends in Lake Salinity in Southwestern Victoria, Australia. *J. Environ. Manag.* **2012**, *112*, 149–159.
64. Yihdego, Y. Engineering and enviro-management value of radius of influence estimate from mining excavation. *J. Appl. Water Eng. Res.* **2017**.

CHAPTER 5

Assessment of the Influence of Rainfall and River Discharge on Sediment Yield in the Upper Tana Catchment in Kenya

Imelda N Njogu[1*] and Johnson U Kitheka[1]

[1]*School of Water Resources Science and Technology, South Eastern Kenya University, Kitui, Kenya*

ABSTRACT

Sediment yield in the Upper Tana Basin in Kenya has implications on the sustainability of Hydro-Electric Power (HEP) dams and water resources development projects. Therefore, a study was undertaken in the basin to establish the extent to which rainfall and river discharges influence the sediment yield in the catchment. The study was based on hydrological data obtained from the Water Resources Management Authority (WARMA) and Kenya Meteorological Department (KMD). The river discharge data was obtained from three RGS Maragua (4BE01), Gura (4AD01) and Tana Sagana (4BC02) and rainfall data was obtained from Sagana Fish Farm and Nyeri Ministry of Works for the of period 1960-2013.The study also applied the Soil Water Assessment Tool (SWAT) Model to determine the extent to which the model can be used to simulate streamflow and sediment yield in the basin. The results of the study showed that there is a significant variability in streamflow and sediment yield in the Upper Tana Basin. In the period between 1960 and 2015, the mean total annual river discharge of Tana Sagana was 128 m3s-1, and the maximum and minimum river discharges were 29.94 m3s-1 and 3.15 m3s-1, respectively. There was an indication of increasing trend in rainfall and subsequently sediment yield in the basin, which may be attributed to alteration of land use and climatic change. The results showed that SWAT model was quite good in simulating the variability of river discharge. The analysis revealed a poor relationship between sediment yield and rainfall. However, the relationship between rainfall and stream flow was strong with r value of 0.9 which is significant at p=0.05. Relationship between simulated and observed river discharge had a R2 of 0.442, r of 0.665 and NSE of -89.43. The relationship between simulated and observed sediment yield had a R2 of 0.733, r of 0.86 and NSE of 0.69. The results of this study showed that SWAT model can be used to predict sediment yield in the Upper Tana catchment. The model had good performance when daily rainfall, stream flow and sediment yield data were used. Thus, the model can be used to establish the relationship between rainfall, discharge, and sediment yield in a highly human-impacted tropical catchment area. The study puts also forward various recommendations on land and water resources management in the basin.

Keywords: SWAT model; Upper tana; Sediment yield; Rainfall; Discharge

1. INTRODUCTION

Sediment yield in the Upper Tana Basin in Kenya is influenced by many factors among them landuse, vegetation cover, river discharge and rainfall variability [1]. The influences of rainfall and river discharge are particularly important but these are usually complicated by changes in land use and vegetation cover, both seasonally and inter-annually. As a result, the amount of sediment delivered by the river to the hydroelectric power dams located downstream is highly variable. While the influence of land use and vegetation cover change has received a lot of attention in past studies, this cannot be said to be so in case of the influence of rainfall and river discharge variability. In this regard, the assessment of the relationship between river discharge and rainfall and sediment yield is important since it is these two variables that shows significant variability as a result of climate change and whose future trends will have major implications on the dams located in the Upper Dam basin. There is still gap in our understanding of the factors influencing the observed sediment yield in the Upper Tana Basin. In particular, the extent to which rainfall variability and hence climate change influences sediment yields in the basin has not been fully established, as is true in many other similar tropical river basins in

Africa. This has been attributed to lack of data. Therefore, to fill the above gaps, we carried out an analysis of the relationship between rainfall, river discharge and sediment yield in some of the key streams draining the Upper Tana Basin in Kenya. This study compliments other studies on sediments yield in the Tana Basin [2-8].

Previous studies carried out on sediment yield in the upper Tana basin have not addressed the relationship between rainfall and discharge and sediment yield. Maingi, et al. [7] examined the hydrologic impacts following construction of dams along the Tana River and Dunne, et al. [3] came up with an approach for estimating the sedimentation in the Upper Tana catchment. Mango, et al. [1] used a calibrated model to explore the potential impacts of continued land use and future climate change in the upper Tana catchment. Dunne T [9] study noted that land use was the main factor that influenced sediment yield. Archer D [10] study in Nairobi region concluded that the sedimentation rates in the reservoirs was high in areas with rainfall amounts ranging between 1000 and 1600 mm and runoff ranging between 350 and 700 mm. Due to geographical differences, land use, climatic and socioeconomic differences, Archer D [10] generalization for Nairobi region further to the south cannot be applied to the Upper Tana catchment. This study on the Upper Tana catchment is important since it provides information on measures that can be used to mitigate the effects of landuse change in specific areas of the basin which are considered to be the main sources of sediments in the Tana river [11]. By providing an analysis on sediment yield in the basin, the study can contribute in the sustainable management of the Upper Tana Basin and the dams that have been constructed downstream e.g Masinga, Kamburu, Gitaru, Kiambere and the planned Grand Falls dam [8].

1.1 Description of the upper Tana basin

The Upper Tana catchment in Central Kenya, is the main catchment area for the Tana river-the largest river in Kenya (**FIGURE 1**). The subbasin covers a surface area of about 12,500 km^2 with elevation ranging from 400 to 5,199 m above sea level. The main rivers in the catchment are Sagana, Thiba, Maragua, Mathioya, Chania, Nyamindi, Chania, Rupingazi, Ena, Tunga and Gura [12]. Five major reservoirs are found in the lower reaches of the sub-basin (**FIGURE 2**), namely Kindaruma (completed in 1968), Kamburu, Gitaru, Masinga and Kiambere. These HEP reservoirs combined provide approximately three quarters of electricity in Kenya. Recent studies have shown that they have modified the flow of the Tana river, particularly in downstream reaches of the river up to the Tana Delta [8]. The Upper Tana catchment experiences two rainy seasons every year as a result of the migration of the Inter- Tropical Convergence Zone (ITCZ). The long rain season lasts from March to June, and the short rains from September to

December. There is a great variance in the rainfall patterns in Central Kenya highlands. The average annual rainfall ranges between 400 mm and 2300 mm [13]. The basin is one of the main water sources of the Tana River basin-the largest and most important river system in Kenya covering basin area of 5,950 km^2. The Tana river basin covering 17% of the total land area in Kenya contributes to 27% of the total mean discharge in Kenya.

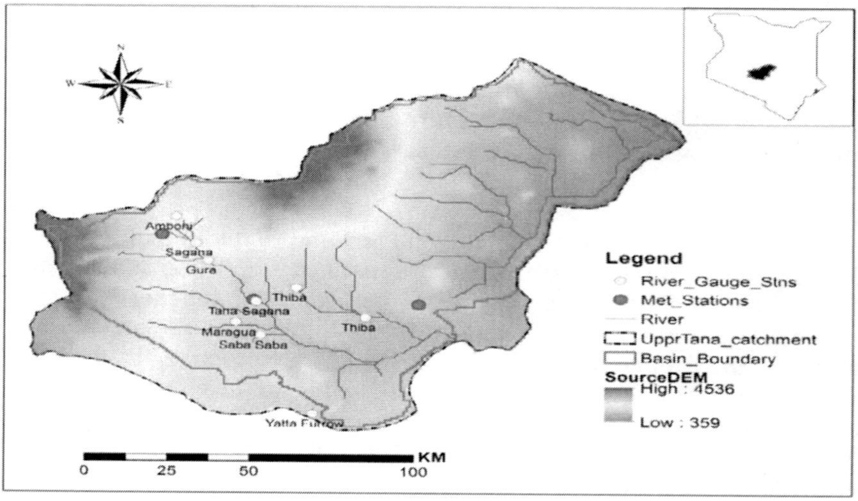

Figure 1: location of the River Gauging Stations (RGS) and rainfall stations in the Upper Tana Basin.

The Upper Tana sub-basin vegetation consists of (i) the forests in the Aberdares conservation area, (ii) the middle zones consisting of farming areas and (iii) the lower drier grazing zone. In the conservation area, vegetation is determined by rainfall distribution and temperature. The forest vegetation in the Aberdares is divided into four categories, namely: the wet evergreen forests; dry evergreen forests; *Juniperus podocarpus*/Olive forests; and low altitude shrubs. The 10 most common species of trees in the three forest reserves of the Aberdares Conservation Area are *Nuxia congesta, Juniperus procera, Olea europaea, Podocarpus latifolius*, and *Neboutonia macrocalyx* [14,15].

The land use in the Upper Tana catchment can be divided into three main classes, namely (i) natural vegetation (forest, grassland, and wetlands), (ii) rain-fed and irrigated agriculture (tea, coffee, maize and cereals) and (iii) Rangeland. The catchment also includes different agro-ecological zones, which corresponds to the different land use types [16]. The population in the Upper Tana catchment is approximately 3.1 million people. The largest urban centres are: Thika, Sagana, Karatina, Murang'a and Nyeri. The population density which is as high as 300 people/km^2 declines with elevation partly due to decreasing rainfall and soil fertility [17]. Most of the people in the basin rely heavily on farming as well as the associated agro-industries which are sources of employment opportunities. The main crops grown include coffee, tea, potatoes, pyrethrum, maize, rice, and bananas.

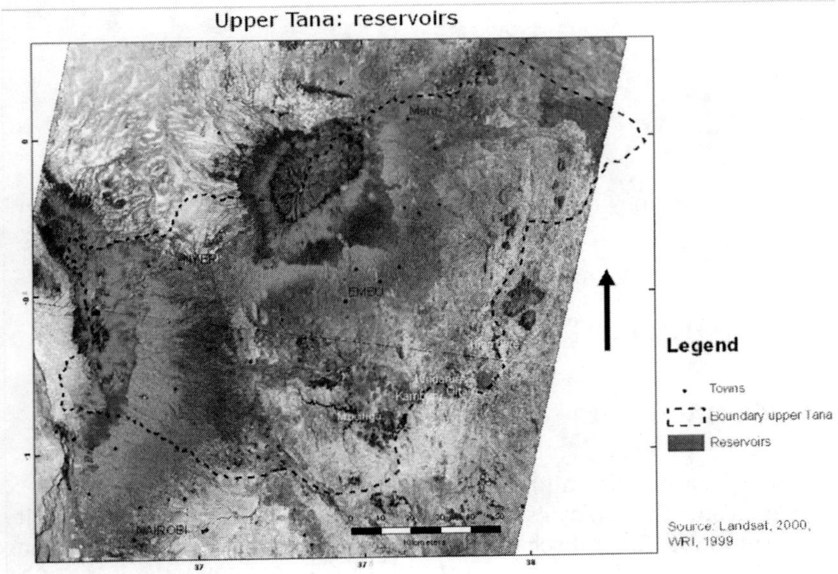

Figure 2: Reservoirs in the Upper Tana catchment.

The soils in the sub-basin are dominated by the humic nitisols which have formed as result of the volcanic deposits on the high-altitude zones. The Nitisol soils are however highly vulnerable to erosion where soil conservation measures are not applied [18]. The other soil types in the catchment include vertisols, cambisols, andosols ferralsols and leptisols. The geology of the area (**FIGURE 3**) is characterized by the volcanic rocks of the Cainozoic era, and metamorphic rocks of the Mozambique belt [13]. Mt. Kenya, an extinct volcano formed between 100-4000 million years ago is located in the west of the catchment and is a source of most of the rivers in the Tana basin [19].

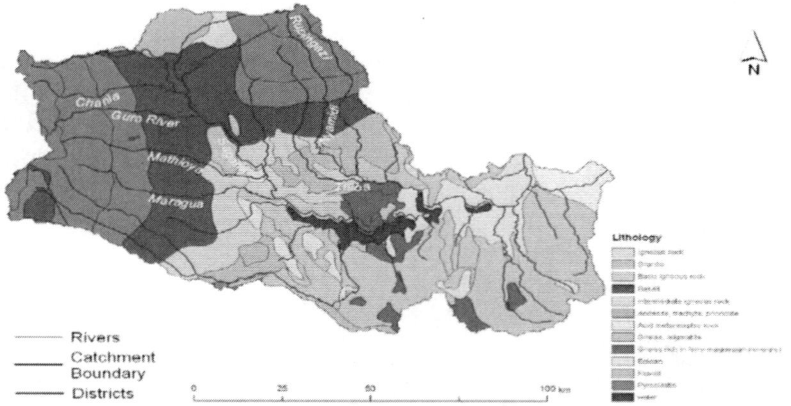

Figure 3: The geology and drainage patterns in the Upper Tana catchment [12].

2. METHODOLOGY

2.1 Streamflow data

Secondary data on the rainfall, river discharge and sediment yield were obtained from the Kenya's Water Resources Management Authority (WRMA) at Embu. River discharge data were obtained from 8 River Gauging stations (RGS), namely Amboni (4AB05), Sagana (4AC03), Gura (4AD01), Tana Sagana (4BC02), Maragua (4BE01), Thiba 1 (4DD02), Saba Saba (4BF02) and Thiba 2 (4DA10) (**FIGURE 1**).

2.2 Rainfall data

The rainfall data was obtained from Sagana Fish Farm and Nyeri Ministry of Works Stations that had more consistent records without gaps. Other stations which data was provided by the WARMA included Kiritiri Chiefs Camp (Embu), Sagana State Lodge, Nyeri Met Station, and Meru Forest Station.

2.3 Sediment data

There is a huge gap in terms of availability of sediment data in the Upper Tana basin. Sediment yield data were therefore not available in most of the

River Gauging Stations with the exception of the Sagana River Gauging station where continuous data was available for the period 1957-1980. Theother sediment yield data for this station were obtained from WRMA for the following RGS- Tana Sagana (4AC03), Gura (4AD01), Maragua (4BE01) and Sagana- Grandfalls. The Sagana- Grandfalls station had continous data for the period 1957 to 1980. The data on the total suspended sediment concentrations (TSSC) was only available for the period between 2010 and 2011 for Maragua (RGS 4BE01), Mathioya (RGS 4BDNEW), Saba Saba (RGS 4BF01) and Sagana (RGS 4BE10).

2.4 Data processing and analysis

The data obtained for this study was subjected to time-series analysis to show the trends in rainfall, sediment yield and river discharge in the period between 1957 and 1990 and in some instances, up to 2015.The flow duration curves and mass curves were plotted for rainfall and river discharge data for Sagana fish farm and Gura RGS, respectively. This was done in order to check the accuracy of the SWAT model and also for detecting shifts in rainfall and river discharges in the basin.

2.5 The soil water assessment tool (SWAT) model

The Soil Water Assessment Tool (SWAT) Model was used to simulate flow rates of the Sagana river which is the main branch emanating from the Upper Tana Basin. The SWAT model has been used in other parts of the world [20-23]. The model dealt with two parameters in the input stage rainfall and discharge of the river. The output of the model was plotted in Excel to show the relationship between measured data and simulated model data on rainfall, sediment yield and discharge. Various quantitative methods used included the measures of central tendency (mean, range), regression analysis, correlation and coefficient of determination and Nash Sutcliffe Efficiency (NSE). The SWAT model was built using the Arc-Map interface (i.e Arc SWAT) which provided the suitable means to enter data into the SWAT code. The SWAT Model deals with soil and water parameters, with primary data inputs being the Digital Elevation Model (DEM), land use and soils shape files. Modelling allowed quantification of soil erosion processes at non-gauged areas and during periods when measurements were absent [24]. The Model was calibrated using data for the period between 1981 and 2010. Model validation was done using datasets from the Gura river gauging station (1980-1985). For modeling purposes, the Upper Tana watershed was partitioned into a number of sub-basins. Input information for each sub-basin was then grouped or organized into the following categories: climate and hydrologic response units (HRUs). The first step in creating a SWAT model involved the delineation of the sub-watersheds in the Upper Tana basin for which each of them is treated as individual units. The sub-basins were further divided into hydrologic response units (HRUs) which had homogenous land use practices, soil type and management practices. The Nash-Sutcliffe efficiency (NSE) was used to as a normalized statistic to determine the relative magnitude of the residual variance compared to the measured data variance [25]. NSE indicates how well the plot of observed versus simulated data fits the 1:1 line. Servat, et al. [26] noted that NSE provides the best objective function for reflecting the overall fit of a hydrograph. The NSE equation used is presented in equation 1:

$$NSE = 1 - \left[\sum_{i=1}^{n} \frac{\left(y_i^{obs} - y_i^{sim}\right)^2}{\left(y_i^{obs} - y_i^{mean}\right)^2} \right] \quad (1)$$

Where,

Y_i^{obs}-i^{th}=observation for the constituent being evaluated

Y_i^{sim}-i^{th}=simulated value for the constituent being evaluated

Y^{mean}=mean of the observed data for the constituent being evaluated

N=total number of observations.

The Nash Sutcliffe Efficiency ranges from -∞ to +1, where the acceptable levels of performance are the values greater or equal to 0.0 to 1. If the NSE number is less than 0.0 it indicates that the mean observed value is a better predictor than the simulated value, which indicates unacceptable performance of the model.

The long-term calibration was validated with inflow data from the Gura River Gauging station that was found to have the least data gaps. A monthly dataset was available from 1981-2010. The erosion rates and sediment yields were calibrated using the model values that were compared with the model predictions to calibrate the soil erosion and sediment routing parameters of the model. The mode fine-tuning was done using the data available from the various RGS and meteorological stations to improve the accuracy of the model parameters and the output. The long-term calibration spurned a 30-year period in which the basin has changed considerably in terms of land use and in terms of infrastructure roads, small-scale hydraulic works, and diversions.

2.6 Statistical data analysis

2.6.1 Multiple regression models presented in equations 2 and 3 were used in this study, respectively.

$$Y=ax_1^2+bx_2+c \tag{2}$$

$$Y=a+bX+e \tag{3}$$

In above equations,

Y=Sediment Yield (tons. $month^{-1}$)

a and b=the coefficients value of X variables

X_1=River discharge (m^3 $month^{-1}$)

X_2=Rainfall (mm.$month^{-1}$)

c=Constant.

The regression equation was deemed to be significant when the p-value is less or equal to 0.05. The correlation analysis was based on the Pearson's correlation coefficient (r) that was used to describe the degree of collinearity between simulated and measured data according to equation 4.

$$r=\frac{n=\left(\sum xy\right)-\left(\sum x\right)\left(\sum y\right)}{\sqrt{\left[n\sum x^2-\left(\sum x\right)^2\right]\left[n\sum y^2-\left(\sum y\right)^2\right]}} \tag{4}$$

In above equation,

r=Correlation coefficient

y=Dependent variable (e.g. Sediment Yield)

x=Independent variable (e.g. River Discharge)

n=Total number of values.

The coefficient of determination R^2 describing the proportion of the variance in measured data explained by the model was computed using equation 5 [27].

$$R^2=\sqrt{r} \tag{5}$$

Where R^2=coefficient of determination

r=Pearson's correlation coefficient

R^2 ranges from 0 to 1, with higher values indicating less error variance, and typically, values greater than 0.5 are considered acceptable [28,29]. Analysis of Variance (ANOVA) was used to determine whether there exist significant differences between two or more variables at a selected probability level. One-way analysis of variance was used in testing hypothesis at 95% confidence level. The null hypothesis was rejected when the critical F>F and when the p-value (x)>0.05.

3. RESULTS

3.1 Seasonal and inter-annual variability of river discharges

The stream flow in the rivers draining the Upper Tana Basin showed significant seasonal and inter-annual variabilities. The high stream flows are experienced during the two rainy seasons, namely long rain season (March, April, and May) and short rain season (October, November, and December). The low stream flows are experienced during the dry seasons (June, July, August, September and January, February). The streamflows also show significant inter-annual variabilities. The maximum river discharges for Amboni (RGS 4AB05), Maragua (RGS 4BE01), Gura (RGS 4AD01) and Tana Sagana (RGS 4BC02) were 4.84 m^3s^{-1}, 33.25 m^3s^{-1}, 29.94 m^3s^{-1} and 59.26 m^3s^{-1}, respectively. The mean river discharges for Amboni (RGS 4AB05), Maragua (RGS 4BE01), Gura (RGS 4AD01) and Tana Sagana (RGS 4BC02) were 1.5 $m^3s^{-1}1$, 14.3 m^3s^{-1}, 19.7 m^3s^{-1} and 21.2 m^3s^{-1}, respectively. Tana Sagana had relatively higher discharge since it represents the main tributary of the Tana draining from the northwest Upper Tana Basin. The flow duration curves of the rivers showed that the high magnitude flows >80 m^3s^{-1} are experienced in <10% of the time, while river discharges <10 m^3s^{-1}occur in 80% of the time (**FIGURE 4**).

3.2 Seasonal and inter-annual variability of sediment yield

The data on the total suspended sediment concentrations (TSSC) for four RGS for the period between 2010 and 2011 showed existence of significant seasonal and inter-annual variability of TSSC. Maragua river (RGS 4BE01) had a minimum, maximum and mean TSSC of 3 mgl^{-1}, 0.517 gl^{-1} and 0.065gl^{-1}, respectively. Mathioya (4BDNEW) had a minimum, maximum and mean of 3 mgl^{-1}, 0.258 gl^{-1} and 0.043 gl^{-1} respectively. Saba Saba (4BF01) had a minimum, maximum and mean of 0.010 gl^{-1}, 1.433 gl^{-1} and 0.243 gl^{-1} respectively. Sagana (4BE10) had a minimum, maximum and mean of 0.03 gl^{-1}, 0.50 gl^{-1} and 0.057 gl^{-1}, respectively. The TSSC ranged from 0.03 to 0.50

gl^{-1} in Sagana river in the period between 2010-2011. The highest TSSC were recorded in the Sagana river which is the main branch of the Tana draining from the northwest Upper Tana Basin. As with TSSC, Sediment yield in the Upper Tana Basin showed both seasonal and inter-annual variabilities. The data for the period between 1955 and 1995, showed that the variability of sediment yield is related to the variability of rainfall and subsequently river discharges (**FIGURE 5**).

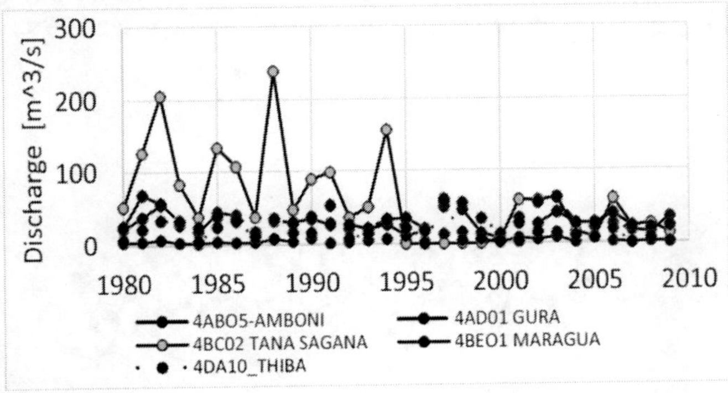

Figure 4: Variability of river discharge at Amboni and Gura River Gauging Stations (RGS) in the Upper Tana Basin in the period between 1980 and 2009.

The computation of sediment yield provided a maximum yield of 39,050 $ton.yr^{-1}$ in 1961 and >50,000 $ton.yr^{-1}$ in 2015. The 1961 sediment yield data are much lower than the 2015 yield, probably pointing to an increase in sediment production in the basin. There is thus an indication of increased sediment yield in the basin. The sediment yields for Maragua, Mathioya, Saba Saba and Sagana were 1,365,603 $ton.yr^{-1}$, 798,365 ton.yr-1, 216,210 $ton.yr^{-1}$ and 2,782,547 $ton.yr^{-1}$, respectively. Previous studies in the basin by Dunne, et al. [3] reported sediment yields ranging from 883,000 $ton.yr^{-1}$ and 2,302,000 $ton.yr^{-1}$, respectively. From the results of sediment yield from different sub-basins, it seems that Maragua and Mathioya are the main sources of sediment in the Upper Tana basin.

3.3 Seasonal and inter-annual variations of rainfall

The rainfall trend shows normal year-to year variability although there is an indication of declining rainfall trend. As with rainfall, the river discharges show normal year to year variations that are related to the variability in rainfall. The sediment yields also showed significant inter-annual variations. Rainfall in the Upper Tana Basin based on data obtained at Nyeri Public Works station and Sagana Fish Farm station showed significant inter-annual variations with an indication of declining trend in the period between 1980 and 2010 (**FIGURE 5**). There is also an indication of an increase in the degree of variability indicating an increase in the frequency and magnitude of extreme rainfall events. It is possible that the declining trend in rainfall observed in the Upper Tana Basin is a result of climate variability or land use change. It is however difficult at this stage to determine exactly what is the main driver. While rainfall shows a declining trend, the degree of variability seems to have increased after 1990.

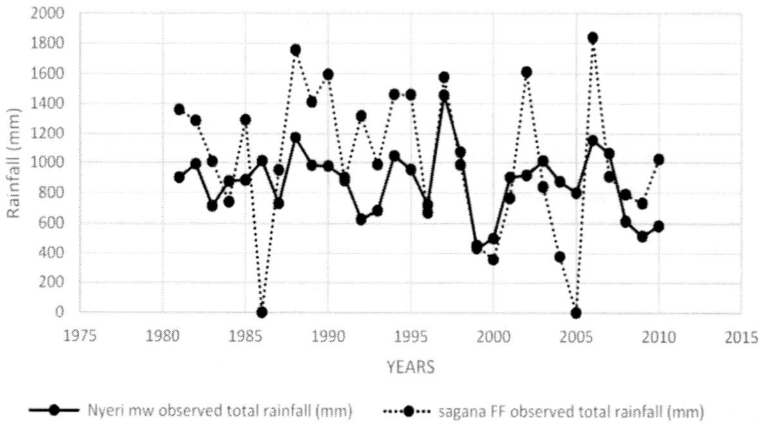

Figure 5: Inter-annual variations of sediment yield, rainfall, and river discharge in the Upper Tana in the period between 1957 and 1990.

The relationship between rainfall and streamflow in the basin was significant with correlation coefficient r of 0.99 and coefficient of determination R^2 of 0.98. This indicates that variations in rainfall explains 98% of the variations in river discharges in the Upper Tana Basin. This would also indicate rapid basin response with limited baseflow contributions to streamflow. This could be a result of land degradation occasioned by increased destruction of catchment areas for agriculture and settlements (**FIGURE 6**).

3.4 Land-use change in the upper tana basin

There has been significant change in land-use in the Upper Tana Basin when one compares the situation during the pre-colonial and post-colonial period. Major significant changes in land-use begun in 1920s following introductions of agriculture in Central Kenya Highlands by European settlers [30]. In the last 100 years, there has been major changes in land use and vegetation cover in the basin. The analysis of satellite images and aerial photographs showed that the forest cover has reduced by more than 50%. Agriculture and settlements have increased by more than 100%. The population in the Upper Tana Basin has increased from <250,000 people in 1969 to >2 million people in 2015 [30]. High population density and intensive cropping in the region leave the soils susceptible to soil erosion. Most of the cultivation is done without the application of appropriate soil erosion protection measures and the cultivation on marginal lands having steep slopes or high erodibility have increased (**FIGURE 7**).

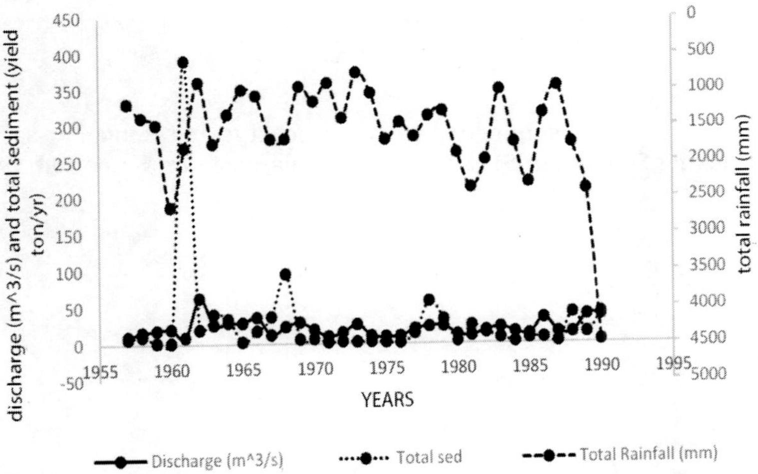

Figure 6: Inter-annual variations of total annual rainfall at Nyeri and Sagana stations in the period between 1981 and 2010.

3.5 Simulation of stream flow and sediment yield

The streamflow in the Upper Tana was modeled using SWAT model using data obtained at Gura River Station for the period between 1980 and 2010. The results of simulation of river discharges are shown in **FIGURE 8**. The relationship between simulated and observed river discharge yielded a correlation coefficient r of 0.67 with R^2 value of 0.44. The NSE index for the relationship was 89 showing the model simulated 89% of the streamflows (**FIGURE 9** and **10**).

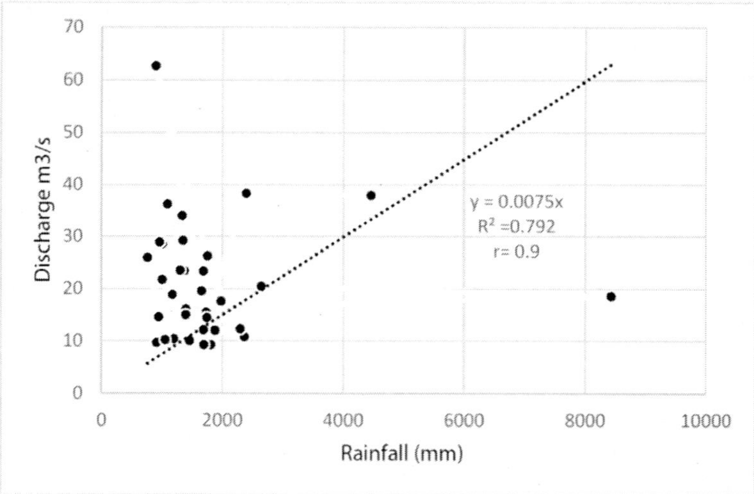

Figure 7: The relationship between river discharge and rainfall at Sagana River Gauging Station (RGS) in the Upper Tana Basin in the period 1981-2010.

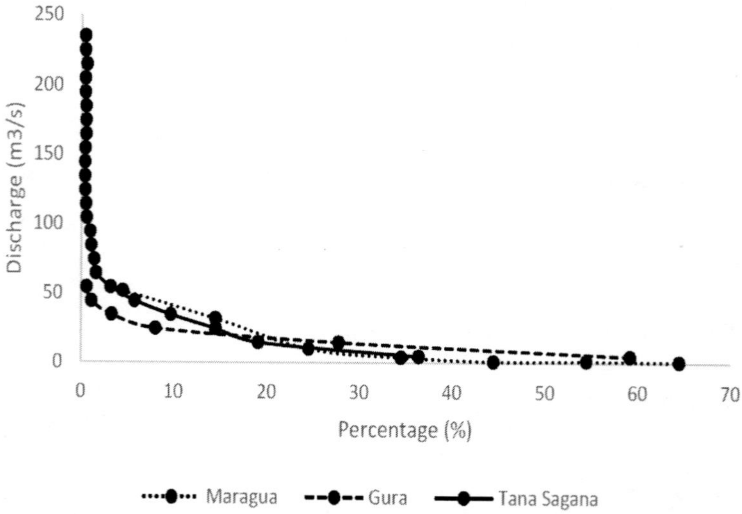

Figure 8: Flow duration curve for Maragua River Gauging Station (RGS) in the Upper Tana Basin based on river discharge data for the period 1981-2010.

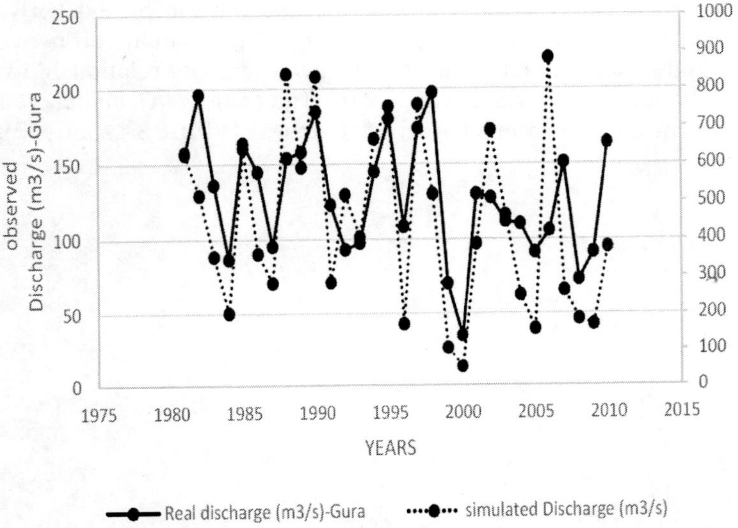

Figure 9: The variations of the simulated and observed total river discharge at Gura River Gauging Station (RGS) in the period 1980-2010.

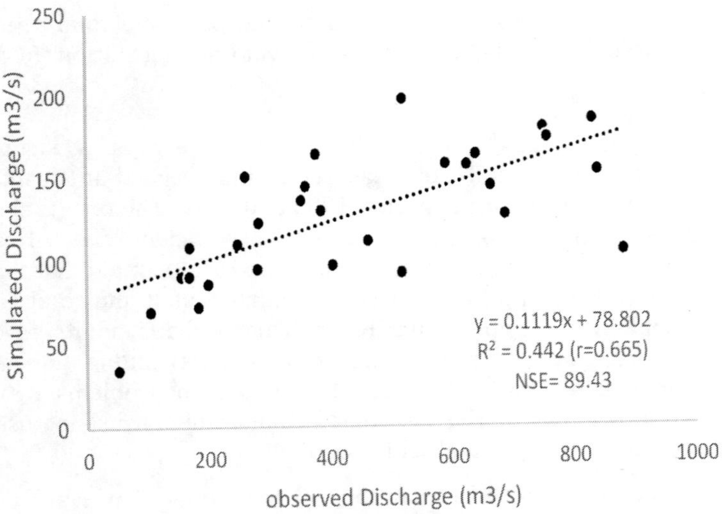

Figure 10: The relationship between the simulated and observed river discharge at Gura River Gauging Station (RGS) in the period between 1981 and 2010.

The sediment yields were simulated using data for the period between 1980 and 2010 at Sagana River Gauging station. The comparison between simulated and observed sediment yield showed that the relationship was weak with r value of 0.02 with R^2value<0.01. Thus, the SWAT model did not simulate the measured sediment yield satisfactorily (**FIGURE 11** and **12**).

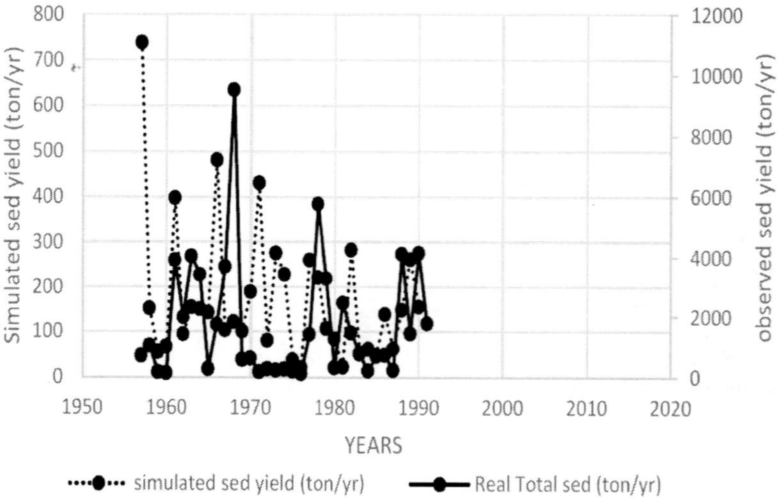

Figure 11: The variations of the simulated and observed total sediment yield for the period of 1981-2010 at Sagana River Gauging Station (RGS).

An attempt was made to establish whether there is a significant relationship between rainfall and measured sediment yield at Sagana RGS in the Upper Tana Basin, for the period 1980 and 2010. The results showed a relatively weak relationship with a correlation coefficient (r) of 0.16 and a coefficient of determination (R^2) of 0.026. These results imply that there is no significant relationship between total annual rainfall and total annual sediment yield in the Upper Tana Basin. This perhaps points to other factors that are playing an important role in determining the annual sediment yield in the basin. It could also be due to problems with the quality of data or due to the fact that actual rainfall was not used to establish the link with the actual sediment yield.

The suspended sediment concentrations in the Upper Tana vary widely with as low as 0.03 gl^{-1} and values greater than 1.40 gl^{-1}. Maragua river (RGS 4BE01) had a minimum, maximum and mean of 0.03 gl^{-1}, 0.517 gl^{-1} and 0.065 gl^{-1}, respectively. Mathioya river (4BDNEW) hada minimum, maximum and mean of 0.03 gl^{-1}, 0.258 gl^{-1} and 0.043 gl^{-1}, respectively. Saba

Saba river (4BF01) had a minimum, maximum and mean of 0.010 gl^{-1}, 1.433 gl^{-1} and 0.243 gl^{-1}, respectively. Sagana river (4BE10) had a minimum, maximum and mean of 0.03 gl^{-1}, 0.500 gl^{-1} and 0,057 gl^{-1}, respectively. These river gauging stations Maragua, Mathioya, Saba Saba and Sagana had sediment yields of 1,365,603 $ton.yr^{-1}$, 798,365 $ton.yr^{-1}$, 216,210 $ton.yr^{-1}$ and 2,782,547 $ton.yr^{-1}$, respectively.

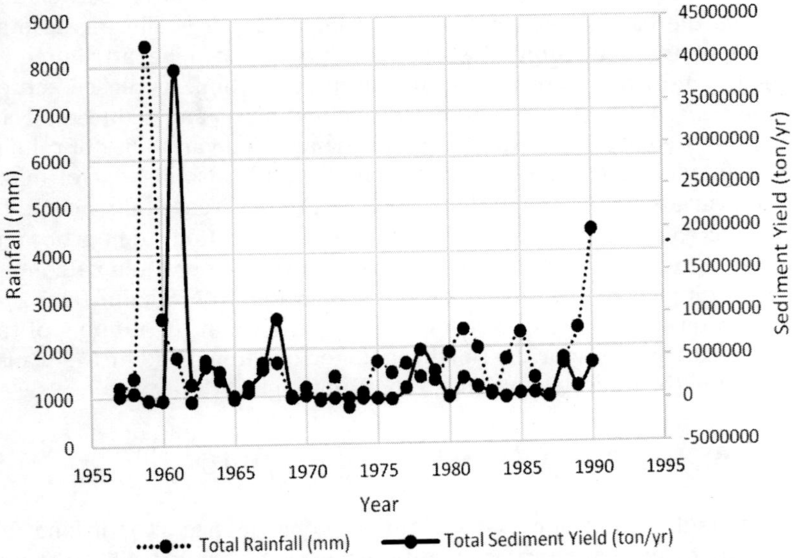

Figure 12: The variations of the observed total rainfall and observed total sediment yield for the period 1955-1995 at Sagana River Gauging Station (RGS).

The results of regression analysis showed that there is no significant relationship between rainfall and river discharge (p=0.9). The same can be said to be true for the relationship between river discharge and sediment concentration, and that between river discharge and sediment yield. In other words, there is no indication that an increase in rainfall will cause a corresponding increase in stream flow. The weak relationship between river discharge and sediment yield and TSSC could be attributed to the time lag between the variables.

4. DISCUSSION

4.1 Stream flow and rainfall

The discharge of the main rivers found in the Upper Tana Basin showed an increase in the range of low and high flows, indicative of large variation in stream flow. For instance, in Tana Sagana river, the high flows reached 239 m^3s^{-1} while the base flows were of the order 0.12 m^3s^{-1}. There is also an indication of the reduction of stream flow which can be attributed to changes in land-use practices (e.g. increased cultivation, increased settled areas), increased deforestation of the catchment areas and reduced glacial cover on Mt. Kenya as a result of climate change. The variability of rainfall in the basin which has showed great variability could also be a factor in the increased variability of stream flow. However, the decreased flow could be due to increased abstraction of water (for irrigation, domestic and municipal water supply). Some of the rivers in the Upper Tana catchment have been tapped for the supply of water of water to Ndakaini Dam that supplies water to the City of Nairobi. It is expected that the tapping of the high flows >80 m^3s^{-1} under the Northern Collector Tunnel will further cause a reduction in the total stream flow of the Tana Sagana [31].

4.2 Sediment yield and rainfall and discharge

Sediment yield in the Upper Tana Basin is a consequence of poor land-use practices that causes high rates of soil erosion. The period 1960-1961 was characterized by high sediment yield which can be attributed to high rainfall events during the same period. There is an indication of a lag between the occurrence of high rainfall event in 1960 and the corresponding high sediment transport rate in early 1961. However, there is no clear indication of an increase in sediment transport rate since 1960, despite high variabilities that are clearly evident in our data. It is not certain whether reforestation and soil conservation programmes in the basin could explain this scenario.

The rainfall in the Upper Tana Basin exhibits significant spatial variations. The maximum total rainfall at Sagana is 1,839 mm while at Nyeri it is 1,456.3 mm. The minimum values for Sagana and Nyeri were 357 mm and 501.3 mm, respectively. The results of simulation of rainfall showed that variabilities are largely due to the difference in the location of the stations. Both Sagana and Nyeri stations showed a negative gradient (-12.49 and -4.93) meaning the amount of rainfall is reducing in the region. This is in agreement with findings of other studies done in the region [8].

4.3 Simulation of river discharges and sediment yield in the upper tana basin

The results of simulation of river discharges in the Upper Tana are based on the results for Gura river. The results had (r) value of 0.665, a R^2 value of 0.442 and NSE of -89.43. These results show a good relationship between the two datasets, although the R^2value is low. This can be explained by the gaps in the available data. The R^2 value showed a small error in the variance of the two datasets. The NSE of -89.43 shows that the mean observed river discharge for the Gura RGS is a better predictor than the mean simulated value and hence an unacceptable performance of the model. The high values for the Gura RGS were over-predicted by the model while in some instances they were under-predicted. The low river discharge values were generally under-predicted which explains the low efficiency in the simulation.

The results obtained for the observed and simulated sediment yield provided a (r) value of 0.02, a R^2 value of 0.0005 and NSE of -0.6. The r and R^2 are low because of relying on the model data. There was no data to input for the model to run as the available data was provided annually and the model required data in daily form. The NSE clearly shows that the mean observed value is a better predictor than the mean simulated value, this means that this is an unacceptable level of performance for the model to simulate sediment yield, and we would rather rely on the observed data as opposed to simulated data. The high values for the sediment yield were highly over-predicted with the low values being highly under-predicted.

4.4 Factors influencing sediment yield

The main objective of this study was to determine how rainfall influences sediment yield in the upper Tana catchment. According to the results of SWAT model, the amount of rainfall had a low significance influence on the amount of sediment yield in the Upper Tana catchment (**FIGURE 12**), with a r value of -0.25 and R^2 value of -0.062. The simulated data on rainfall was used for the plotting because the model produced a good simulation between the observed and simulated data for the Sagana fish culture farm. The results obtained for the relationship between rainfall and sediment yield is quite low on the r and R^2 which can be attributed to low performance of the model to simulate sediment yield. The data available was in annual form while the model required data in daily format. It would be expected from the graph that an increase in rainfall would result to an increase in the amount of sediment yield but this is not the case in many years. This shows that there are other factors apart from rainfall that influence the amount of sediment yield in the Upper Tana catchment.

According to the results provided by the model, the amount of rainfall and sediment yield has no threat to the future of the Masinga dam reservoir. However, land use and the type of soil are the major determinant of the sediment supply to the reservoir. This is in agreement with other studies carried out by Kitheka, et.al. [8], that found a correlation of 0. 67. Mutiso (1980) found a correlation coefficient r of 0.8 which is much higher than the one reported in this study which is 0.5. The difference is attributed to the approaches used to determine the relationship between the parameters and the difference in geographical locations. This can also be attributed to the use of different data sets. The multiple regression analysis on the extent to which sediment yield is influenced by stream flow and rainfall variations yielded r value of 0.973 and R^2 of 0.94 at 95% confidence level (p=0.05). This shows the combined influences of rainfall and river discharge influence sediment yield. This study established that there is a good relationship between river discharge and sediment yield in the Upper Tana basin. This is expected since increased river discharge is associated with the increased capacity to transport the detached sediments [32].

4.5 Variability of sediment yield

The production of sediments in the Upper Tana Basin is a consequence of soil erosion processes in the sub-basin. Soil erosion is influenced by numerous factors among them rainfall, runoff, landuse, vegetation cover, and soil erodibility. This study has examined the influence of river runoff and rainfall. The extent to which land-use change and soil erodibility influences sediment production in the basin has not been examined in detail due to lack of data. However, we based our arguments on their contributions from the results of other studies.

The computation of sediment yield in the basin was done using the river discharge data and suspended sediment concentrations (TSSC) data for Sagana river. This was based on the mean, maximum and minimum TSSC values of 0.384 gl^{-1}, 0.032 gl^{-1} and 0.0042 gl^{-1} respectively and the corresponding mean, maximum and minimum river discharges of 20. 53 m^3s^{-1}, 62.56 m^3s^{-1} and 9.23 m^3s^{-1}, respectively. These data yielded the mean, maximum and minimum sediment yield of 31,062,960 ton.km^2/yr, 781,777,440 ton.km^2/yr and 157,680 ton. km^2/yr respectively.

The Upper Tana catchment has changed considerably in terms of land use with increased extent of Cultivated areas due to increased population, increased livestock keeping and in terms of infrastructure roads, small-scale hydraulic works, and diversions [9,33]. There exists a strong relationship between the land-use and sediment yield in the Upper Tana catchment. The consequence of high sediment yield is the siltation of the Masinga dam reservoir [4,5,7-9]. The primary concern of high sediment discharge in the Masinga reservoir is the loss of storage capacity which affects the economic life of the reservoir. Loss of storage capacity hampers a reservoir's ability to fulfill its main functions which includes; store water, which impacts on other functions, e.g. power generation, irrigation supply, flood control, navigation, and domestic water supply. Due to the nonlinear relationship between water yield and reservoir capacity, even a small loss of reservoir capacity can result in substantial reductions in reservoir yield [34].

4.6 Sustainable land-use and protection of reservoirs in the upper tana basin

Sustainable land use practices are seen as one of the best approaches of reducing sediment yield in the Upper Tana Basin. This can be done through application of various management practices which have the potential to benefit both upstream as downstream stakeholders. These include strip cropping, gully erosion control [9], and reforestation. There is a need for a comprehensive reforestation programme in the Upper Tana catchment. There is also a need for eco-hydrological studies to establish the relationship and feedbacks between ecological and the hydrological processes [35]. In addition, there is need for comprehensive monitoring of river discharges, rainfall and sediment yield by the WRMA. Data collection programme should be consistent to avoid gaps in hydrological and climatological data [36-41].

REFERENCES

1. Mango LM, Melesse AM, McClain ME, Gann D, Setegn SG (2011) Land use and climate change impacts on the hydrology of the upper Mara River Basin, Kenya: results of a modeling study to support better resource management. Hydrology and Earth System Sciences 15: 2245-2258.
2. Kitheka JU, Nthenge P, Obiero M (2003) Dynamics of sediment transport and exchange in the Tana estuary in Kenya. School of Water Resources Science and Technology.
3. Dunne T, Ongweny GS (1976) A new estimate of sedimentation rates on the upper Tana River. The Kenyan Geographer 2: 109-126.
4. Ongwenyi GSO (1978) Erosion and sediment transport in the Upper Tana catchment with special reference to the Thiba basin. Unpublished PhD thesis, University of Nairobi.
5. Ongwenyi GSO (1979) Patterns of sediment production in the Upper Tana basin in Eastern Kenya. In: Hydrology of Areas of Low Precipitation. Proceedings of Canberra Symposium, pp: 447-457.
6. Maingi SM (1991) Sedimentation in Masinga Reservoir. Unpublished MSc Thesis, Appropriate Technology Centre, Kenyatta University, Nairobi.
7. Maingi JK, Marsh SE (2002) Quantifying hydrologic impacts following dam construction along the Tana River, Kenya. Journal of Arid Environments 50: 53-79.
8. Kitheka JU (2014) Assessment of modification of the Tana River runoff due to developments in the Upper Tana Basin. In: The Proceedings of the 2nd Hydrological

Society of Kenya (HSK) Workshop: "Hydrology in Water Cooperation and Security for Sustainable Economic Development". Sub- Theme: Hydrology and Sustainable Development, Nairobi, Kenya, pp: 178-191.

9. Dunne T (1979) Sediment yield and land use in tropical catchments. Journal of hydrology 42: 281-300.
10. Archer D (1996) Suspended sediment yields in the Nairobi area of Kenya and environmental controls. IAHS Publications-Series of Proceedings and Reports- Intern Assoc Hydrological Sciences 236: 37-48.
11. Ongwenyi GS (1983) Development of water resources. In the Kenyan Geographer-special issue on the Proceedings of the Nairobi workshop: Strategies for developing the resources of the semi-arid areas of Kenya, pp. 36-47.
12. Geertsma R, Wilschut L, Kauffman JH (2010) Review for the Green Water Credits Pilot Operations in Kenya. Green Water Credits Report 8 / ISRIC Report 2010/02, ISRIC- World Soil Information, Wageningen.
13. Notter B, MacMillan L, Viviroli D, Weingartner R, Liniger HP (2007) Impacts of environmental change on water resources in the Mt. Kenya region. Journal of Hydrology 343: 266-278.
14. Mutangah JG, Mwangangi O, Mwaura PK, Shem OM, Abiero L, et al. (1992) Kenya Indigenous Forest Conservation Programme.
15. Lamond G (2007) Local knowledge of biodiversity and ecosystem services in smallholder coffee farms in Central Province, Kenya (Doctoral dissertation, Master Thesis, Bangor. School of the Environment and Natural Resources, University of Wales, Bangor).
16. Jaetzold R, Helmut SH (1983) Farm Management Handbook of Kenya: Natural Conditions and Farm Management Information, Central Kenya: (Rift Valley and Central Provinces)/ Ministry of Agriculture, Kenya, pp: 3-29.
17. Vahabi J, Nikkami D (2008) Assessing dominant factors affecting soil erosion using a portable rainfall simulator. International Journal of Sediment Research 23: 376-386.
18. Nachtergaele FO, van Velthuizen H, Verelst L, Batjes NH, Dijkshoorn JA, et al. (2008). Harmonized world soil database (Version 1.0). Food and Agric Organization of the UN (FAO); International Inst. for Applied Systems Analysis (IIASA); ISRIC-World Soil Information; Inst of Soil Science-Chinese Acad of Sciences (ISS-CAS); EC-Joint Research Centre (JRC).
19. Baker BH (1967) Geology of the Mount Kenya area. Geological Survey of Kenya, pp: 78-79.
20. Tripathi MP, Panda RK, Raghuwanshi NS (2003) Identification and prioritisation of critical sub-watersheds for soil conservation management using the SWAT model. Biosystems Engineering 85: 365-379.
21. Parajuli PB, Mankin KR, Barnes PL (2008) Applicability of targeting vegetative filter strips to abate fecal bacteria and sediment yield using SWAT. Agricultural water management 95: 1189-1200.
22. Levesque E, Anctil F, Van Griensven ANN, Beauchamp N (2008) Evaluation of streamflow simulation by SWAT model for two small watersheds under snowmelt and rainfall. Hydrological sciences journal 53: 961-976.
23. Rostamian R, Jaleh A, Afyuni M, Mousavi SF, Heidarpour M, et al. (2008) Application of a SWAT model for estimating runoff and sediment in two mountainous basins in central Iran. Hydrological Sciences Journal 53: 977-988.
24. Neitsch SL, Arnold JG, Srinivasan R (2002) Pesticides fate and transport predicted by the soil and water assessment tool (SWAT). Final report submitted to Office of Pesticide Programs. Washington, DC: USEPA.
25. Nash JE, Sutcliffe JV (1970) River flow forecasting through conceptual models part I- A discussion of principles. Journal of hydrology 10: 282-290.
26. Servat E, Dezetter A (1991) Selection of calibration objective functions in the context of rainfall-runoff modelling in a Sudanese savannah area. Hydrological Sciences Journal 36: 307-330.
27. Legates DR, McCabe GJ (1999) Evaluating the use of "goodness-of-fit" measures in hydrologic and hydroclimatic model validation. Water resources research 35: 233-241.

28. Santhi C, Arnold JG, Williams JR, Dugas WA, Srinivasan R, et al. (2001) Validation of the SWAT model on a large water river basin with point and nonpoint sources. JAWRA Journal of the American Water Resources Association, 37: 1169-1188.
29. Van Liew MW, Arnold JG, Garbrecht JD (2003) Hydrologic simulation on agricultural watersheds: Choosing between two models. Transactions of the ASAE, 46: 1539-1551.
30. Jacobs J, Angerer J, Vitale J, Srinivasen R, Kaitho R, et al. (2004) Exploring the potential impact of reforestation on the hydrology of the upper Tana River catchment and the Masinga Dam, Kenya. Impact Assessment Group, Cent. For Natural Resource Information Technol., Texas A and M Univ., College Station, pp: 11-13.
31. Runo KS (2015) Evaluation of geological conditions along the proposed northern collector tunnel using electrical resistivity tomography method in Gatanga, Kenya, Doctoral dissertation, University of Nairobi.
32. Kitheka JU, Mavuti KM, Nthenge P, Obiero M (2014) The dynamics of the turbidity maximum zone in a tropical Sabaki estuary in Kenya.
33. McCully P (1996) Rivers no more: the environmental effects of dams, Zed Books, pp: 29-64.
34. Annandale GW (1987) Reservoir sedimentation. Elsevier 99-116.
35. Zalewski M (2002) Eco-hydrology-The use of cological and hydrological processes for sustainable management of water resources. Hydrological Sciences Journal 47: 823-832.
36. Palmieri A, Shah F, Dinar A (2001) Economics of reservoir sedimentation and sustainable management of dams. Journal of Environmental Management 61: 149-163.
37. Mati BM, Morgan RP, Gichuki FN, Quinton JN, Brewer TR, et al. (2000) Assessment of erosion hazard with the USLE and GIS: A case study of the Upper EwasoNg'iro North basin of Kenya. International Journal of Applied Earth Observation and Geoinformation 2: 78-86.
38. Kitheka JU, Obiero M, Nthenge P (2005) River discharge, sediment transport and exchange in the Tana Estuary, Kenya. Estuarine, Coastal and Shelf Science 63: 455-468.
39. Kitheka JU, Mavuti KM (2016) Tana Delta and Sabaki Estuaries of Kenya: Freshwater and Sediment Input, Upstream Threats and Management Challenges. In Estuaries: A Lifeline of Ecosystem Services in the Western Indian Ocean. Springer International Publishing, pp: 89-109.
40. Baker TJ, Miller SN (2013) Using the Soil and Water Assessment Tool (SWAT) to assess land use impact on water resources in an East African watershed. Journal of Hydrology 486: 100-111.
41. Fao I (2002) WFP (2002) Reducing poverty and hunger: the critical role of financing for food, agriculture and rural development, pp: 18-22.

CHAPTER 6

Field Analysis of Stepwise Effective Thermal Conductivity along a Borehole Heat Exchanger under Artificial Conditions of Groundwater Flow

Yoshitaka Sakata [1,*], Takao Katsura [1], Katsunori Nagano [1] and Manabu Ishizuka [2]

[1] *Division of Human Environmental Systems, Faculty of Engineering, Hokkaido University, Sapporo 060-8628, Japankatsura@eng.hokudai.ac.jp (T.K.); nagano@eng.hokudai.ac.jp (K.N.)*

[2] *AquaGeoTechono Co., Ltd., Sapporo 003-0025, Japan*

AUTHOR CONTRIBUTIONS

Y. Sakata performed the field tests, analyzed the data, and wrote the paper as the corresponding author; T. Katsura supported the tests and the analysis. K. Nagano contributed as the supervisor for this study; M. Ishiduka performed the pumping test and analyzed the data.

ABSTRACT

Heat advection caused by groundwater flow can potentially improve the performance of a borehole heat exchanger. However, the required flow velocity is not achieved under most natural conditions. This study focuses on artificial groundwater flow generated by pumping and investigates the associated effect in a lowland area near the Toyohira River alluvial fan, Sapporo, Japan. Thermal response test results are compared under natural and artificial groundwater flow conditions. A pumping well is constructed one meter from the borehole. Temperature profiles are measured in the U-tube during testing, using a pair of optic fiber distributed temperature sensors. The effective thermal conductivity is calculated from the profiles obtained in each 10-m sub-layer; this thermal conductivity is termed the stepwise thermal conductivity. Additionally, the upward flow velocity in the pumping well is measured to estimate the mean groundwater flow velocity at the borehole. The results show that effective thermal conductivity increases at depths less than 50 m, where the pumping creates mean velocities greater than 0.1 m d^{-1} in each sub-layer (1.5 md^{-1} on average). Thus, a borehole length of 50 m is more reasonable at the test site for its efficiency in a ground source heat pump system coupled with the pumping well than that used.

Keywords: artificial groundwater flow; effective thermal conductivity; borehole heat exchanger; thermal response test; optic fiber distributed temperature sensor; groundwater flow velocity

1. INTRODUCTION

Ground source heat pump systems (GSHPs) have been increasingly established for heating and air conditioning in residences and commercial buildings owing to their energy efficiency, ubiquitous availability, and environmental friendliness [1,2]. Among a variety of ground heat exchangers, closed loop borehole heat exchangers (BHEs) are the most widely used. BHEs consist of a high-density polyethylene U-tube(s) in a vertical borehole of a depth of approximately 100 to 300 m. The performance of a BHE in a GSHP is dependent on the effective thermal conductivity around the BHE. The effective thermal conductivity is variable, with ranges between <1 W m^{-1} K^{-1} in unsaturated fine sediments and >3 W m^{-1} K^{-1} in dense consolidated rocks [3]. In addition, the effective thermal conductivity may increase in aquifers with groundwater flow velocities sufficiently high for heat advection relative to the conduction, i.e., the Peclet number is of order 1 in principle [4]. Several numerical studies have indicated that the threshold of the groundwater flow velocity for the advection effect is commonly on the order of 0.1 md^{-1} [5]. The threshold velocity was also evaluated on the same order by other numerical studies [6,7,8]. Such high velocities of groundwater flow are limited to cases in which both hydraulic gradients and conductivities are relatively high, such as in unconsolidated gravelly aquifers beneath steep slopes [4]. Therefore, the design and planning of GSHPs typically do not take into account the groundwater flow effect on the performance of a BHE.

This study focuses on the artificial groundwater flow effect generated by pumping water from a well adjacent to a BHE. The pumping causes radial groundwater flow, and the flow velocity increases as the pumping rate increases. In addition, when the BHE is close to the pumping well and the groundwater flow velocity exceeds the threshold at the BHE, the effective thermal conductivity around the BHE increases; thus, the performance of the GSHP improves. The effect of artificial groundwater flow can differ depending on the pumping conditions (i.e., well depth and pumping rate) and on the hydrogeologic conditions (i.e., sedimentary structure and hydraulic conductivity in each layer). Thus, field measurements and analyses are required to evaluate the effect of artificial groundwater flow for the design and planning of GSHPs at each site.

The objective of this study is to obtain field evidence of the effect of artificial groundwater flow on the effective thermal conductivity around a BHE, based on thermal response tests (TRTs). TRT was originally developed in Sweden [9] to determine the average effective thermal conductivity along a BHE. The standard analysis is valuable for practitioners in terms of its simplicity under the assumption of an infinite line heat source. The methodology is summarized in texts such as ASHRAE [10] and IEA [11]. The analysis method has been also developed by using analytical solutions of finite line heat source [12,13], inversion techniques [14], Monte-Carlo method [15] and numerical simulations [5,6,7,8,16]. Especially in this study, the standard TRT is limited because the averaged effective thermal conductivity is insensitive to the geologic structure under the ground [16]. In order to address the limitation, our pervious study [17] measured and analyzed temperatures in a U-tube during the TRT, resulting in effective thermal conductivity in each 10-m thickness sub-layer; this variation is termed the stepwise effective thermal conductivity.

This study performs the multi-layer concept TRTs under two different conditions of groundwater flow. One is the natural condition without pumping, and the other is the artificial condition with pumping from a water well. The test site is located in a lowland area near the Toyohira River alluvial fan, Sapporo, Japan. The natural flow velocity is moderate on the flat slope, but pumping from the gravelly aquifers increases the groundwater flow velocity. The water well is located one meter from the BHE. The temperature of heat transfer fluid is measured in the U-tube during the testing using a pair of fiber optic distributed temperature sensors (DTSs). The effective thermal conductivity is estimated on the line heat source analysis, indicating apparent increases of the stepwise effective thermal conductivity due to the advection effect. This study also measures the upward flow velocity in the pumping well and estimates the mean groundwater flow at the BHE for comparison with the stepwise effective thermal conductivity. This study shows the actual effect of artificial groundwater flow and discusses a reasonable BHE length in terms of efficiency for GSHPs under the artificial conditions.

2. MATERIALS AND METHODS

2.1. Site Description

Figure 1a,b illustrate maps of the test site at regional and local scales, respectively. The test site is located at 43°03′06″ N and 141°15′25″ E at Hokkaido University, Sapporo, Japan. Sapporo is a capital city in the northern island of Japan, Hokkaido, and has been developed on and around the Toyohira River alluvial fan. Figure 1b shows that the alluvial fan is surrounded topographically by mountains to the northwest, hills to the northeast, and lowlands to the south. The radius of the fan is approximately 7 km, and the area is approximately 31 km^2. The fan consists of a western Holocene fan and an eastern Pleistocene fan [18]. The Holocene fan contains the subsurface groundwater reservoir. The test site is located on the lowlands near the alluvial fan. The topographic slope of the lowland area is less than 1%; thus, the groundwater flow velocity is moderate for BHE performance.

Figure 1c shows a geologic cross section along a longitudinal line (the black line in Figure 1b). Tertiary rocks can be observed beneath the riverbed at the fan head, but the fan basement suddenly inclines northward to a depth of several hundreds of meters. Because of this depression, various Quaternary sediments underlie the central to distal parts of the fan, with a thickness of hundreds of meters. The unconsolidated formation is classified hydrogeologically into Aquifers I–IV [19]. The deepest aquifer, Aquifer IV, is composed of alternating clay, sand, and gravelly layers and formed before the alluvial fan. The Pleistocene, Aquifer III, and Holocene, Aquifer II, aquifers consist mainly of alluvial fan gravel deposits, i.e., poorly sorted sandy gravel sediments. The total thickness of the gravel aquifers is more than 100 m at the middle of the fan. The boundary of Aquifers III and II is generally not obvious from the available borehole data. Aquifer I is distributed only near the ground level in the fan toe, and it consists of fine materials related to the fan recession. Figure 1c also shows the water table level and the potentiometric head contours in the cross section. In the upper fan (<6 km in distance), the hydraulic head decreases in the downstream and downward directions, reflecting the three-dimensional groundwater flow system.

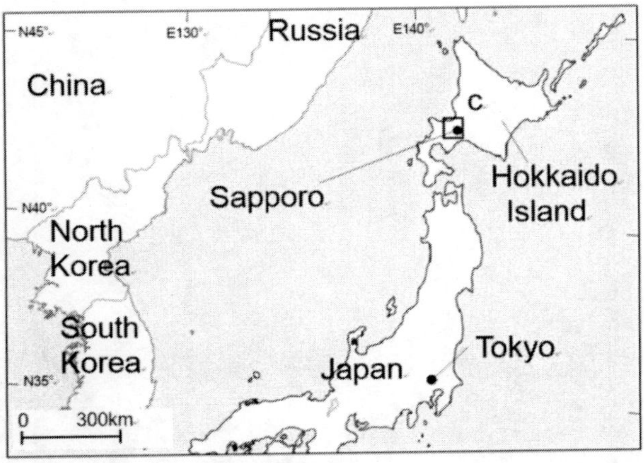

(a)

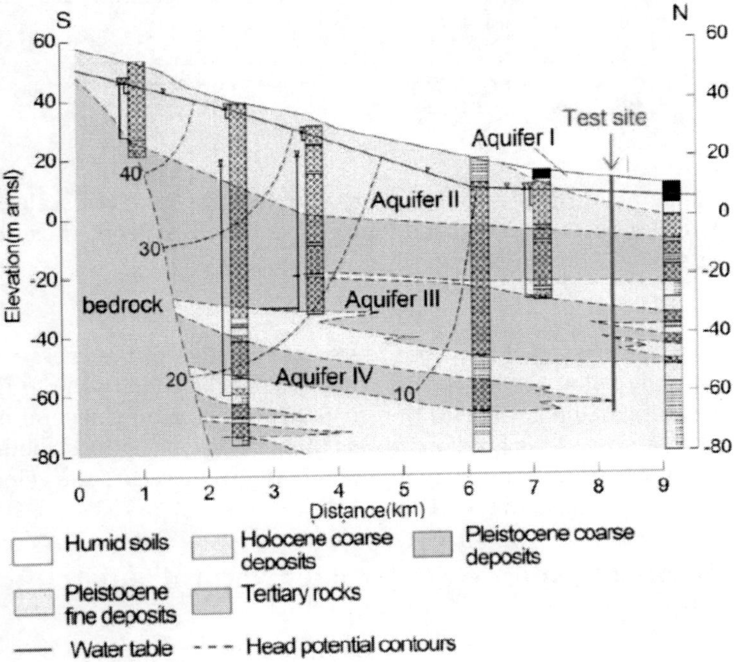

(b)

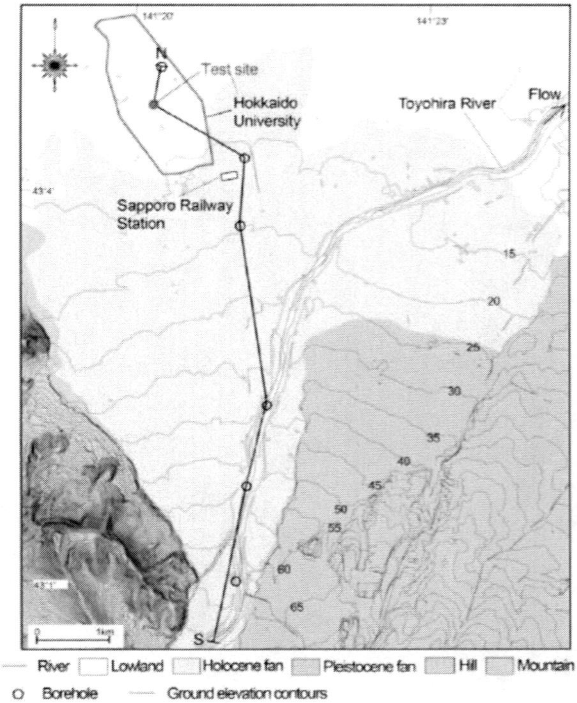

(c)

Figure 1. Maps of the study site at regional (**a**) and local (**b**) scales and a geologic cross section (**c**). These figures are modified from those of our previous study [20] with permission of the journal.

The test site is located 8 km from the southern edge of the cross section. Figure 1c shows that the formation at the test site is composed of Aquifers I–IV. The water table is close to the surface (approximately several meters deep), and the hydraulic gradient is less than 1% under natural conditions. However, pumping from gravelly Aquifers II and III increases the velocity of groundwater flow around the BHE.

2.2. Thermal Response Tests under Natural and Artificial Conditions of Groundwater Flow

This study performed standard TRTs as follows: (1) a BHE with configurations (diameter, length, and U-tube(s)) similar to those planned is constructed for testing; (2) the U-tube contains the heat carrier fluid (water or brine); (3) the fluid is circulated at a constant rate in the U-tube; (4) the heat is injected into the fluid at a constant rate; (5) the fluid temperatures are measured during the heating at both ends of the U-tube; and (6) the temperature measurements are analyzed to determine the average effective thermal conductivity along the BHE and the thermal resistance within the BHE.

In addition to the standard TRT, this study measured the temperature of the heat transfer fluid in the U-tube by installing a pair of fiber optic DTSs. One DTS is installed in the inlet tube, and the other DTS is installed in the outlet tube. The DTSs were used to measure temperatures at multiple points along a distance of several kilometers based on the Raman scattering effect of the laser pulse of the optic fibers [21]. Notably, the diameter of each fiber sensor was sufficiently small to avoid disturbances due to flow circulation; thus, the DTSs could be used to monitor and investigate the GSHP system [22]. The temperature measurements by the DTSs were also used to match the analytical solutions by the numerical inversion techniques, resulting in the depth-varying effective thermal conductivity of soils [23] and the thermal resistances of BHEs [24].

The test configuration is illustrated in Figure 2. A photo of the TRT is also shown in Figure 3. The BHE was constructed at a depth of L = 80 m using a double-casing driller in February 2015. The BHE was composed of a single U-tube (ID = 25 mm and OD = 32 mm), which was buried with silica sand as the grout material. Tap water was enclosed in the U-tube under a pressure greater than 0.1 MPa. The water well was constructed with the same depth and at a distance of R = 1 m from the BHE in May 2016. The water well included a plastic screen (ID = 75 mm) and silica sand between the screen and the ground. After construction, the water well was flushed via upward pumping until the discharge water was sufficiently clear.

This study performed TRTs at the BHE without pumping from the water well (the first TRT) and with pumping at a constant rate (the second TRT). In both TRTs, the configurations were the same, excluding pumping and non-pumping. The total heating rate via the electric heater was Q = 5.2 kW during the period of 60 h. The flow rate for circulating the heat transfer fluid was constant at v = 0.00033 m^3 s^{-1} (20 L/min). The temperatures of the fluid at the ends of U-tube were observed using PT100 platinum resistance thermometers, which have an accuracy of 0.1 K. The temperatures of the fluid in the U-tube $\theta_1(z, t)$ and $\theta_2(z, t)$ were observed using the DTSs. The DTS specifications are shown in Table 1. The sensors were able to reduce the random errors by alternating the laser pulse emitted in the forward and backward directions. In addition, offset calibration and a moving average were performed to reduce the standard deviations of temperature fluctuations within 0.2 K [17]. The first TRT was performed from 28 April to 1 May 2016. The second TRT was performed during 10–20 May 2016 at a constant pumping rate of Q_v = 0.0053 m³/s (320 L/min) using a centrifugal vacuum pump. The extraction mouth for pumping was set at a depth of 6 m (2.3 m below the water table). The total displacement of the water table was 1.7 m at the end of pumping (t = 60 h).

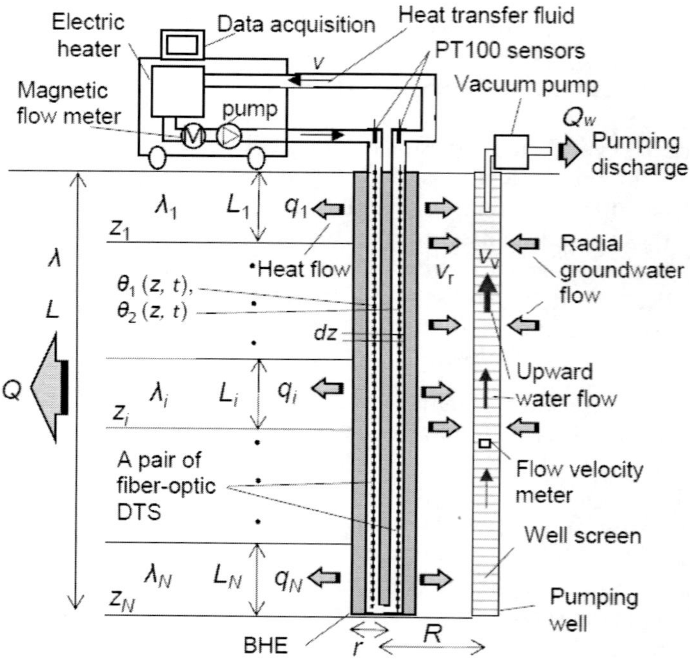

Figure 2. Cross-sectional illustration of the test arrangement of the borehole heat exchanger (BHE) and the pumping well.

This study also measured the upward flow velocity toward the extraction mouth in the pumping well during the second TRT. The measurement was performed every 0.5 m in depth using an in-hole micro-flowmeter, Model 3216, Oyo Co., Ltd., Tokyo, Japan. The device measured the number of propeller rotations in the pumping well, which was translated to the flow velocity. The groundwater discharge per unit depth was based on the upward flow through the meter. This study calculated the mean groundwater flow velocity at the location of the BHE based on the assumption of a circular flow field. First, the radial flow discharge into the well per unit depth was calculated as a product of the vertical slope of the upward velocity measurements and the cross-sectional area of the well

(0.0088 m^2). Next, the mean groundwater flow velocity was calculated as the discharge divided by $2\pi r$ (the distance r at the BHE is one meter).

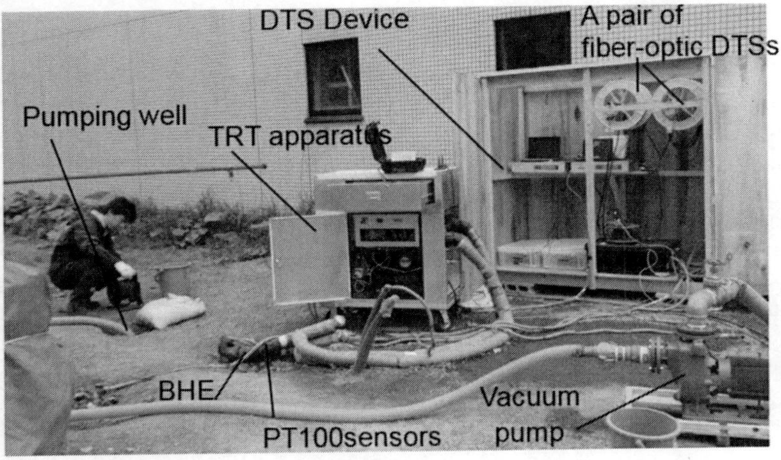

Figure 3. Photo of the field test on 2 May 2016.

Table 1. Distributed temperature sensor (DTS) device and sensor specifications.

Instrument	N4385A-008 (A.P. Sensing, Ltd)
Maximum distance range	8 km
Spatial resolution	0.5 m
Measurement interval	1 min
Accuracy (Uncertainty)	+/− 0.2 K (moving average per 20 min)
Optic fiber	SKF-VP13L404CC140 (NK Systems, Ltd.)
Structure	Dual fibers in a single water-proofed tube
Diameter	4 mm (18 mm at the top)
Total length	140 m
Temperature range	−10 °C to 85 °C

2.3 Analysis of Temperature Profiles to Determine Stepwise Thermal Conductivity

In the standard TRT, the temperature measurements of the heat transfer fluid at the ends of the U-tube were analyzed based on the assumption that the BHE was the line heat source of uniform heat injection along the length of a finite region of homogeneity. The approximate solution of the temperature around the BHE due to the heating process was obtained as follows [25]:

$$\overline{\theta} - \theta_0 \approx \frac{q}{4\pi\lambda}\left(-\ln\frac{r^2}{4\alpha\, t} - \gamma\right) + qR_b \tag{1}$$

where θ is average temperature surrounding the BHE (K), θ_0 is the initial temperature of the ground (K), q is the heat exchange rate per unit depth (W m^{-1}), r is the radius of the BHE (m), λ is the depth-averaged ground thermal conductivity (W m^{-1} K^{-1}), α is the thermal diffusivity (m^2 h^{-1}), γ is Euler's constant (≃ 0.5772), t is the elapsed time since the start of heating (h), and R_b is the thermal resistance of the BHE (m KW^{-1}). Equation (1) shows that the temperature displacement increases linearly on a log scale of elapsed time t during the testing. When the temporal temperature slope is assumed constant, the effective thermal conductivity λ is calculated using two factors: the heat exchange rate per unit depth q and the temporal temperature slope:

$$\lambda = \frac{q}{4\pi k_t} \tag{2}$$

where k_t is the temporal slope of the temperature on a log scale of elapsed time t (K). The slope is estimated from the average temperatures of the circulating fluids via manual fitting upon visual inspection or linear regression fitting. q is also calculated by dividing the total heat exchange rate Q (W) by the total length of the BHE. The total heat exchange rate was calculated as follows:

$$Q = C_s \rho v \Delta\theta \tag{3}$$

where C_s is the specific heat of the heat transfer fluids (J kg^{-1}), ρ is the fluid density (kg m^{-3}), and v is the flow rate (m^3 s^{-1}). λ is determined in Equation (2), and R_b is calculated based on Equation (1) as follows:

$$R_b = \frac{\theta_{intersect} - \theta_0}{q} - \frac{1}{4\pi\lambda}\left(\ln\frac{4\alpha}{r^2} - \gamma\right) \tag{4}$$

where $\theta_{intersect}$ is the intersect of the linear fitting line of temperature displacements at the elapsed time t = 1 h on a logarithmic scale. It should be considered that the estimates are influenced by uncertainty in related to measurement and parameterization errors, resulting in the theoretical errors in TRTs are on the order of 5% for effective thermal conductivity and 10-15% for the thermal resistance [26].

This study extended the analytical approach based on the line heat source theory to a horizontally stratified formation, as shown in Figure 2. The formation consisted of *N* sub-layers. The effective thermal conductivity was constant at λ_i within each layer but different among layers. The heat exchange rate was also constant at q_i throughout each layer during the TRT. L_i and z_i are the thickness and the boundary depth, respectively, of the ith layer. The depth-averaged conductivity λ_i was assumed equal to λ in Equation (2) [27]. Thus, Equation (2) was transformed as follows:

$$\lambda_i = \frac{q_i}{4\pi k_{ti}}, \ i = 1, 2, \cdots, N \tag{5}$$

where λ_i is the individual effective thermal conductivity in the ith layer (W m^{-1} K^{-1}), qi is the heat exchange rate per unit depth in the ith layer (W m^{-1}), and k_{ti} is the temporal temperature slope of elapsed time on a logarithm scale (K).

In each sub-layer, the heat exchange rate was estimated from the sum of the temperature displacements in the inflow and downflow tubes. Although the displacements were directly calculated in the standard TRT using the temperature displacement between the temperature measurements in the ends of the U-tube, as shown in Equation (3), the proposed method calculates the temperature displacements as the vertical slope of the sequential temperature data:

$$q_i = q_i^{(in)} + q_i^{(out)} = c\rho v(\Delta\theta_i^{(in)} + \Delta\theta_i^{(out)})/L_i \cong c\rho v(-k_{zi}^{(in)} + k_{zi}^{(out)}) \tag{6}$$

where $k_{zi}^{(in)}$ and $k_{zi}^{(out)}$ are the vertical slopes of sequential temperature data in the inflow and outflow paths, respectively, of the ith layer. The temperature measurements recorded by the DTSs fluctuated more and were less accurate than those recorded by conventional temperature sensors, such as PT100 sensors. Thus, the vertical slopes of the sequential temperature data effectively reduced the measurement uncertainty in a statistical sense. The vertical temperature slopes were calculated at each elapsed time based on a linear regression analysis of the sequential data within each vertical segment. The heat exchange rate was calculated using Equation (6) at each measurement time (every one minute in this study) and was averaged during the heating period in each sub-layer. The average rates were used as inputs into Equation (5) to calculate the individual conductivities λ_i. This study also calculated the standard deviations of q_i and the ranges of λ_i to evaluate the uncertainty in this analysis.

The line heat source analysis was based on the minimum elapsed time required to transform the original error function into a logarithmic function and the maximum time required to approach steady state rather than radial heat conduction [11]. This study analyzed the temperature data during the heating period between t = 12 h and 60 h, as recommended in [28]. Additionally, in this analysis, the thickness in each sub-layer should be assigned to validate the line heat source assumption. Although the BHE in each sub-layer has a limited length, the temperatures based on the finite line source model were practically the same with those based on the infinite line source when the vertical length was about 10 m (40 feet) [25]. As a field study [17], our previous study conducted in Asahikawa City, Hokkaido, Japan, indicated that the effective thermal conductivity was approximately constant when the thickness was at least 10 m. Thus, this study assumed that the thickness required to validate the line heat source assumption was constant at 10 m, and the formation around the BHE (L = 80 m) was assumed to consist of 8 sub-layers of 10 m in thickness. The DTS measurement interval was d_z = 0.5 m. Temperature data recorded at 21 depths were used for the regression analysis of q_i and k_{ti} in each sub-layer. This study also calculated the total heat exchange rate Q^* from q_i and the average effective thermal conductivity λ^* from λ_i. Additionally, Q^* and λ^* were compared with the standard test results based on Q in Equation (3) and λ in Equation (2).

3. RESULTS AND DISCUSSION

3.1. Standard Test Results

Figure 4 shows the average fluid temperatures $\overline{\theta}$ measured by the PT100 sensors and the fitted lines for the determination of k_t based on Equation (2). The measurement data set is included in the excel file as the supplementary material. The temperatures in the first and second TRTs increased during the heating process. The temperature increase was almost the same during both TRTs until the elapsed time of approximately t = 10 h because heating was dominant in the borehole. After approximately t = 10 h, the temperature increase was smaller in the second TRT than that in the first TRT. The relatively small increase in the second TRT was due to the heat advection caused by artificial groundwater flow. The average effective thermal conductivity λ was calculated using Equation (2). In the first TRT, λ = 2.00 W m^{-1} K^{-1}, which is that of 'Light sand and 15% water' in Table 5 of Chapter 34 in [10]. In the second TRT, average effective thermal conductivity was λ = 3.19 W m^{-1} K^{-1}, which was almost 1.6 times larger than that during the first TRT. Additionally, the thermal resistance value of the

BHE was calculated using Equation (4). In the first TRT, R_b = 0.114 m KW^{-1}, and R_b = 0.128 m KW^{-1} during the second TRT, were almost the same in terms of an error criterion of R_b as described above (10%–15%), and were reasonable based on Table 6 in [10].

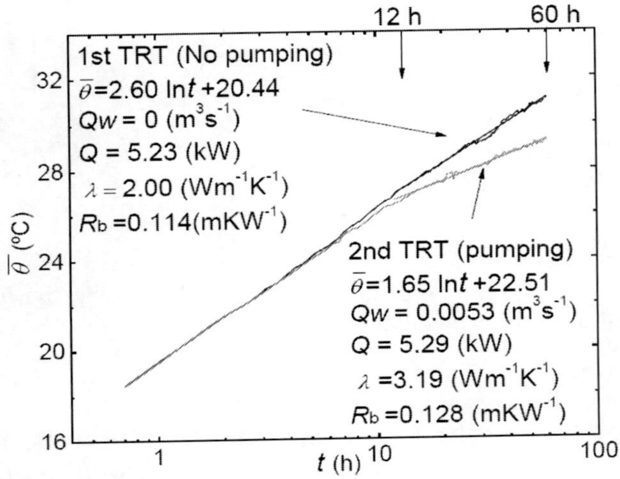

Figure 4. Plots of the average temperature of the fluid during the first (black plots) and second (red plots) thermal response tests (TRTs).

3.2. Profiles of Temperature in the U-Tube

Figure 5 shows the temperature profiles in the U-tube at extracted elapsed times (t = 0, 1, 24, and 60 h) during the first and the second TRTs for examples of the DTS measurements. The profile data are also included in the supplementary material. At the initial time (t = 0 h), the temperatures were constant at 12 °C in both the inlet and outlet tubes as a result of fluid circulation without heating for approximately 0.5 h before the tests. The temperature profiles were almost the same between the first and second TRTs after heating at t = 1 h because the heating domain was limited within the borehole, as described above. At t = 24 h and 60 h, there were differences in the temperature profiles between the first and second TRTs. The average temperature differences along the BHE were one kelvin at t = 12 h and two kelvin at t = 60 h. These differences were due to the effect of artificial groundwater flow into the pumping well.

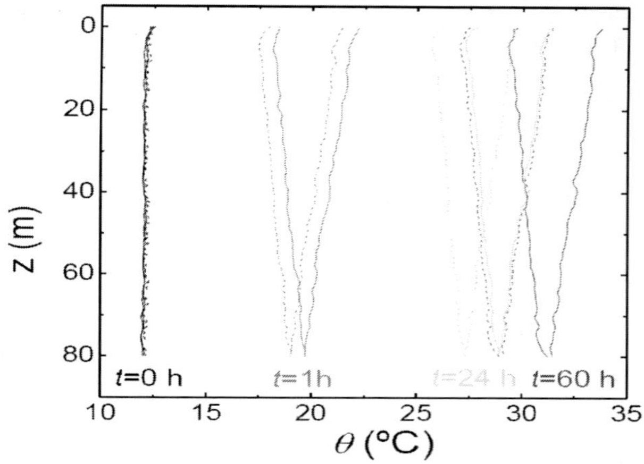

Figure 5. Fluid temperature profiles in the U-tube during the first (solid lines) and the second (dashed lines) TRTs.

3.3. Comparison of Stepwise Thermal Conductivity under Natural and Artificial Conditions of Groundwater Flow

Table 2 summarizes the measurement and analysis results of the first and second TRTs. In the upper rows, the total heat exchange rates, λ^*, and the average thermal conductivity along the BHE, Q^*, almost agreed with λ and Q based on the standard method, and the differences were less than 5%. The average conductivity λ^* was different between the first and second TRTs. In the first TRT, $\lambda^* = 2.08$, and $\lambda^* = 3.12$ in the second TRT, although the heat exchange rate Q^* was almost constant at about 5.2 kW. The increase was due to groundwater advection caused by pumping. In the lower rows, the heat exchange rate varied between 36.4 W m^{-1} and 84.9 W m^{-1} in the first TRT and between 48.6 W m^{-1} and 81.4 W m^{-1} in the second TRT. In the first to fifth layers, the rates during the second TRT increased relative to the rates in the first TRT. The temporal slope of temperature was almost constant in the sub-layers at approximately 2.5 K in the first TRT and at approximately 1.6 to 1.7 K in the second TRT. The slope in the second TRT was approximately 0.6 times the slope in the first TRT. The effective thermal conductivity in each sub-layer ranged between 1.18 and 2.71 W m^{-1} K^{-1} in the first TRT and between 2.35 and 3.92 W m^{-1} K^{-1} in the second TRT. On the right side of Table 2, the mean groundwater velocity at the BHE distance (r = 1 m) ranged from 10^{-2} to 10^{0} m d^{-1} orders of magnitude.

Table 2. Summary of the TRT results under natural (first) and artificial (second) conditions.

	First TRT (no pumping)				Second TRT (pumping)				
$^{1}Q_w$	0				0.0053				
^{2}Q	5.23				5.29				
$^{3}Q^*$	5.20				5.16				
$^{4}\lambda$	2.00				3.19				
$^{5}\lambda^*$	2.08				3.12				
^{6}i	$^{7}z_i$	$^{8}q_i$	$^{9}k_{ti}$	$^{10}\lambda_i$	$^{7}z_i$	$^{8}q_i$	$^{9}k_{ti}$	$^{10}\lambda_i$	$^{11}v_r$
1	10	36.4	2.46	1.18	10	55.4	1.65	2.67	1.57
2	20	56.2	2.48	1.80	20	77.8	1.66	3.73	2.70
3	30	73.5	2.48	2.36	30	76.2	1.65	3.67	0.17
4	40	66.9	2.50	2.13	40	81.4	1.65	3.92	1.06
5	50	80.5	2.50	2.56	50	73.0	1.61	3.61	0.35
6	60	68.9	2.48	2.21	60	54.3	1.62	2.66	0.01
7	70	52.7	2.48	1.69	70	48.6	1.66	2.33	0.06
8	80	84.9	2.49	2.71	80	48.9	1.65	2.35	0.08

[1] Q_w is water discharge due to pumping (m^3 s^{-1}); [2] Q and [3] Q^* are total heating rates (kW) based on the standard method and the proposed method, respectively; [4] λ and [5] λ^* are average effective thermal conductivities (W m^{-1} K^{-1}) based on the standard analysis and the proposed analysis, respectively. In each [6] ith layer, [7] z_i, [8] q_i, [9] k_{ti}, and [10] λ_i are the bottom depth (m), heat exchange rate per unit depth (W m^{-1}), temporal slope of temperature on a logarithmic time scale (K), and effective thermal conductivity (W m^{-1} K^{-1}), respectively. [11] v_r is the mean groundwater velocity estimate (m d^{-1}) based on the upward flow measurements.

Figure 6 summarizes the measurement and analysis results of the TRTs. The profile of k_{ti} was omitted due to the minimal variations in the vertical direction. The geologic column at the test site indicated that the formation at this site was composed mainly of Aquifer I (unsaturated fine deposits; z = 0 to 5 m), Aquifers II and III (upper aquifers of sandy gravel deposits; z = 5 to 45 m), and Aquifer IV (alternating fine and coarse sediments; z = 45 to 80 m). This composition was similar to that shown in Figure 1c. A comparison between Figure 6b,c indicates that the layers of relatively high q_i values were potentially related to the layers of relatively high λ_i values in both TRTs. The standard deviations of qi and λ_i were within 10%, respectively. In Figure 6c, during the first TRT (black lines), the effective thermal conductivity in the first layer (z = 0 to 10 m) was relatively small (λ_1 = 1.18 W m^{-1} K^{-1}) due to the unsaturated conditions above the water table (z = 6 m). The effective thermal conductivities in the second to eighth layers fluctuated, but the difference was within 15% of the average in the layers (2.21 W m^{-1} K^{-1}). During the second TRT (red lines), the effective thermal conductivities were larger than those during the first TRT, except for the value in the eighth layer. In the second to fifth layers, the effective thermal conductivities were commonly greater than 3.5 W m^{-1} K^{-1}. The average conductivity in the upper layers was 3.73 W m^{-1} K^{-1}, which was almost double the average conductivity along the BHE (λ = 2.00 W m^{-1} K^{-1}).

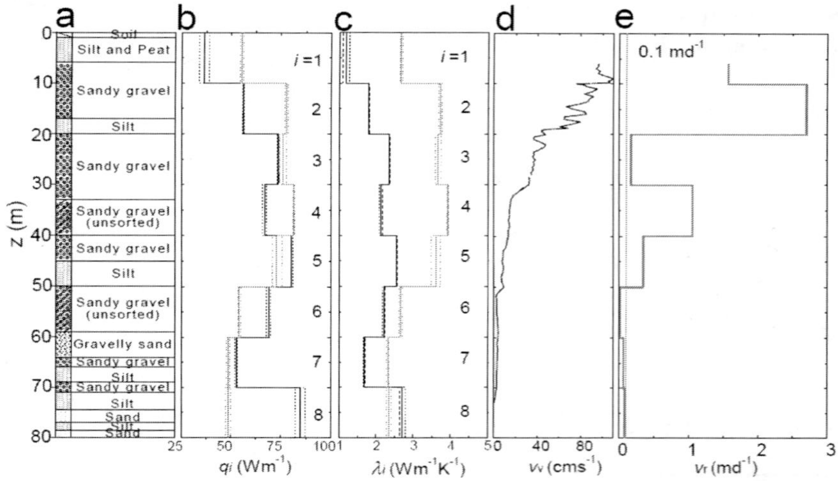

Figure 6. Measurement and analysis results of the TRTs: (**a**) geologic column near the test site; (**b**) heat exchange rate per unit length; (**c**) effective thermal conductivity (black and red solid lines represent natural and artificial conditions of groundwater flow and their dashed lines represent the ranges in related to the standard deviations of heat exchange rate, respectively); (**d**) upward flow velocity measurement in the pumping well; (**e**) mean groundwater velocity estimate at the BHE distance.

Figure 6d shows the upward flow velocity measurement v_v within the well. The velocity measurement data are included in the supplementary material. The flow velocity in the well was small below a depth of 50 m and increased above 50 m toward the extraction mouth of the vacuum pump. Thus, most of the drainage water was supplied from the upper gravelly aquifers above 50 m in depth according to the high permeability. The depth of the increase in the upward velocity corresponded to the depth of the increase in the effective thermal conductivity shown in Figure 6c. Additionally, the mean groundwater velocity estimates v_r at the BHE distance were more than 0.1 m d^{-1} above 50 m in depth, as shown in Figure 6e. The velocity measurement results confirm that the increase in the effective thermal conductivity above 50 m in depth was due to the artificial groundwater flow condition caused by pumping near the well. In actuality, the groundwater flowed not only in the radial direction but also in the vertical direction toward the extraction mouth. Thus, the average velocity of 1.5 m d^{-1}, i.e., the total discharge Q_w (0.0053 m^3 s^{-1}) divided by $2\pi r$L (r = 1 m, L = 50 m), was likely the threshold velocity for the heat advection effect in this case.

The comparison of the stepwise effective thermal conductivities showed not only an increase in the effective thermal conductivity, which was similar to that of the standard TRT, but also the depth at which the effective thermal conductivity typically increased or remained relatively constant. As described above, the preliminary determination of the artificial groundwater flow effect is difficult due to data limitations and the unique hydrogeologic conditions. However, we believe that the analysis of the stepwise effective thermal conductivity was able to effectively evaluate the artificial groundwater flow effect for the design and planning of GSHPs. In this study, an analytic approach based on the line heat source assumption was applied. The approach was clear and straightforward, especially relative to other methods of numerical inversion [23,24]. However, this approach contained uncertainty associated with the thickness of each sub-layer due to the line heat source assumption. Future studies should improve this analysis to determine not only the effective thermal conductivity but also the thickness of each sub-layer.

The limitations of this study and its future approaches are discussed. First, this study showed only apparent increases in the stepwise effective thermal conductivity. The groundwater flow velocity at the BHE was estimated not from temperature, but from the flow velocity meter within the well. The moving line source theory [29] is potentially available to determine both effective thermal conductivity and groundwater flow velocity, although the approach is not directly used in such radial flows into the well. Second, the approach was based on the assumption of the infinite line heat source. The finite line heat source model [12,13] was required to analyze high-resolution stepwise effective thermal conductivities. Finally, the analytic solutions of temperature profiles in a U-tube [30] would aid the analysis especially in the relatively short heating time to obtain more reasonable estimates of thermal resistance of the BHE. Numerical simulations might be required when the boundary conditions are complex and variable, for example when actual operation schedules are considered.

The in-situ test results were used for designers to determine reasonable length and number of BHEs at this site. Based on the same pumping condition, it is recommended that the length of the BHE is 50 m, and BHEs should be installed around the pumping well. Thus, the total number and length of BHEs can be effectively reduced compared with the number and length under natural conditions. However, the effect of artificial groundwater flow is likely variable based on the actual conditions such as the pumping discharges, the operation schedule and the heating loads of GSHPs. Thus, numerical simulations are needed to determine the optimized arrangement of BHEs and pumping wells. In the near future, we will investigate the stepwise effective thermal conductivity under different pumping conditions, and actual field data will be used in numerical simulations.

4. CONCLUSIONS

This study performed TRTs under natural and artificial conditions of groundwater flow in a lowland area near the Toyohira River alluvial fan, Sapporo, Japan. The artificial groundwater flow was produced through the pumping of water from a well located one meter from the BHE. The temperature of the heat transfer fluid was measured in the U-tube during the tests using a pair of optic fiber DTSs. The upward flow velocity was also measured in the pumping well to estimate the mean groundwater flow velocity at the BHE distance. As a result, the temperature profiles in the U-tube were approximately 1 to 2 K lower under artificial conditions than those under natural conditions. The standard TRT provides the difference in average effective thermal conductivity between natural (2.00 W m^{-1} K^{-1}) and artificial conditions (3.19 W m^{-1} K^{-1}). This study also analyzed the temperature profiles to determine the effective thermal conductivity in each 10-m sub-layer. The resulting values of stepwise thermal conductivity showed that the conductivity increased at depths less than 50 m, where the mean groundwater flow velocity was more than 0.1 m d^{-1} (1.5 m d^{-1} on average). The average conductivity under the artificial condition was 3.73 W m^{-1} K^{-1}, which was almost double the average conductivity along the BHE under the natural condition. In the near future, TRTs will be performed again at this site to investigate the stepwise effective thermal conductivity under different pumping conditions. Additionally, numerical simulations will be performed to determine the optimal arrangement of BHEs for a GSHP coupled with pumping wells.

ACKNOWLEDGMENTS

This study is based on results obtained from a "Renewable energy heat utilization technology development" project commissioned by the New Energy and Industrial Technology Development Organization (NEDO), a Japanese national agency. This study also greatly benefited from the thermal response tests performed with the assistance of our technical staff, Mr. Makoto Nakamura, and our graduate students.

REFERENCES

1. Spitler, J.D. Ground-source heat pump system re-search—Past, present and future. HVAC & R Res. 2005, 11, 165–167.
2. Nagano, K. The progress of GSHP in Japan. IEA Heat Pump Cent. Newsl. 2015, 33, 21–25.

3. Ingersoll, L.R.; Zobel, O.J.; Ingersoll, A.C. Heat Conduction with Engineering and Geological Applications; McGraw-Hill Book: New York, NY, USA, 1954; pp. 241–245.
4. Chiasson, A.C.; Rees, S.J.; Spitler, J.D. A preliminary assessment of the effects of ground-water flow on closed-loop ground-source heat pump systems. ASHRAE Trans. 2000, 106, 380–393.
5. Signorelli, S.; Bassetti, S.; Pahud, D.; Kohl, T. Numerical evaluation of thermal response tests. Geothermics 2007, 36, 141–166.
6. Angelotti, A.; Alberti, L.; Licata, I.L.; Antelmi, M. Borehole Heat Exchangers: Heat transfer simulation in the presence of a groundwater flow. J. Phys. Conf. Ser. 2014, 501, 1–9.
7. Raymond, J.; Therrien, R.; Gosselin, L.; Lefebvre, R. Numerical analysis of thermal response tests with a groundwater flow and heat transfer model. Renew. Energy 2011, 36, 315–324.
8. Katsura, T.; Nagano, K.; Nakamura, Y. Development of simulation tool for ground heat exchanger considering multiple geological layer and its Application. Trans. Jpn. Soc. Refrig. Air Cond. Eng. 2015, 32, 335–344.
9. Mogensen, P. Fluid to duct wall heat transfer in duct system heat storages. In Proceeding of the International Conference on Subsurface Heat Storage in Theory and Practice, Stockholm, Sweden, 6–8 June 1983; Swedish Council for Building Research: Stockholm, Sweden, 1983; pp. 1652–6571.
10. American Society of Heating and Air-Conditioning Engineers (ASHRAE). ASHRAE Handbook Heating, Ventilating and Air-Conditioning Applications; ASHRAE: Atlanta, GA, USA, 2011; p. 34.
11. IEA ECES Annex 21 Sub-task 4. Standard TRT procedure. In Thermal Response Test Final Report; IEA ECES Annex 21; IEA: Garching, Deutsch, 2013; pp. 1–28. Available online: http://www.iea-eces.org/files/a4.1_iea_eces_annex_21_final_report_1.pdf (accessed on 11 June 2015).
12. Bandos, T.V.; Montero, Á.; Fernández, E.; Santander, J.L.G.; Isidro, J.M.; Pérez, J.; Fernández de Córdoba, P.J.; Urchueguía, J.F. Finite line-source model for borehole heat exchangers: Effect of vertical temperature variations. Geothermics 2009, 38, 263–270.
13. Bandos, T.V.; Campos-Celador, Á.; López-González, L.M.; Sala-Lizarraga, J.M. Finite cylinder-source model for energy pile heat exchangers: Effects of thermal storage and vertical temperature variations. Energy 2014, 78, 639–648.
14. Zarrella, A.; Emmi, G.; Zecchin, R.; De Carli, M. An appropriate use of the thermal response test for the design of energy foundation piles with U-tube circuits. Energy Build. 2017, 134, 259–270.
15. Pasquier, P. Stochastic interpretation of thermal response test with TRT-SInterp. Comput. Geosci. 2015, 75, 73–87.
16. Lee, C.K. Effects of multiple ground layers on thermal response test analysis and ground-source heat pump simulation. Appl. Energy 2011, 88, 4405–4410.
17. Sakata, Y.; Katsura, T.; Zhai, J.; Nagano, K. Estimation of effective thermal conductivities according to multi-layers by thermal response test with a set of fiber optics in a U-tube. J. Jpn. Soc. Civ. Eng. Ser. G 2016, 72, 50–60.
18. Daimaru, H. Holocene evolution of the Toyohira River alluvial fan and distal floodplain, Hokkaido, Japan. Geogr. Rev. Jpn. Ser. A 1989, 62, 589–603.
19. Hu, S.G.; Miyajima, S.; Nagaoka, D.; Koizumi, K.; Mukai, K. Study on the relation between groundwater and surface water in Toyohira-gawa alluvial fan, Hokkaido, Japan. In Groundwater Response to Changing Climate; Taniguchi, M., Holman, I.P., Eds.; CRC Press: London, UK, 2010; pp. 141–158.
20. Sakata, Y.; Ikeda, R. Regional mapping of vertical hydraulic gradient using uncertain well data: A case study of the Toyohira River alluvial fan, Japan. J. Water Resour. Prot. 2013, 5, 823–834.
21. Grattan, K.T.V.; Sun, T. Fiber optic sensor technology: An overview. Sens. Actuators 2000, 82, 40–61.
22. Kallio, J.; Leppäharju, N.; Martinkauppi, I.; Nousiainen, M. Geoenergy research and its utilization in Finland. Geol. Surv. Finl. Spec. Pap. 2011, 49, 179–185.

23. Fujii, H.; Okubo, H.; Nishia, K.; Itoi, R.; Ohyama, K.; Shibata, K. An improved thermal response test for U-tube ground heat exchanger based on optical fiber thermometers. Geothermics 2009, 38, 399–406.
24. Acuña, J.; Mogensen, P.; Palm, B. Distributed thermal response test on a U-pipe borehole heat exchanger. In Proceedings of the Effstock—The 11th International Conference on Energy Storage, Stockholm, Sweden, 14–17 June 2009.
25. Ingersoll, L.R.; Plass, H.J. Theory of the ground pipe heat source for the heat pump. ASHVE Trans. 1948, 54, 167–188.
26. Witte, H.J.L. Error analysis of thermal response tests. Appl. Energy 2013, 109, 302–311.
27. Banks, D. An Introduction to Thermogeology: Ground Source Heating and Cooling, 2nd ed.; Wiley-Blackwell: Oxford, UK, 2008; pp. 285–286.
28. Nagano, K. Standard Procedure of Standard TRT, version 2.0; Heat Pump and Thermal Storage Technology Center of Japan: Tokyo, Japan, 2011; pp. 1–13.
29. Diao, N.; Li, Q.; Fang, Z. Heat transfer in ground heat exchangers with groundwater advection. Int. J. Therm. Sci. 2004, 43, 1203–1211.
30. Beier, R. Transient heat transfer in a U-tube borehole heat exchanger. Appl. Therm. Eng. 2013, 62, 256–266.

CHAPTER 7

Groundwater Potentials Estimation of a Basement Terrain Using Pumping Test Data for Parts of Sanga Local Government Area, Kaduna State, Northwestern Nigeria

Hamidu Hassan[1], Sidi Muhammad Waru[2], Garga Ali Bukar[3], Kana Muhammad Abdullahi[4]

[1]*Department of Geology, Usmanu Danfodiyo University, Sokoto, Nigeria*

[2]*Department of Geology, University of Maiduguri, Maiduguri, Nigeria*

[3]*Department of Civil and Water Resources Engineering, University of Maiduguri, Maiduguri, Nigeria*

[4]*Department of Geology, Ahmadu Bello University, Samaru Zaria, Nigeria*

ABSTRACT

Aquifers derived from the crystalline basement rocks for parts of Sanga Local Government area of Kaduna State, Nigeria consist of clay, silt, sand, gravel and laterite materials which may be in various proportions. Using Cooper-Jacobs non-equilibrium graphical method, the hydraulic properties were estimated from pumping test data of 18 boreholes. Conductivity varies from 1.02×10^{-2} m/s to 4.07×10^{-2} m/s and transmissivity varies from 1.14×10^{-1} m^2/s to 4.40×10^{-1} m^2/s. The values of specific capacity range between 1.03×10^{-1} m^2/s/m and 9.00×10^{-2} m^2/s/m, and these values indicate that the aquifers in the area have low to moderate ground water potentials. The low yield range of between 0.45 l/s and 1.00 l/s recorded in the area shows the heterogeneous and anisotropy nature of the basement aquifer system in terms of groundwater discharge. On the average, the boreholes assessed have potentials to sustain local to regional supply provided the best drilling method and materials are used and well completion properly done.

Keywords: Pumping Test, Heterogeneous, Anisotropy, Conductivity, Discharge, Sanga

1. INTRODUCTION

Sanga local government is located at the extreme southern part of Kaduna state. It shears boundary with Nasarawa State Southward and covers a total land area of about 1402 Km^2 with a population of 149,333 [1]. Being a Rural Local Government with no pipe borne water, the populace depends, to a greater extent, on surface water sources made of streams with sources from nearby spring, perennial rivers that sustain flow throughout the year and streams and rivers whose flow cease at the peak of the dry season. The surface sources are prone to pollution and contamination. The groundwater sources are mainly Dug wells and Boreholes. Based on the statistic given by Kaduna State ministry of water resources [2] , Sanga Local Government has a total of 128 boreholes; this serves a population of about 18,300 while projected population of 179,354 by the year 2015 is to be without enough portable water supplies, with majority of its population relying on surface water that is not protected from pollution or through the direct harvesting of rainwater during the wet season, a process that is not sustained for a longer period and unreliable. This water deficit can only be made up by the groundwater sources. [3] pointed out that the low yield of boreholes in the crystalline complex was attributed to thinness of the weathered regolith, interconnectivity of the fractures in fractured layers and the geology of the underlying rock materials. This research work is aimed at assessing the groundwater potentials of the area through the use of hydraulic properties determined from borehole pumping test data in the area of study.

The area investigated is located between Latitudes 9°15'N to 9°30'N and Longitude 8°8'E to 8°40'E at the southern part of Kaduna State, Nigeria. The area falls within the Guinea Savannah climatic belt of West Africa which is characterized by two distinct seasons, the wet and dry season. The wet period lasts for a period of 7 months commencing in March/April and terminates in October. The Eastern portion of the area receives more rainfalls due to its close proximity to the Jos Plateau region [4] (Figure 1).

The area consists of rocks that range in age from Pre-Cambrian to Lower Palaeozoic and Quaternary period. Four groups of rocks can be distinguished for the Basement Complex Terrain in the area [4] - [9]. The crystalline basement rocks which consist of gneisses and migmatites with different varieties of the gneisses like the banded gneiss, granite gneiss, biotite gneiss, hornblende gneiss and ortho genessis. The second group consists of the metasediments in the area and is represented by the quartzite. The third group is the intrusive rock consisting mainly of granites. The fourth group of rock is the basalt which is Quaternary in age. The ground water in the area contains in the soft overburden, fractured bedrock, and alluvium aquifers [10]. Parkman international and Parkman Nigerian limited (1997) reported the presence of three hydrogeological units in Kaduna state which includes: a modern alluvium of present day river channels and ancient alluvium of the Fadama under silts, and abnormally thick regoliths overlying Basement Complex, Granite and Metasedimentary rocks and fractures in the fresh granitic rocks, regoliths and ancient alluvium underlying basaltic flows (Figure 2).

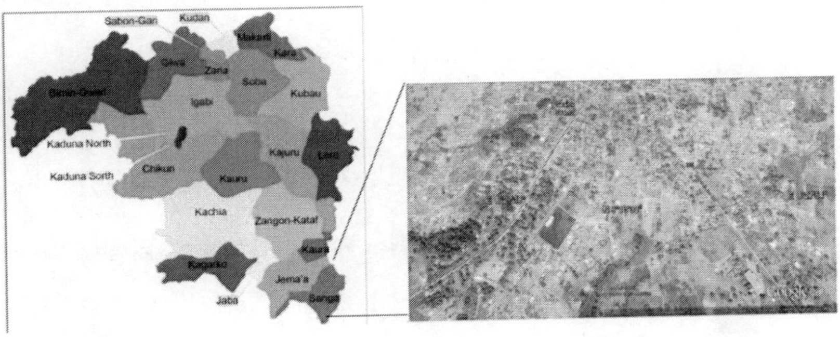

Figure 1. Map of Kaduna state showing the study area and satellite imagery showing part of Sanga local government area.

2. DATA AND METHODS

2.1. Data

Pumping test data for the 18 boreholes tested were obtained from the rural water supply unit of Kaduna state ministry of water resources. The boreholes were drilled under the MDGs (millennium development gold) program for the year 2009 by Carling Earth Resources Nigerian Limited Kaduna Nigeria were used for this study [11]. The pumping phase data were used from the study. Table 1 contains information on the boreholes used for the hydraulic properties assessment.

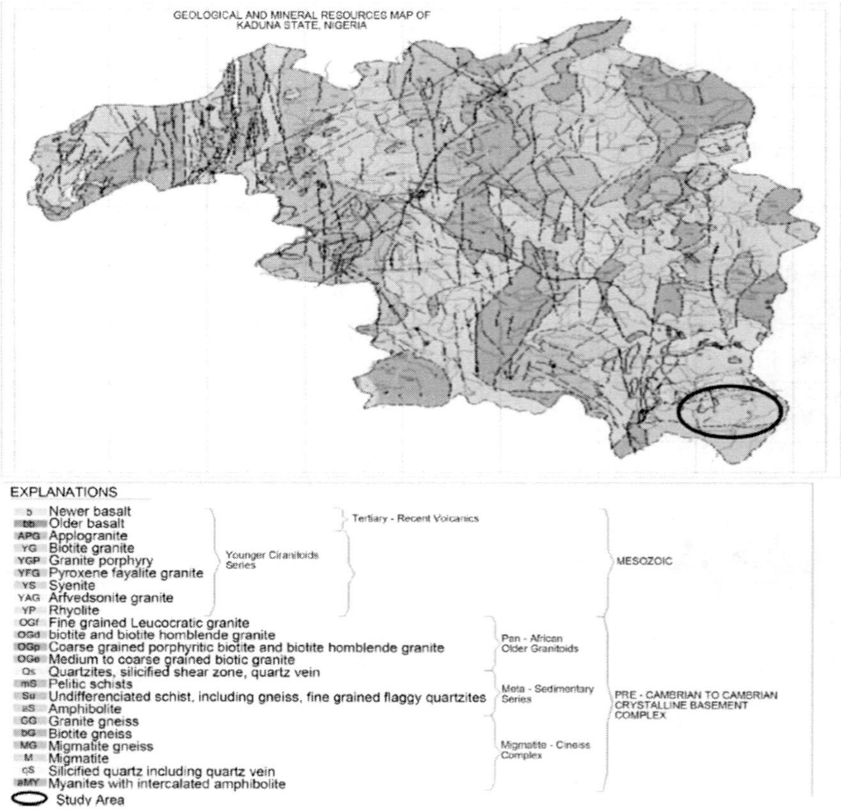

Figure 2. Geologic map of Kaduna state showing the study area (adopted from the geologic and minerals map of Kaduna state of the Nigerian geological survey agency 2006).

2.2. Methods

Pumping test data from 18 boreholes were used. The aquifer was tested for periods of between 60 and 180 minutes depending on the time equilibrium was reached in the individual well. The is' non-equilibrium graphical method is given bellow

$$s = \frac{Q}{4\pi T}\int_{U}^{\infty}\frac{e^{-u}\,du}{u} \tag{1}$$

where s = drawdown,

Q = constant well discharge,

u = well function,

T = Transmissivity,

r = radius of well.

Table 1. Pumping test data of borehole used in the study area.

S/NO	Community	Borehole depth (m)	SWL (m)	Screen length (m)	Final drawdown (m)
1	Agamati Market	30	7.00	9.00	16.20
2	Fadan Karshi	43	7.00	12.00	9.94
3	Kufai Aboro	43	13.48	12.00	6.89
4	Ankara	40	11.74	9.00	13.42
5	Maitozo Distric House	29	6.20	9.00	20.70
6	Kwagiri Village	32	6.20	9.00	16.70
7	LGEA Landa	27	5.1	9.00	10.70
8	Dorowa	30	14.43	9.00	3.80
9	Kwassu Ung. Dutse	33	13.74	12.00	7.03
10	FadanAyu	28	4.00	9.00	12.50
11	Dangam	22	6.50	12.00	9.90
12	LGEA Ancha	24	4.20	9.00	6.80
13	LGEA Aboro	27	7.50	12.00	15.30
14	PHC Kutal	25	5.60	9.00	8.20
15	LGEA Amantu	23	4.00	9.00	12.50
16	Aboro Village	31	7.30	9.00	11.00
17	LGEA SabonGida	30	6.90	9.00	5.10
18	Wasa Village	31	9.20	9.00	13.30

This was used for the estimation of transmissivity (T) and Storage (S) but assuming a long pumping time, Equation (1) above was simplified [12] to

$$T = \frac{2.30Q}{4\pi\Delta s} \text{ Measured in } (m^2/s) \tag{2}$$

where Δs = drawdown over one long cycle.

In determining the slope Δs early times were excluded because for early time approximation is not valid due to casing/wellbore storage, true aquifer response to pumping is masked until storage is exhausted. Only the medium and late times were considered, these represent a transition period during which matrix blocks feed their water to an increasing rate to fractures resulting in partly stabilizing drawdown and pumped water coming from storage in the fractures and matrix block respectively [13]. The pumping phase data were plotted on a semi-log graph sheet against the corresponding time; a straight line was then drawn through the field data points. The values of the drawdown per log cycle of time were determined from the slope of the graph as Δs. The above Cooper-Jacob equation of (1946) was used for the estimation of the transmissivity of the aquifers in the area, single well model was utilized for the studies since no observations wells was available [14]. Meaning the pumping well also serves as the piezometer, the straight line obtained for the semi-log plot of drawdown against time indicates the radial flow of water into the pumped well (Figure 3).

The Hydraulic Conductivity K is computed from the relation:

$$K = T/B \text{ measured in (m/s)} \tag{3}$$

where B = aquifer thickness or length of screen used,

T = Transmissivity.

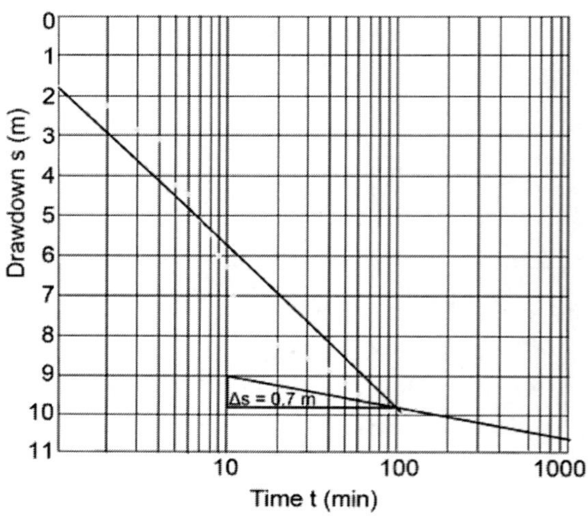

Figure 3. A semilog time-drawdown curve plot of pumped-tested borehole at Dangam.

While Specific capacity was computed from the equation

$$Q/s \text{ Measured in } m^3/s \tag{4}$$

where Q = borehole yield or discharge,

s = total drawdown recorded in pumped well.

3. RESULTS AND DISCUSSION

Transmissivity recorded in the area ranges between 1.14×10^{-1} and 4.40×10^{-1} with an average of 2.18×10^{-1} while the hydraulic conductivity ranges between 1.02×10^{-2} and 4.07×10^{-2} with an average of 2.18×10^{-2}. Hydraulic conductivity obtained falls within the range given in Table 2 and Table 3, this shows that the composition of materials of the aquifer is heterogeneous in nature comprising of finer clay, silt, gravel and fractured basement rocks with average to low yield.

Reduction as well as break in discharge experience during pumping, is an indication of the heterogenic and anisotropic nature of the basement aquifer in the study area. The low to average Transmissivity value of 2.18×10^{-2} m^2/s of the aquifer is attributed to the decrease in aquifer thickness and saturation. High average Drawdown recorded values obtained shows the aquifer to be poor in terms of both recharge and discharge of groundwater; this is also an indication of the inefficiency of the aquifer as a transmitting medium. It is evidently clear also that the low yields range of between 0.45 to 1.00 l/s recorded, high level of drawdown recorded and poor aquifer properties obtained for the area, indicated that the aquifers in this area are predominantly made of fine clayey or Silty materials which are derived from the in situ chemical decomposition of the basement rock which are generally characterised with low permeability and poor water yield. The value range of conductivity obtained in this study fall within the range of $10^{-3}$to 10^{-1} as given by [12] for clay, sand and gravel mixtures (till). Similarly poor performance of the fractured aquifer in the area can be attributed to low fracture density with poor conduit connectivity which characterise most basement aquifers. Although they yields obtained can sustain water abstraction by hand pump for small population, it will not be sufficient for a higher populated community with greater water demand.

Table 2. Aquifer characteristic in the study area.

S/N	Community	Discharge Q (l/s)	Transmissivity T (m^2/s)	Hydraulic conductivity K (m/s)	Specific capacity $m^2/s/m$	Drawdown per log cycle Δs (m)
1	Agamati Market	0.50	1.31×10^{-1}	1.10×10^{-2}	3.10×10^{-2}	0.70
2	Fadan Karshi	0.86	1.43×10	1.20×10^{-1}	9.00×10^{-2}	0.11
3	Kufai Aboro	0.84	4.40×10^{-1}	3.66×10^{-2}	4.00×10^{-2}	0.20
4	Ankara	0.68	2.07×10^{-2}	2.30×10^{-2}	5.10×10^{-2}	0.60
5	Maitozo District Head House	0.60	3.04×10^{-1}	3.05×10^{-2}	3.00×10^{-2}	0.40
6	Kwagiri Village	1.00	2.30×10^{-1}	2.54×10^{-2}	6.00×10^{-2}	0.80
7	LGEA Landa	0.75	2.30×10^{-1}	2.54×10^{-2}	7.01×10^{-2}	0.60
8	Dorowa	0.70	3.00×10^{-1}	3.20×10^{-2}	1.84×10^{-1}	0.45
9	Kwassu Ung. Dutse	0.50	2.61×10^{-1}	2.20×10^{-2}	7.11×10^{-2}	0.35
10	Fadan Ayu	0.64	1.60×10^{-1}	1.74×10^{-2}	5.12×10^{-2}	0.70
11	Dangam	0.70	1.83×10^{-1}	1.53×10^{-2}	7.07×10^{-2}	0.70
12	LGEA Ancha	0.70	3.22×10^{-1}	3.60×10^{-2}	1.03×10^{-1}	0.40
13	LGEA Aboro	0.60	1.22×10^{-1}	1.02×10^{-2}	4.00×10^{-2}	0.70
14	PHC Kutal	0.45	2.10×10^{-1}	2.30×10^{-2}	5.50×10^{-2}	0.40
15	LGEA Amantu	0.50	1.14×10^{-1}	1.30×10^{-2}	4.00×10^{-2}	0.80
16	Aboro Village	0.70	2.60×10^{-1}	2.85×10^{-2}	4.02×10^{-1}	0.50
17	LGEA SabonGida	0.80	4.00×10^{-1}	4.07×10^{-2}	2.00×10^{-1}	0.40
18	Wasa Village	0.50	1.14×10^{-1}	1.30×10^{-2}	4.00×10^{-2}	0.80

Table 3. Summary of aquifer properties for the 18 boreholes in the area of study.

Aquifer property	Transmissivity T (m^2/s)	Hydraulic conductivity k (/m/s)
Minimum	1.14×10^{-1}	1.02×10^{-2}
Maximum	4.40×10^{-1}	4.07×10^{-2}
Mean	2.18×10^{-1}	2.18×10^{-2}

4. CONCLUSION

The area is characterised by low to moderate groundwater resources, since this is the readily available portable source of water available, there is a need for this resource to be properly harness by carrying out an in-depth investigation on the groundwater resources reserve in the area. Similarly it is strongly recommended that a longer duration of pumping test period of about 24 hours should be adopted in the future so that a better assessment of the aquifer productivity and water potentials can be fully properly determined. The use of modern technology and scientific methods such as groundwater modelling will assist in giving the clear picture of the subsurface and understanding of the groundwater resources of the area. It will also give the chance to water managers to properly plan for the present and future water demand for the fast growing population in the area before the provision of pipe borne water is realised in the area.

ACKNOWLEDGEMENTS

They authors wish to thanks the staffs management of Kaduna state ministry of Water resources for making available the pumping test results and other relevant inform that has assisted in the facilitating this study.

REFERENCES

1. Nigerian Population Commission (2007) Official Gazette 2006 Population Census, Nigeria. Vol. 94, Abuja, Nigeria.
2. Kaduna State Ministry of Water Resources (2008) Baseline Survey and Inventory of Boreholes in Kaduna State. Teamwork Global Associates, Kaduna.
3. Hamidu, H., Garba, M.L., Kana, M.A. and Yelwa, N.A. (2014) Causes of Low Groundwater Yield of Boreholes in Crystalline Basement Complex: a Case Study of Gwantu Town and Environs Southern Part of Kaduna State Nigeria. Water Resources Journal of the Nigerian Association of Hydrogeologists, 24, 20-33.

4. Jacobson, R. and Webb, J.S. (1946) The Pegmatite of Central Nigeria Geol. Survey of Nigeria Bulletin No. 17S.
5. McCurry, P. (1976) The Geology of the Precambrian to Lower Paleozoic Rocks of Northern Nigeria. A Review. In: Kogbe, C.A., Ed., Geology of Nigeria, Elizabeth Pub. Co., Ibadan, Nigeria, 15-38.
6. Okezie, C.N. (1970) Interim Report on the Geology of the Basement Complex of the 1:100.000 Sheet 167 Kafanchan S.W. Unpublished GSN Report No. 1347.
7. Nehikhare, J.I. (1971) Geology of Western half of 1:100.000 Sheet 188 (Jema'a) Geology Survey of Nigeria. Unpublished Report No. 1472.
8. Macleod, W.N. (1971) The Geology of the Jos Plateau with 1:100,000 Sheets 147, 148, 168, 169, 189 and 190. GSN, Bulletin No. 32.
9. Young, B. (1962) Report on the Basement Complex Geology of the Eastern Half of the 1:100,000 Sheet 189 (Kurra). Unpublished GSN Report No. 1323.
10. Parkman International and Parkman Nigeria Limited (1997) State-Wide Water Resources Master Plan, Kaduna State.
11. Kaduna State Water Board (2009) Analysis of Meteorological Data. Hydrological Department, Kaduna State Water Board, Kaduna.
12. Bouwer, H. (1978) Groundwater Hydrology, McGraw-Hill Book, New York, 480.
13. Kruseman, G.P. and de Ridder, N.A. (2000) Analysis and Evaluation of Pumping Test Data. 2nd Edition, International Institute for Land Reclamation and Improvement, 372.
14. Cooper, H.H. and Jacob, C.E. (1946) A Generalized Graphical Method for Evaluating Formation Constants and Summarizing Well-field History. Trans-America Geophysical Union, 27, 526-534.

CHAPTER 8

Assessment of Climate Change Impacts on the Water Resources of Megech River Catchment, Abbay Basin, Ethiopia

Endalkachew Abebe[1*], Asfaw Kebede[2]

[1]*Department of Soil and Water Resource Management, Woldia University, Woldia, Ethiopia*

[2]*School of Natural Resource and Environmental Engineering, Institute of Technology, Haramaya University, Dire Dawa, Ethiopia*

ABSTRACT

This study was aimed to assess the impact of climate change on the water resource of Megech river catchment. In this study, large scale regional climate model (REMO) output was downscaled statistically to metrological variables at a daily resolution using SDSM model version 5.11. We noticed that statistical downscaling smooth out the bias between REMO output and observed data. According to the projected climate data, the maximum temperature is likely to have an increasing trend +0.57°C while the minimum temperature shows a decreasing trends −0.61°C. There is no clear trend for precipitation, both increasing and decreasing trend observed in the catchment. The HBV- Light hydrological model was successfully calibrated (1991-1995) and validated (1998-2000) using current climatic inputs and observed river flows. The overall performances of the model was good at monthly time scale both on calibration (NSE = 0.91) and validation (NSE = 0.86). Future discharge (2015- 2050) was simulated using statistically downscaled 20 ensembles climate scenario data for both A1B and B1 scenarios. HBV-Light model simulation results showed a reduction of the peak discharge in August and September.

Keywords: HBV-Light Model, Megech, REMO, SDSM

1. INTRODUCTION

Climate change refers to a change in the state of the climate that can be identified by changes in the mean and/or the variability of its properties and that persists for an extended period, typically decade or more [1]. Climate model projections show an increase in the global mean near-surface air temperature [2].

It is now widely acknowledged that climate change will have impacts on water resources availability and management throughout the world, in the near and longer terms. Some of the sectors under concern include urban water supply, irrigated agriculture and hydropower production [1] [3].

According to [4], the current climate variability is already imposing a significant challenge to Ethiopia by affecting food security, water and energy supply, poverty reduction and sustainable development efforts, as well as by causing na- tural resource degradation and natural disasters. Therefore, assessing the impact of climate change on the water resource of Megech river catchment will be expected to have importance to be considered in development plans in water resources, agriculture and to overcome the impacts of intensifying recurrent droughts. This gives an opportunity to plan appropriate adaptation of measure s that must be taken ahead of time based on the projected climate change.

Several studies have attempted to evaluate the impacts of climate change in Ethiopia [5] [6] [7] . Most of this studies used Global Climate Model (GCM) which has a coarser resolution and some degree of uncertainty. In addition, these studies did not use model ensemble predictions, which provide a means of forecasting robust simulations for weather and climate prediction uncertainties. No Impact assessment of climate change study had been performed for the Megech catchment of the Abbay basin, even though there have been studies conducted on other part of the Abay basin [6] [7] [8]. The objective of this study was to assess the impact of climate change on the water resource of Megech River catchment.

2. MATERIALS AND METHODS

2.1. Study Area

Megech catchment is located in the Northwest part of Ethiopia between 12°29'00" to 12°43'40" latitude and 37°23'53" to 37°33'35" longitude with an approximation altitude range between 1849 and 2942 m.a.s.l. The catchment covers a total surface of 513 km^2 as it shown in Figure 1. The annual rainfall ranges between 896mm and 1592 mm. The monthly maximum temperature is between 21°C and 27°C, and the monthly minimum temperature is between 10°C and 13°C. The study area had been selected due to the reason that it is one of the major tributaries of Lake Tana and no impact assessment studies had been done on the catchment.

2.2. Methodology

In this work, the output variables from REMO (Regional Climate Model) for both emission scenarios of A1B and B1 were statistically downscaled by using Statistical Downscaling Model (SDSM) version 5.11. HBV-Light hydrological model was calibrated and validated using historical climate data of three stations (Ambagiorgis, Gondar and Maksengit) and observed discharge data of Megech river. The downscaled future scenario 20 ensembles data of both REMO A1B and B1 scenario were used as an input for HBV-Light model to assess catchment hydrological response to climate change.

2.2.1. Climate Model and Downscaling

For this study the climate scenario data (A1B and B1) were extracted from REMO model based on longitude and latitude that had grid resolution of 50 km (0.5° latitude by 0.5° longitude grid size).The coarser climate data (REMO output) further downscaled in to station level by using statistical downscaling model (SDSM version 5.1.1) and these downscaled data was taken directly as an input to the hydrological model to assess the future climate change impact on hydrology of the catchment.

2.2.2. REMO Model

The regional climate model REMO is a hydrostatic regional climate model developed at the Max Planck Institute for Meteorology on the basis of the former operational weather forecast model Europa-model of the German Weather Service (DWD) [9] . In addition, the physical parameterization package of the general circulation model ECHAM4 has been implemented in REMO. REMO quantification of the IPCC A1B and B1 (Medium-Low

Emission) scenarios for the years 2015?2050 was used in this study after being statistically downscaled to the local scale using the SDSM 5.11.

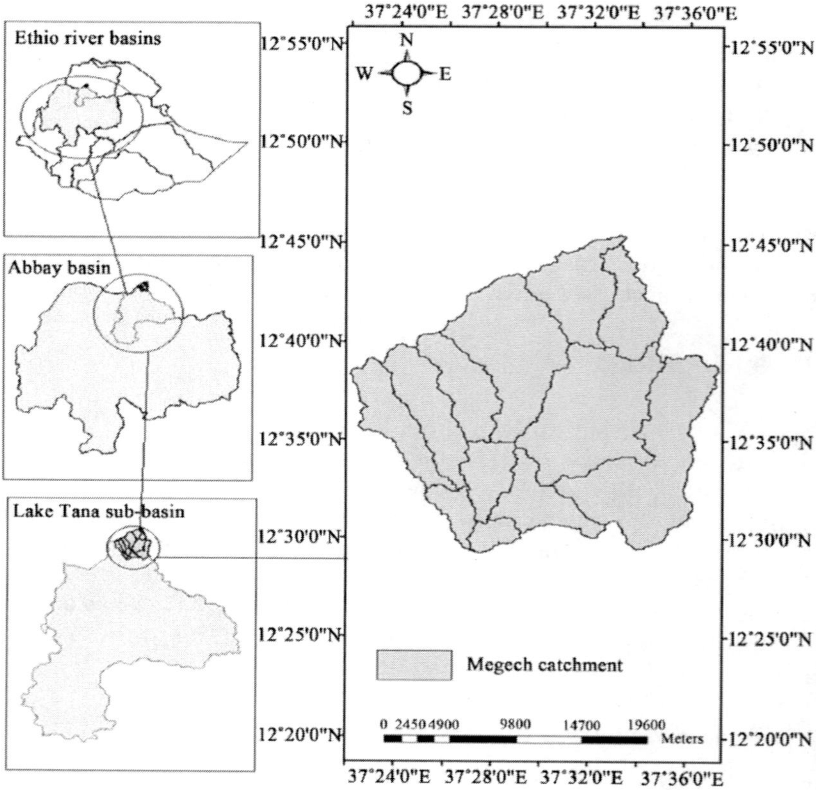

Figure 1. Location of Megech catchment.

2.2.3. Statistical Downscaling Model

According to Wilby et al. [10] empirical downscaling involves developing a quantitative relationship between large-scale atmospheric variables (predictors) and local surface variables (predictands). For this study the REMO grid nodes data were takes as predictors and the station data were taken as predictands. The catchment was covered by 8 REMO raster nodes. Among those 3 grids nodes were used for downscaling. The base line data for the base period were from 3 stations in and around the Megech catchment within the range of 30 to 21 years period from 1971-2000 and

1980-2000 respectively. The first 20 (1971-1990) to 14 (1980-1993) years of data were considered for calibrating SDSM while the remaining 10 (1991-2000) to 7 years (1994-2000) respectively, were used for validation. After calibrating SDSM model, the future climate scenarios (2015-2050) was generated based on the calibrated parameter and large scale predictor (REMO predictor) based on the mean of 20 ensembles for both A1B and B1 scenarios.

2.2.4. Grid Selection for Statistical Downscaling Model (SDSM)

The REMO model which covers the area of interest (12°29'00''N to 12°43'40''N and 37°23'53''E to 37°33'35''E) was used in this study. REMO grid outputs data were classified based on their grid location (Latitude and Longitude) with 0.5° horizontal interval. The catchment was covered by 8 REMO raster nodes as shown in Figure 2. Among those 3 grids nodes were used for downscaling. The grid data were selected as a predictor for a given metrological station and the station data were taken as predictand.

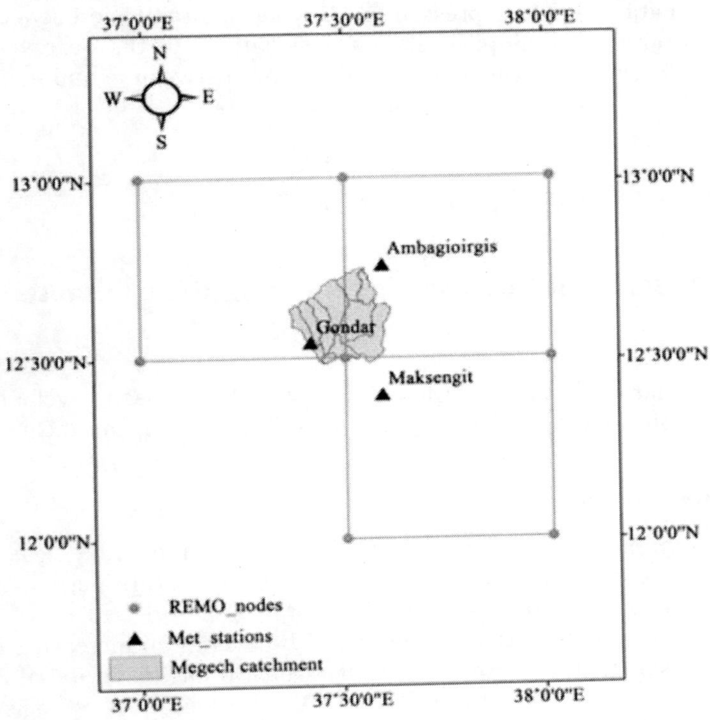

Figure 2. REMO grids selected for the study area.

2.2.5. Hydrological Modeling

To simulate the water balance components of the catchment, HBV-Light model was used. The HBV (Hydrologiska Byråns Vattenbalansavdelning)-Light hydro- logy model is a widely used conceptual model [11] . It computes runoff from observed daily rainfall, daily temperature, long-term monthly potential evapotranspiration and runoff data. The daily areal rainfall (1991-2000) was calculated by Thiessen polygon method. Potential evapotranspiration for the study area was estimated by the Hargreaves method. The total period of the data that was used for this specific study was 8 years. From this period by using split sample technique 2/3rd of the data (1991-1995) were used for calibration and the remaining 1/3rd of the data (1998-2000) were applied for validation.

2.2.6. Model Performance Criteria

For this particular study, three performance criteria namely Nash and Sutcliffe efficiency (NSE), Coefficient of determination (R^2) and Relative volume error (RVE) were used. Nash and Sutcliffe efficiency (NSE) was proposed by [12] , its value lies between 1.0 (perfect fit) and $-\infty$. Coefficient of determination (R^2) is expressed as [13] the squared ratio between the covariance and the multiplied standard deviations of the observed and predicted. Relative volume error (RVE) can vary between ∞ and $-\infty$ but it performs best when a value of 0 (zero) is gene- rated.

3. RESULTS AND DISCUSSION

3.1. REMO output Downscaling for the Future Climate Scenarios

Climate scenarios for future periods (2015-2050) have been developed for two emission scenarios of REMO A1B and B1 based on the mean of 20 ensembles.

1) Maximum temperature

As it shown in Figure 3 and Figure 4, the mean monthly maximum temperature shown generally an increase trend for future period (2015-2050) for both A1B and B1 scenario in the range of +0.1°C to +0.51°C and +0.12°C to +0.57°C respectively. There will be overall an increasing trend for Gondar sta- tion for both A1B and B1 scenario for the period of 2015-2050 in the range of +0.17°C to +0.8°C. However Ambagirogis and Maksengit stations show almost no change for both A1B and B1 scenario.

2) Minimum Temperature

As it revealed in Figure 5 and Figure 6 generally there will be a decreasing trend for mean monthly minimum temperature for Both A1B and B1 scenario in the range of −0.11°C to −0.61°C and −0.12°C to −0.60°C respectively for future window (2015-2050).

3) Precipitation

As it depicted in Figure 7 and Figure 8, the mean monthly precipitation shows both increasing and decreasing trend for both A1B and d B1 scenario for future period (2015-2050). It is likely to be increased in the range of +6.7% to +34.5% and +11.0% to +38.89% while it is expected to be decreased in the range between −1.14% to −31.88% and −1.6% to −36.42% for A1B and B1 scenarios respectively.

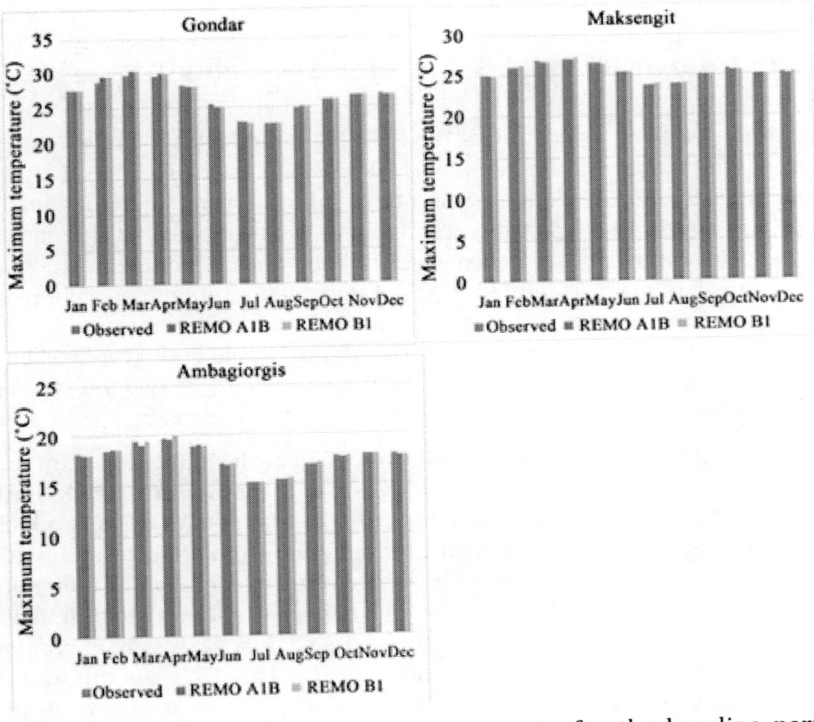

Figure 3. Monthly mean maximum temperature for the baseline period (1971-2000) and REMO A1B and B1 scenarios of selected stations.

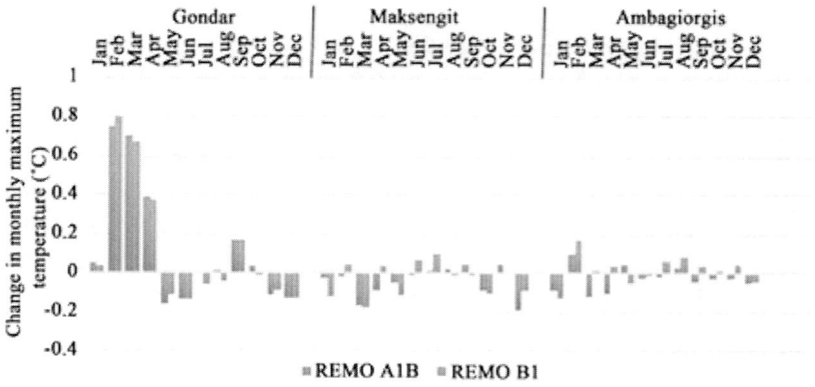

Figure 4. Change anomalies of mean monthly maximum temperature of selected stations for future period (2015-2050) for both A1B and B1 scenarios.

3.2. Hydrological Model Calibration and Validation

The calibration and verification of the HBV-Light model were implemented by using split sample technique (1991-1995) data for calibration and the remaining (1998-2000) data for validation. Calibration was done manually by optimizing the model parameters that gave the best possible agreement between observed and simulated discharge of a catchment. Due to that several runs had been attempted to select the most optimum parameter set that shows good agreement between observed and simulated discharge. The performance of the model has been evaluated using both graphical and statistical criteria. On graphical analysis the agreement between observed and simulated discharge was evaluated through visual inspection.

The statistical criteria's selected for showed good performance for daily and monthly calibration (with NSE = 0.91 and R^2 = 0.92 for monthly simulation). In addition, the models were validated using independent data set, which shows good agreement for both daily and monthly simulation results (with NSE = 0.86 and R^2 = 0.87 for monthly simulation). Generally speaking, the results show that the HBV-Light model can reproduce historical daily discharge with an acceptable accuracy. Previous impact assessment studies done using HBV model showed similar result [6] [14] [15] . The calibration and validation result of the HBV- Light model is shown in Figure 9 and Figure 10.

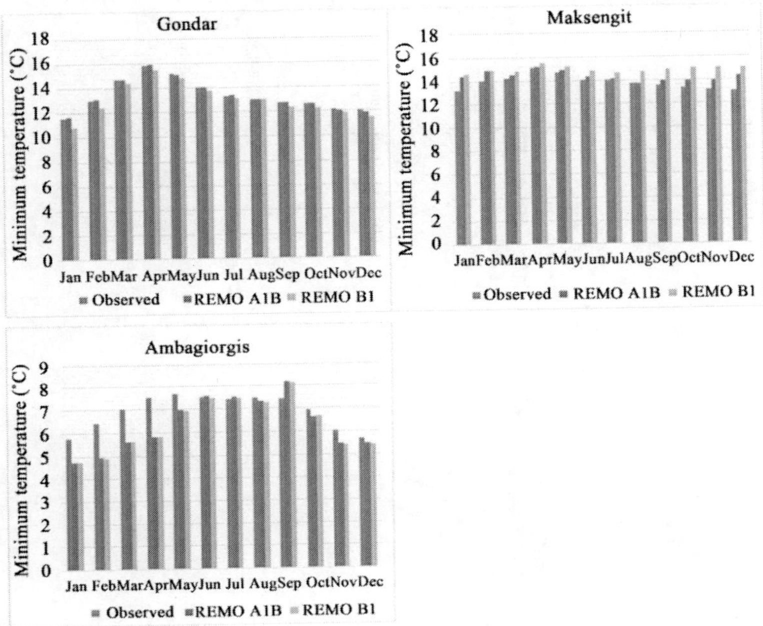

Figure 5. Monthly mean minimum temperature for the baseline period (1971-2000) and REMO A1B and B1 scenarios of selected stations.

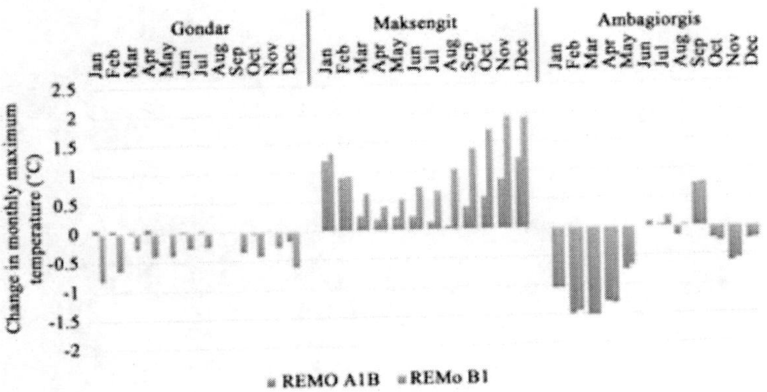

Figure 6. Change anomalies of mean monthly minimum temperature of selected stations for future period (2015-2050) for both A1B and B1 scenarios.

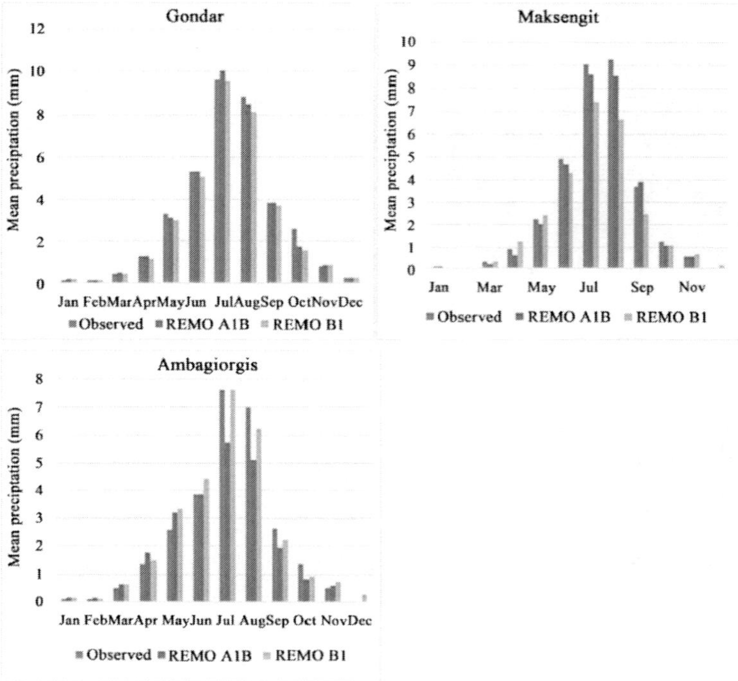

Figure 7. Monthly mean Preciptation for the baseline period and REMO A1B and B1 scenario of selected stations.

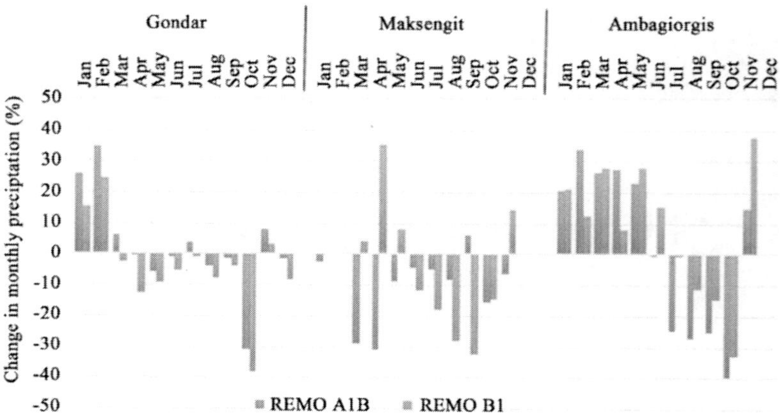

Figure 8. Change anomalies of mean monthly precipitation of selected stations for future period (2015-2050) for both A1B and B1 scenarios.

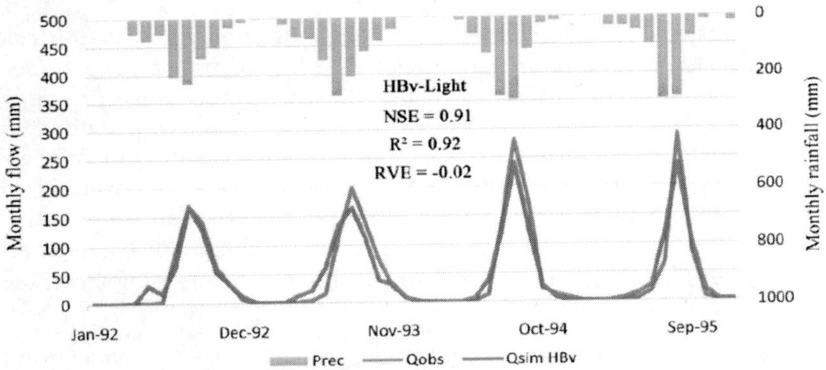

Figure 9. Monthly calibration result of Megech catchment (1992-1995).

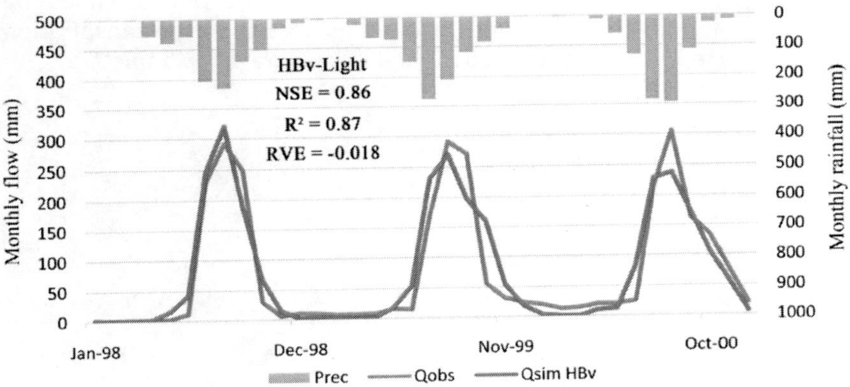

Figure 10. Monthly validation of Megech catchment (1998-2000).

3.3. Catchment Predicting Hydrological Response of Changed Climate

For analyzing the catchment response of climate change, 20 ensembles of climate variables (i.e. Precipitation and temperature) for both REMO A1B and B1 scenario were developed). Each 20 ensembles of climate data have been used as an input for simulating the discharge of Megech catchment for future period (2015- 2050). Among the simulated 20 ensembles the maximum and minimum one has been selected to find the range of discharge Change for future period (2015- 2050). As displayed on Figure 11 and Figure 12 the HBV-Light hydrological model showed a reduction in peak discharge (August & September) for future period (2015-2050. Simulation result showed that there will be a shift in wet season towards May to July. There will a reduction of peak flow for August and September (−17.47% to −30.58%) this is mainly because for those periods temperature shows no change while Precipitation expected to be fall down up to −10%. For the low flow (October to December) HBV-Light models shows a decreasing trend. For low flow of the dry season (November to April), since there will be an increment of both maximum temperature (+0.57°C) and precipitation (+38.39%), simulation result showed almost no change except for the month of April which shows an increasing trend for both A1B and B1 scenarios.

Generally the mean annual simulated water balance for the observed period and future climate change scenarios is portrayed in Table 1. The table results are the averages of the 20 ensemble run for each climate scenario of potential evapotraspiration (ETP), actual evapotraspiration (ETR) and total discharge (QT). Change in percentage (Δ%) had been determined by comparing the reference observed period discharge and evpotranspiration from the future scenarios (2015-2050) and it is portrayed in Table 1.

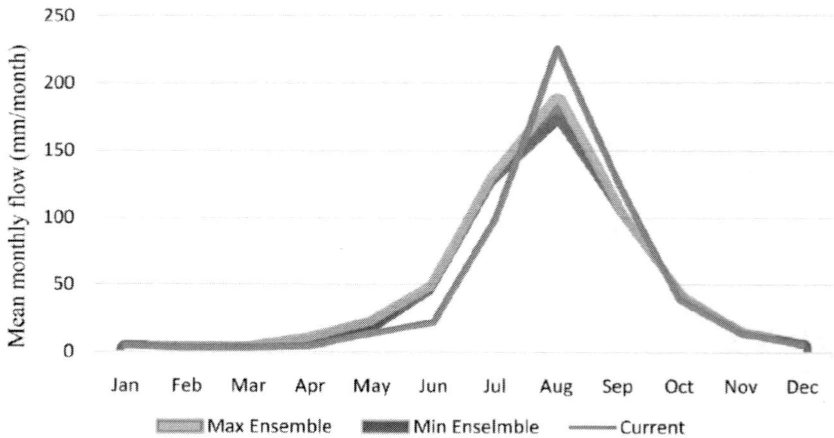

Figure 11. Comparison of current and future period discharge (2015-2050) for maximum and minimum ensembles of REMO A1B Scenario as it simulated by HBV-Light model.

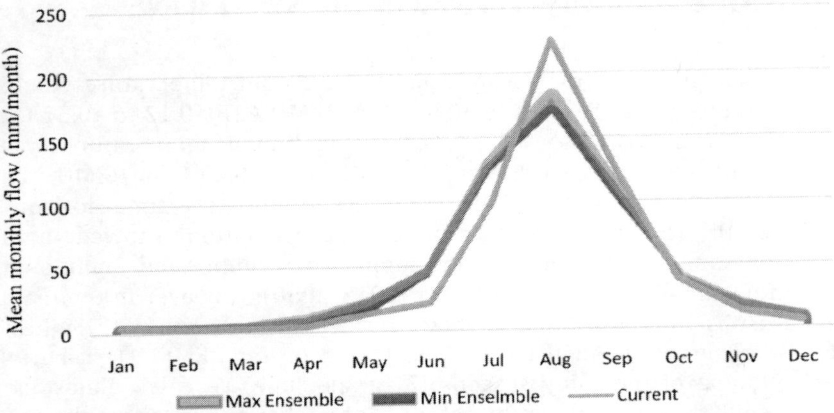

Figure 12. Comparison of current and future period discharge (2015-2050) for maximum and minimum ensembles of REMO B1 Scenario as it simulated by HBV-Light model.

Table 1. Mean simulated annual water balance for the observed period and future climate change REMO A1B and B1 scenarios in the Megech River catchment.

		Q_T (mm)	ETP (mm)	ETR (mm)	Q_T (Δ%)	ETP (Δ%)	ETR (Δ%)
HBV-Light	Observed (1991-1995, 1998-2000)	559	1424	433	0	0	0
	REMO A1B						
	2015-2024	545.76	1459.13	485.62	−2.37	+2.46	+12.15
	2025-2034	543.78	1460.0	477.58	−2.72	+2.52	+10.29
	2035-2044	531.08	1463.76	475.53	−4.99	+2.79	+9.82
	2045-2050	534.60	1462.49	477.08	−4.36	+2.70	+10.18
	REMO B1						
	2015-2024	539.62	1435.6	487.16	−3.46	+0.81	+12.50
	2025-2034	540.0	1439.8	478.20	−3.39	+1.10	+10.43
	2035-2044	531.80	1438.71	478.11	−4.86	+1.03	+10.41
	2045-2050	527.20	1446.25	481.95	−5.68	+1.56	+11.30

4. CONCLUSIONS AND RECOMMENDATIONS

As the result indicates the mean monthly maximum temperature would increase in the range of +0.1°C to +0.51°C for REMO A1B +0.12 to +0.57 for REMO B1 scenarios. The mean monthly minimum temperature was expected to be decreased in the range of −0.11°C to −0.25°C for REMO A1B and −0.12 to −0.57 for REMO B1 scenarios in all future time horizons (2015-2050). The result of downscaled precipitation showed both increasing and decreasing trend unlike maximum and minimum temperature in all future time horizon (2015-2050). However in the main rainy seasons (July and August) showed a decreasing trend with highest decrease observed in August with 10% reduction for REMO B1 scenario. The beginning of the rainy seasons (May and June) revealed almost no change while the end of the rainy season (September and October) showed a decreasing trend for both A1B and B1 scenarios in all future time horizon.

The calibration and validation results of HBV-Light shows that the model is able to reproduce discharge with good performance (R^2 = 0.92, NSE = 0.91 RVE = 0.02). The performances of the model in simulating future discharge (2015- 2050) were tested using downscaled REMO climate scenario data. The hydrological models showed that there will be a reduction of peak flow for August and September (−17.47% to −30.58%). For the low flow (October to December) HBV- Light models shows a decreasing trend while for low flow of the dry season (November to April) almost no change will be for both A1B and B1 scenarios.

The study showed that there would be a reduction in precipitation in major rainy season (Kiremit) and an increment of precipitation in Bega season. Coping mechanisms like adaptation actions should be taken in accordance with the present and future scientific studies include: changing the cropping calendar, use of moisture stress resistant crops, adopt different irrigation methods, crop diversification, mixed crop-livestock farming systems, and scaling up the soil and water conservation practices. The catchment water management system should be in accordance with the future trends of rainfall peaks as the temporal shift in peak rainfall showed a direct impact on the flow of Megech river catchment.

According to the study, there would be a reduction in precipitation in major rainy season (Kiremit) and an increment of precipitation in Bega season. Therefore, soil and water conservation activities should be adopted by the community as well as water harvesting structure should be properly designed and applied on the catchment since for the future period temperature and evapotrasnpiration would increase and would cause a reduction in the flow of Megech River.

REFERENCES

1. IPCC (Intergovernmental Panel on Climate Change) (2007) Climate Change 2007: the Physical Science Basis. Contribution of Working Group II to the Fourth Assessment Report of the Intergovernmental Panel on Climate Change, Cambridge University Press, Cambridge, UK
2. IPCC (Intergovernmental Panel on Climate Change) (2001) Climate Change: The Scientific Basis. In: Houghton, J.T., Ding, Y., Griggs, D.J., Noguer, M., van der Linden, P.J., Dai, X., Maskell, K. and Johnson, C.A., Eds., Technical Summary of the Working Group I Report, Cambridge University Press, Cambridge, UK
3. Jasper, K., Calanca, P., Gyalistras, D. and Fuhrer, J. (2004) Differential Impacts of Climate Change on the Hydrology of Two Alpine River Basins. Climate Research, 26, 113-129. https://doi.org/10.3354/cr026113
4. Tadege, A., Ed. (2007) Climate Change National Adaptation Programme of Action (NAPA) of Ethiopia. Ministry of Water Resources and National Meteorological Agency, Addis Ababa.
5. Beyene, T., Lettenmaier, D.P. and Kabat, P. (2010) Hydrological Impact of Climate Change on the Nile River Basin: Implication of the 2007 IPCC Scenarios. Climate Change, 100, 433-461. https://doi.org/10.1007/s10584-009-9693-0
6. Abdo, K.S., Fiseha, B.M., Rientjes, T.H., Gieske, A.S. and Haile, A.T. (2009) Assessment of Climate Change Impacts on the Hydrology of Gilgel Abay Catchment in Lake Tana Basin, Ethiopia. Hydrological Processes, 23, 3661-3669.
7. Belay, H. (2011) Evaluation of Climate Change Impacts on Hydrology on Selected Catchments of Abbay Basin. MSc Thesis, Institute of Technology, Addis Ababa University, Addis Ababa, Ethiopia.
8. Mulushewa, Z. (2013) Hydrological Impacts of Climate Change on Lake Tana's Water Balance. MSc Thesis, Faculty of Geo-Information Science and Earth Observation, University of Twenty, The Netherlands.
9. Jacob, D. (2001) A Note to the Simulation of the Annual and Inter-Annual Variability of the Water Budget over the Baltic Sea Drainage Basin. Meteorology and Atmospheric Physics, 77, 61-73https://doi.org/10.1007/s007030170017
10. Wilby, L.R., Zorita, E., Timbal, B., Whetton, P., Mearns, L.O. and Charles, S.P. (2004) Guidelines for Use of Climate Scenario Developed from Statistical Downscaling Methods. Environmental Agency of England and Wales, UK.
11. Seibert, J. (2005) HBV Light Version 2. User's Manual. Department of Physical Geography and Quaternary Geology, Stockholm University, Stockholm.
12. Nash, J.E. and Sutcliffe, J.V. (1970) River Flow Forecasting through Conceptual Models, Part I—A Discussion of Principles. Journal of Hydrology, 10, 282-290.
13. Krause, P., Boyle, D.P. and Base, F. (2005) Comparison of Different Efficiency Criteria for Hydrological Model Assessment. Advances in Geosciences, 5, 89-97. https://doi.org/10.5194/adgeo-5-89-2005
14. Gebreyesus, K. (2014) Impact Assessment of Climate Change on the Hydrology of Gojeb River Catchment in Western Omogibe River Basin, Ethiopia. MSc Thesis, Haramaya University, Haramaya, Ethiopia.
15. Kebede, A. (2013) Downscaling Climate Model Outputs for Estimating the Impact of Climate Change on Water Availability over the Baro-Akobo River Basin, Ethiopia. Doctoral Dissertation, Bonn University, Germany.

CHAPTER 9

Integrating Geotechnical Investigation with Hydrological Modeling for Mitigation of Expansive Soil Hazards in Tabuk City, Saudi Arabia

Abd-Alrahman Embaby[1], Ayman Abu Halawa[2], Medhat Ramadan[1]

[1]*Geology Department, Faculty of Science, Damietta University, Damietta, Egypt*

[2]*Nuclear Material Authority, Cairo, Egypt*

ABSTRACT

Due to global climatic changes, flash floods are followed as a yearly disaster with high magnitude of influence. During the years 1981, 1988, 2010, 2012 and in January 2013, Tabuk city, northwest of Saudi Arabia suffered huge flash floods. These are major factors affecting on the swelling behavior of expansive Tabuk shale. The examined geotechnical properties of the surface and subsurface lithology of the sedimentary deposits distinguished Tabuk city into three zones. The expansive zone is spread in the middle and the non-expansive zones are distributed in the east and west of the city. The Watershed Modeling System (WMS) and Hydrologic Engineering Center (HEC-1) models were used to delineate and identify the drainage system and basin morphometry, where flash floods and accumulation of water might take place. Integration between geotechnical distribution maps of the expansion soil and surface hydrological data in terms of runoff maps was done. It has been identified the whereabouts the soils which have expansion characteristics and areas prone to flooding and surface runoff. They are helpful in defining the hazard zones map. Based on this map, it can be avoided constructions on the risk neighborhoods such as Al Qadsiyah, Al Maseif, Arrwdah, Al Nakhil and Al Rajhi. Also, it can suggest that the western side of Tabuk city is suitable for future urban extension. These results will help planners and citizens to create alternative development scenarios and determine their impact on the future urbanization patterns. Moreover, the direction of surface runoff flow or storm water discharge should be away from the expansion soil areas. Therefore, constructing dams on the outlet of the high-risk basins, south of Tabuk city is an important solution to control flash flood events, as well as increase groundwater recharge.

Keywords: Expansion Soil Hazards, Flash Floods, Geotechnical Investigation, Saudi Arabia, Tabuk City, Watershed Modeling System

1. INTRODUCTION

Soils and soft rock that tends to swell or shrink due to changes in moisture content are commonly known as expansive soils [1]. Damages to infrastructure caused by expansive soil movement have been widely reported in many countries such as Australia, China, United States, India and South Arica [2]. In America, the economic loss of the expansive soil question approximately amounts to above $15 billion every year [3] [4] [5] more than twice the damage from natural disasters.

One of the major geotechnical problems in Kingdom of Saudi Arabia (KSA), as has recently been noticed, is the swelling potential of the soils upon change in moisture content. Tabuk city, northwest KSA, is underlain by a shale; it is the primary source of expansive soils [6] [7] [8] [9] [10]. The problem in Tabuk is a terrifying for the administrative, municipality officials and legal authorities; luxury villas, modern built, commercial buildings, schools, hospital, mosques and roads are subjected to upheaval forces. It is resulted in serious cracks, tilting, twisting and sticking doors and windows. Damages seen in some districts are estimated in the hundreds of millions US dollars [11] [12] [13]. Conducted com- prehensive studies covering greater parts of the Tabukcity are included in [6] [10] [12] [14] [15] [16].

The hazard posed by expansive soils is greatest in regions with pronounced wet and dry seasons [17]. Therefore, building and infrastructure built in arid to semi-arid regions will be predisposed to expansive soil problems when compared with humid regions that maintain a rather uniform soil moisture condition throughout the year [18]. In years of extreme temperatures or rainfall, the damage is the most severe [19]. Factors that control the behavior of expansive soils have been thoroughly investigated by many researchers [1] [20] [21] [22]. The primary factor is a change in water content. Pronounced changes in water content in soils seasonally occur in many areas of the world. The potential change in water content is generally attributed to the environmental conditions which influence volume change of soil is seasonal moisture variations due to rainfall, runoff and groundwater table.

Water is introduced to the sub-soil through infiltration of rainfall and runoff, rise in the groundwater level, and human activities (leaking from subsurface utilities i.e., water or sewage systems) could adversely affect ambient moisture conditions [23] [24]. Surface drainage leads to moisture accumulation or ponding, which can provide a source of moisture for expansive soil by infiltration.

Tabuk city is hardship surface water scarcity compensated by flash flooding, which has been affecting the entire city for many years during the winter season. The catchment experienced major floods during the years 1981, 1988, 2010, 2012 in January 2013, causing unfortunately the death of many people and serious damage to residential buildings [25] [26]. Hydrological processes must play a dominant role in the swelling of the soil and rocks. Surface hydrology deals with movement of water along the watershed because of precipitation. Drainage basin morphometry plays a significant role in delineating the areas prone to runoff or flooding [27]. Poor surface drainage is a frequent problem associated with constructions on expansive soils. It leads to water accumulation or ponding, which can provide a source of moisture for expansive soil by infiltration. The extent of the infiltration is a function of the transverse and longitudinal gradients in the drainage tributaries, lag time of runoff, volume and peak discharge of runoff and the hydrologic soil groups.

The main objective of the current study is to investigate the runoff hazards on the expanded shale of Tabuk residential neighborhoods and to identify basin tributaries; surface runoff, flash floods location, and depressions allow accumulation or standing water, which activates the expansive shale in a localized manner leading to differential heaving of expansive soil. In addition, it will be carried an integration of geotechnical investigation and surface hydrology to construct a hazard expansive soil map of Tabuk city. It is helpful in determining the suitable future extension and development of urban areas.

2. LOCATION OF STUDY AREA

Tabukcity lies in the northwestern part of KSA, between latitudes 28°20'N and 28°28'N and longitudes 36°26'E and 36°36'E. It is situated on an alluvial plateau with an elevation of about 760 m [28]. Tabuk area is almost flat; it gently slopes downward to the north and northeast with gradually moderate relief. The plateau is frequently cut with several wadis, which drain northwards for some 20 km to the Qa'a Sharawa dry lakebed (Figure 1). This area has the lowest relief and forms an internal drainage for the wadis.

3. METHODOLOGY AND TECHNIQUE

3.1. Geotechnical Properties

The soil samples (disturbed and undisturbed) were collected from the drilled 72 boreholes to depth varied from 10m to 25 m distributed in the Tabuk city (Figure 1). Laboratory tests are carried out to attest the potential expansiveness of the soils by grain size distribution, soil moisture content, Atterberg limits, and swell tests. All of these tests were performed in conformity with the American Standards, ASTM [29].

3.2. Hydrological Methodology

The used rainfall data is collected from the Hydrology Division of the Ministry of Water and Electricity [30]. The available rainfall records cover a period of 49 years (1965 to 2004). HyFrAn PLUS (Hydrological Frequency Analysis) program [31] had been used for the available rainfall data to obtain the corresponding rainfall depth values of the probabilities distribution (Gumbel, General Extreme value GEV, Pearson Type III and Log-Pearson Type III) for different return periods. The Chi-square test comparing computed values with observed values was carried out to identify the best-fit method.

Delineating drainage networks is essential to predict flow of surface water runoff, and consequently the flash flood hazards zones [32] [33]. Basin delineation was condu- cted using the Shuttle Radar Topography Mission (SRTM v.3, 1 arc) 30-m digital elevation model "DEM" (Projected Coordinate System: WGS_1984_UTM_Zone_37 N) in the Watershed Modeling System (WMS, v.10). WMS and Global Mapper (v.17) were used to provide most of the morphometric parameters of the studied basins and their stream attributes [34].

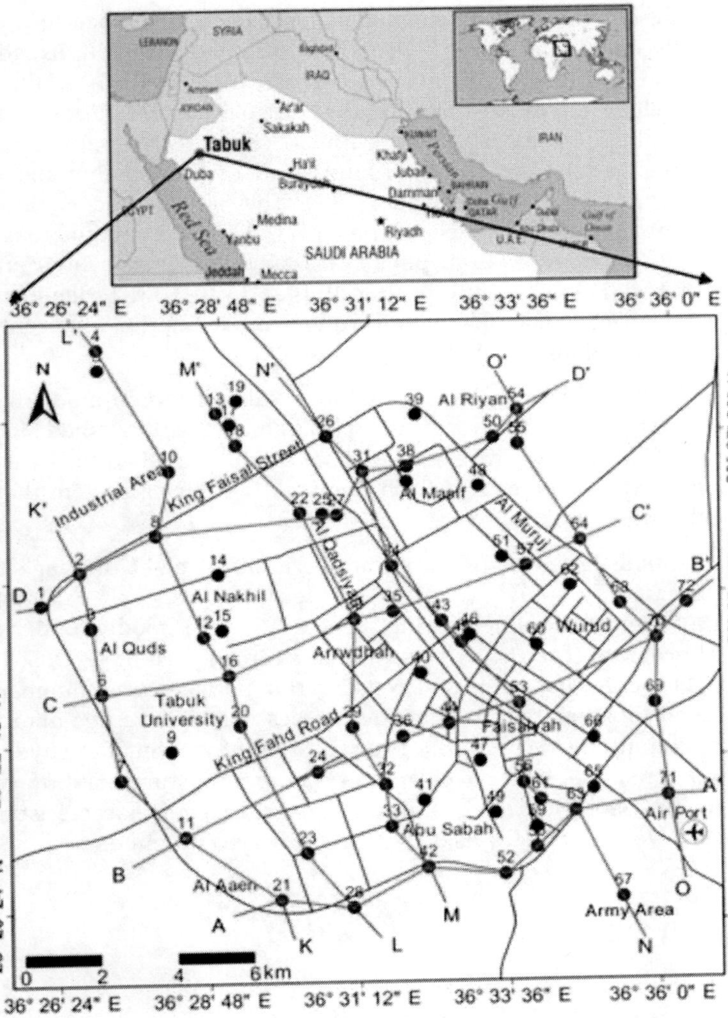

Figure 1. Location map showing distribution of drilled boreholes and cross sections, Tabuk City, KSA

The time of concentration constitutes an important factor in flood assessment studies, since it is the time required by runoff to travel from the most distant point to the basin's outlet point and lag time as the time from the center of mass of excess rainfall to the hydrograph peak [35]. For ungagged watersheds, the Soil Conservation Service (SCS) suggests that the UH lag time (t_L) may be estimated as 0.6 times the t_c ($t_L = 0.6\ t_c$).

A number of detailed hydrologic models have been developed to estimate the peak discharge and runoff hydrograph for a given rainfall distribution, such as the Hydrologic Engineering Center model (HEC-1), the SCS's Technical Release-20 model (TR- 20), the SCS's Technical Release-55 model (TR-55). Sonbol [36] used HEC-1 with SCS Unit Hydrograph option to simulate peak discharges in Sinai, Egypt and concluded that the model reasonably predicted observed peak flow. The SCS method is suited for humid, semi-arid, and arid conditions [37]. The SCS method has been successfully applied to several ephemeral watersheds in arid environments similar to Saudi Arabia [38] [39] [40] [41] [42]. HEC-1 demonstrated acceptable performance in rainfall runoff modeling for different case studies in arid regions [43] [44] [45].

In the present study, the selected computer Rainfall-Runoff model was the Hydrologic Engineering Center (HEC-1) [32] under the Watershed Modeling System (WMS v. 10) software as it was successfully applied for hydrologic modeling of watercourses and flash floods of the daring system that flow into Tabuk city.

The peak flood discharge and volume of runoff of eight drainage basins were estimated by applying the Soil Conservation Services (SCS) method using the Hydrologic Engineering Center (HEC-1) model under WMS software program. The HEC-1 was run for 7 return periods (2, 3, 5, 10, 20, 50 and 100 years) using the soil conservation service curve number (SCS CN) method to estimate the infiltration losses with an average antecedent moisture condition (AMC II) [37]. The SCS CN is a function of the hydrologic soil group, land use, land cover and antecedent moisture conditions [46]. Based on the geological, soil, land use and land cover characteristics, the weighed curve number (CN) is 80 for the study area catchments.

4. RESULTS

4.1. Geological and Geotechnical Setting

Geology of the study area was based on the geology of Tabuk Quadrangle mapped (GM-137 Sheet No. 28B) by Janjou [47] (Figure 2), field investigation and description and analyses of the drilling boreholes samples; in addition to the general geology and subsurface hydrology [48] - [56].

4.1.1. Surface Geology

Tabuk area is located within the sedimentary cover in the northwestern part of the Arabian Peninsula Shelf. The outcrop rock units in the Tabuk city belong to Paleozoic and Quaternary sediments (Figure 2).

Qasim and Sarah Formations represent the exposed Paleozoic sediments in Tabuk city, where it outcrops in several hills within the town limits (Figure 3). Qasim and Sarah Formations are belonging to Tayma and Tabuk groups, respectively [47]. It composed of a thick sequence of shale, siltstone, and sandstone units that belong to marine and continental facies.

It rests conformably but with a sharp contact on the sandstones of the Saq Formation. Age of Qasim Formation is Early to Late Ordovician. Its maximum thickness is about 1070 m but diminishes gradually in the southeast direction. The Formation is subdivided into four members, from bottom to top: Hanadir, Kahfah, Ra'an and Quwarah. In Tabuk city, Ra'an and Quwarah members are present. Ra'an Member consists of gray and dark-red, finely laminated silty clay to greenish gypsiferous clay (Figure 2 and Figure 3(a)). Quwarah Member is overlies Ra'an member. It is late Ordovician age [57]. Quwarah Member consists of medium to fine-grained micaceous sandstone and silty clay (Figure 2).

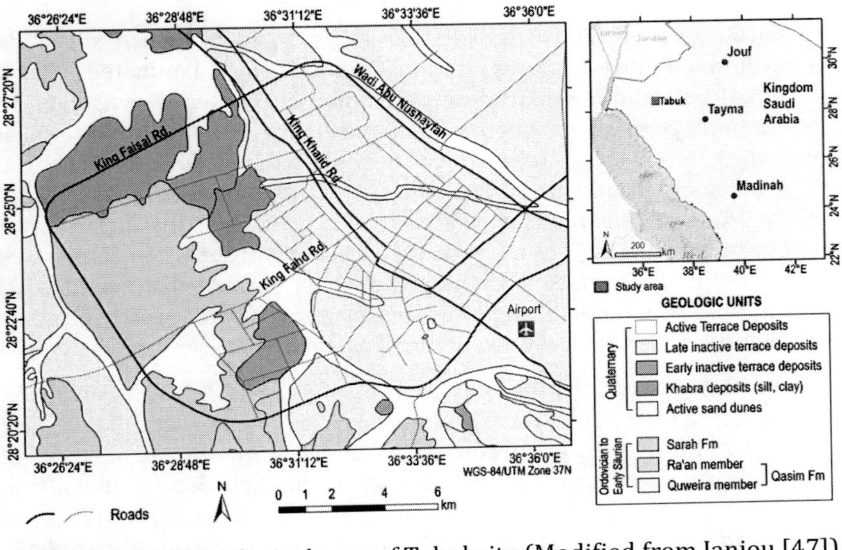

Figure 2. General geological map of Tabukcity (Modified from Janjou [47]).

Sarah Formation is belonging to Tabuk Group. It is overlies Qasim Formation of Tayma Group. The age of Sarah Formation is of late

Ordovician to early Silurian. Sarah Formation consists of medium to coarse-grained sandstone (Figure 2 and Figure 3(b)).

Quaternary sedimentary deposits overly Paleozoic sedimentary rocks, covered most of the study area (Figure 2). It represented by alluvial deposits (active and inactive terraces), khabra and sabkhah deposits (Figure 3(c) & Figure 3(d)). They consist mainly of silt, clay and blown sand with variable proportions of gravel and rock fragments of basement complex and vesicular basalt (scoria), its thickness ranges from 0.5 m to 6 m.

4.1.2. Subsurface Lithology

Through the nine lithological cross sections (Figure 1 & Figure 4), it can be divide Tabuk city according to the subsurface lithology into three zones; these are western, central, and eastern zones.

The western zone consists of yellowish to brownish color, very fine to medium grain size, moderately sorted sandstone rock. It is found from surface to the extending depth of the borings (Figure 4). It is interbedded with thin lamination of shale. In some location, sandstone overlies by thin sandy clay, clayey sand or sand blown layers, ranges in thickness from 0.5 m to 1 m. Sandstone represents the subsurface layers of the Al Aaen, Tabuk University, Al Quds Districts.

The central zone of Tabuk city consists of green to brownish silt shale layers often abundantly micaceous, oriented and laminated, flaky, interbedded with thin lamination of sandstone and lenses of gypsum. The depths of the layers range from 0 m to 20 m (Figure 4). Lateral and vertical extend of shale layers increase towards the north. Shale founds on the surface and some localities overlies by clayey sand, calcareous clayey sand, sandy clay and/or calcareous clayey sand with rock fragments layers, with thickness range from 0.5 - 3 m (Figure 4). Shale represents the subsurface layer of the center districts of Tabuk city such as Al Qadsiyah, Arrwdah, Al Masief, El Nakil, Annahdhah, and buildings are constructed directly on shale layers (Figure 3(a)) which represent the expansive soils.

While the eastern zone of Tabuk city consists of clayey sand with gravel and rock fragments (alluvial terraces) and sandy clay layers. These layers are occasionally thicknesses calcareous interbedded with thin lamination of gypsum or gravel coating by thin layer of gypsum. Their depths range from 0 m to 15 m (Figure 4). It represents the subsurface layers of the most of districts east King Khaled Street such as Al Muruj, Al Wurud, Al Faisalyah, Ulaya, Sultanah, Asulimaniyah, Al Mahrajan, etc.

Figure 3. Exposed Tabuk rocks and sedimentary deposits: (A) shale, (B) sandstone, (C) sabkha and (D) alluvial terraces, building built direct up on it, Tabuk, KSA.

4.1.3. Geotechnical Properties of Expansive Soil

Shale is the primary source of expansive soils in Tabuk due to the presence of clay minerals (smectite and illite) that are derived from shale [6] [7] [8] [9] [58]. Sabtan [10] found that, the degree of soil expansion in Tabuk ranges from low to very high.

From the geotechnical properties and mineralogical composition of Tabuk shale, it had low to mediums welling potential, according to the criteria proposed by [1] [59] [60] [61] [62]. The mineralogical composition of clay constituents of the shale, investigated using X-ray diffraction analysis reveals that the predominant clay minerals are kaolinite with minor amounts of illite [62]. The expansive lattice-type clay mineral i.e., montmorillonite) cannot be traced in the shale. Both kaolinite and illite minerals are known for being low to moderately expansive materials.

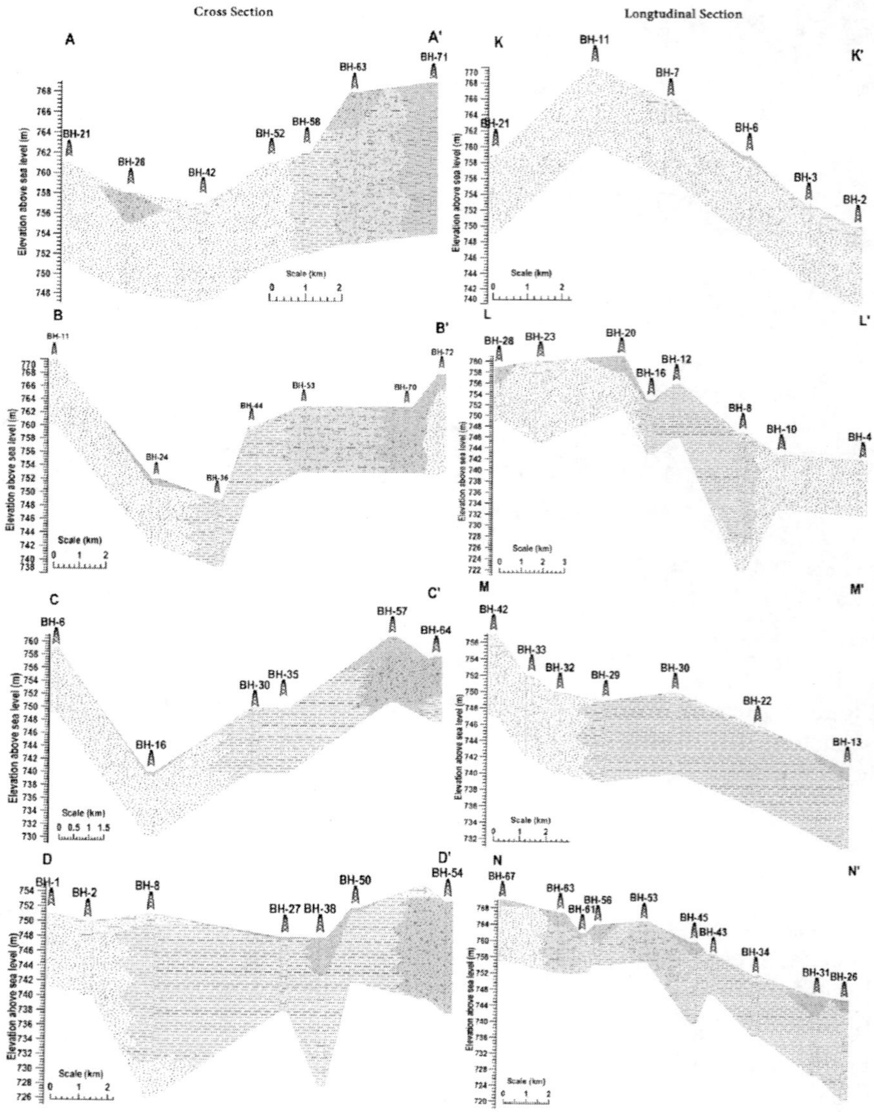

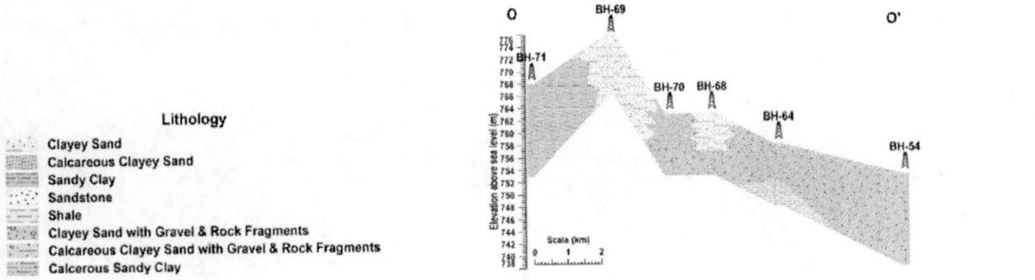

Figure 4. Lithological cross-sections (A-A', B-B', C-C', D-D', K-K', L-L', M-M', N-N', O-O') of Tabuk city, Tabuk, KSA.

The geotechnical properties of the surface and subsurface lithology of the sedimentary deposits and/orrocks showed that, shale layers of (Ra'an member) have expansive characteristic, whatever sandstone (Sarah Formation) and clayey sand with gravel and rock fragments (Quaternary alluvial deposits) layers are non-expansive.

4.1.4. Distribution of Expansive Soil

Based on the field geology, subsurface lithology, and geotechnical properties of the investigated soil samples of Tabuk city, it can be distinguished Tabuk city into three zones (Figure 5) from geotechnical view as follow.

The expansive zone is spread in the middle of the city, composed of shale. This zone is narrow in the south and become wider in the north of the city. The neighborhoods of Arrwadah, Al Farabi, Annahdhah, Al Nakhil, Rahil, Al Rajhi, Al Qadsiyah and Al Masef are situated in the expansive zone and suffered significant damage to constructions.

The non-expansive zones are distributed in the east and west of Tabuk city, represented by alluvial deposits (silt, clay and blown sand with variable proportions of gravel and rockfragments) and sandstone eastwards and westwards zones. The neighborhoods of the east, west, and southern parts of Tabuk city are underlain by non-ex- pansive deposits and/or rocks and did not show problems to constructions.

4.2. Surface Hydrology Analysis

4.2.1. Rainfall Analysis

Climate conditions play an important role in defining the hydraulic response of the watersheds existing in Tabuk city and it's environ. The most important factor affecting the hydraulic behavior of the wadi basins is rainfall. Its duration, intensity, distribution, and return periods are major influences.

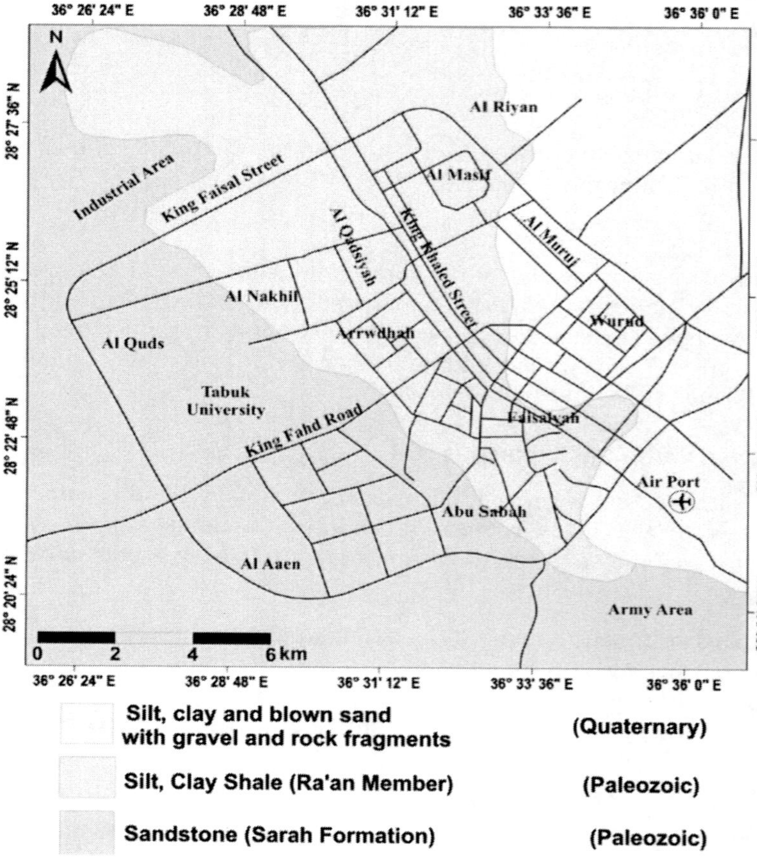

Figure 5. Distribution of the expansive and non-expansive soils in Tabuk city, KSA.

In Tabuk city, the amount of rainfall is irregular through the years and through the months (Table 1, Figure 6). Monthly rainfall distributions are uneven, with little or no rainfall in the May, June, July, August and September. The maximum rainfall occurs in January, November, October and December (Figure 6). The daily maximum-recorded rainfall in the region was 96.2 mm on the 19thof December 1985 [30].

The computed values and visual inspection revealed that Pearson Type III pdf produced the best fit in most of cases (Figure 7). The predicated maximal daily rainfall depth for 2, 3, 5, 10, 20, 50 and 100 return periods of Pearson Type 3 distribution is presented in Table 2.

Table 1. Statistical analysis of daily rainfall data for the period from 1965 to 2004 at Tabuk Weather Station 403,750 Latitude: 28.38, Longitude: 36.63, Altitude: 778.

Parameters	Value
Number of years	49.00
Minimum	1.00
Maximum	96.20
Average	24.20
Median	20.00
Standard deviation	21.80
Coefficient of variation (C_v)	0.90
Skewness coefficient of (C_s)	1.53
Kurtosis coefficient of (C_k)	4.74

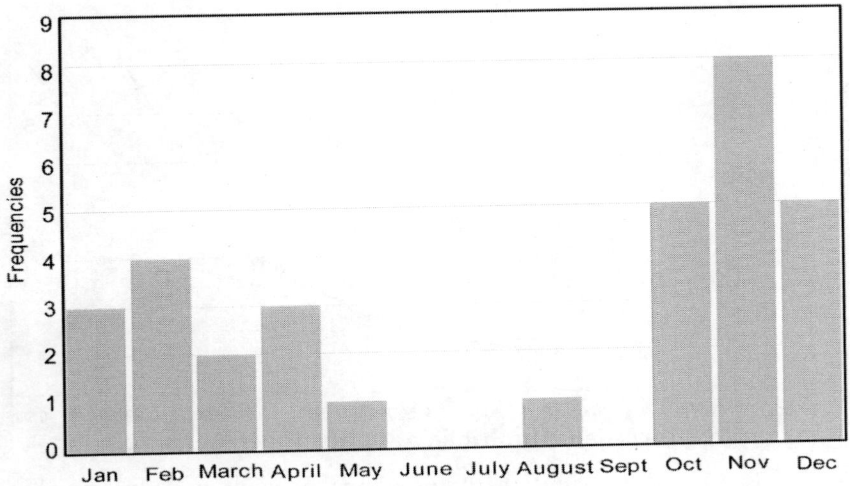

Figure 6. Statistical distribution of daily rainfall depth records over months of the year at Tabuk Weather Station 403,750.

4.2.2. Delineation of Drainage System

Analysis DEM (Figure 8) of the area shows that the high lands occur towards the south, east, and the west of Tabuk area (Figure 9); the northern part occupies the relatively lower elevations towards Qa'a Sharawra. The general slope is towards the north (0.13 m/m). Flatness terrain present in the last eight kilometers of Tabuk city, starting from Arrawdah district to the northern border of Tabuk city, the slope is nearly 0.0002 m/m (Figure 9).

WMS program was determined mainstream network and delineated the watershed of the Tabuk city and its environ (Figure 10(a)), also define and calculate the drainage area upstream of the outlet points of interest (selected outlet at the northern borders of Tabuk city). The wadis are generated from Nahamah Mountain in the southern part of the catchment. Generally, Tabuk city represents the lowland and consider the outlet of 403,750.

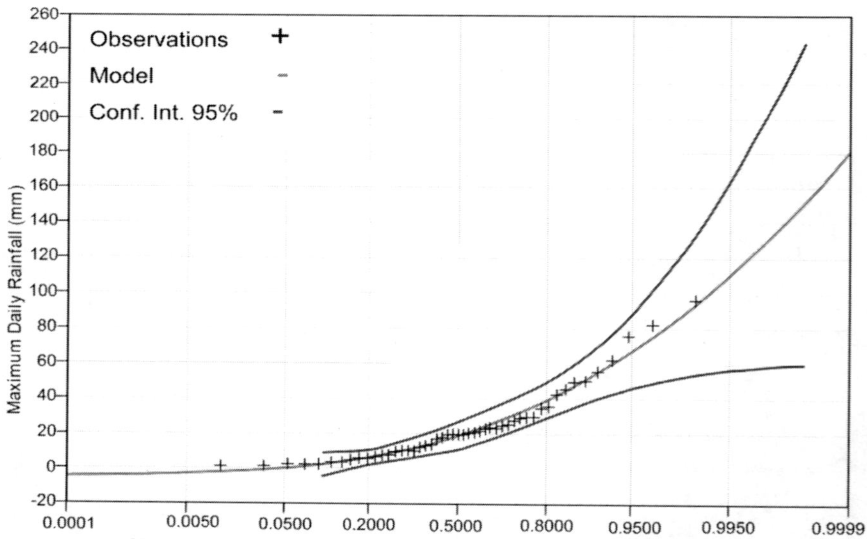

Figure 7. Gumbel distribution fit for annual maximum daily rainfall at Tabuk weather station

Table 2. Estimated return periods for maximum daily rainfall depth (Pearson Type III) for Tabuk weather station 403,750.

Return period (years)	Annual chance of exceedance	Rainfall depth (mm)	SD	Confidence interval
100	0.9900	103.0	26.70	50.2 - 155
50	0.9800	89.4	20.50	49.2 - 130
20	0.9500	71.6	13.30	45.5 - 97.7
10	0.9000	57.6	9.06	39.8 - 75.4
5	0.8000	42.9	6.63	29.9 - 55.9
3	0.6667	31.7	5.88	20.2 - 43.3
2	0.5000	21.3	5.25	11.0 - 31.6

Eight drainage basins (B-1, B-2, B-3, B-4, B-5, B-6, B-7 and B-8) affect on Tabuk city (Figure 10(b)). The upstream of all drainage basins have their source from outside Tabuk city, except the Basins 2 and 4 within the city. The downstream is towards the north at Qa'a Sharawra depression (Figure 10(b)). The main different morphological parameters of these basins indicated that most of the basins are elongated, with low relief, and moderate drainage density (Table 3).

Tabuk city lies in the downstream of the middle basin Number 3 (Figure 10(b)). It is a bigger and more effect basin; it occupies an area of about 2860 km^2. The basin comprises Wadi Na'am and Al Baqqar in the south, these were connected together to form Wadi Dab'an within Tabuk city. It is passing cross the center of the city and influential coordinate system: WGS_1984_UTM_Zone_37N) of Tabuk, KSA.

4.2.3. Peak Discharge

The results of the peak discharge and volume of runoff estimated for return periods of 2, 3, 5, 10, 20, 50 and 100 years from the run of the HEC-1 model of the drainage basins at the outlet of Tabuk area are represented in Table 4. From the hydrograph (Figure 11), it is noticed that the discharge rate of the drainage basins is highly variable from one basin to another, according to the basin area. Three basins (B-3, B-1 and B-5) have a high peak discharge and volume of runoff. The maximum peak discharge for the lowest return periods of 2 year of the Basins 3, 1 and 5 are 41.19, 20.39 and 15.5 m^3/s, respectively, whatever for the highest return period 100 year are 2201, 1085 and 824 m^3/s, respectively (Table 4 and Figure 11).

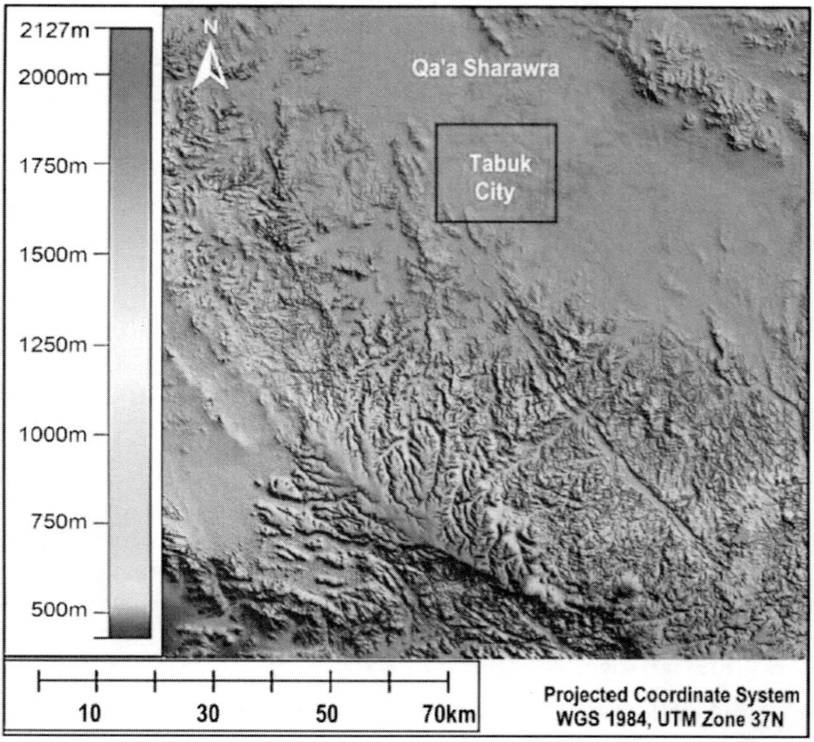

Figure 8. Shuttle radar topography mission (SRTM v.3, 1 arc) 30-m DEM (projected on the most of Tabuk districts. Basin 5, which includes Wadi Abu Nushayfah and Ghudayy, is the second dangerous basin on Tabuk city districts from the east. The northwest part of Tabuk city is affected by basin 1 of Wadi Ghuwaul; the basin has area of about 1520 km^2. Whatever, Basins 2, 4, 7 and 8 are affecting the farthest northern, eastern, and northwestern borders of the city. Due to the large areas of the Basins 3, 1 and 5 (2860, 1520 & 1185 km^2, respectively), the maximum surface runoff values are expected to be high. Therefore, the flood and surface runoff behavior in these wadis are important for any flood protection projects.

5. DISCUSSION

The hazards posed by expansive soils have significant relation with water runoff. It leads to water accumulation or ponding, which can provide a source of moisture for expansive soil by infiltration.

Table 3. The main basins geometric parameters of Tabuk area, KSA.

Parameter	Unit	B-1	B-2	B-3	B-4	B-5	B-6	B-7	B-8
Basin area	km^2	1519.67	24.46	2860.11	8.53	1185.08	206.31	57.33	44.07
Basin length	km	66.45	13.98	95.87	6.40	87.71	54.85	22.02	12.65
Basin slopes	m/m	0.0811	0.0219	0.1384	0.2200	0.0910	0.0255	0.0268	0.0244
Periphery	km	324.26	56.12	434.45	25.73	380.05	230.62	95.55	52.15
Shape		2.910	7.990	3.210	4.800	6.490	14.580	8.460	3.630
Sinusitis	msl/l	1.810	0.960	1.560	0.940	1.540	1.470	1.290	1.400
Av. overland flow	m	1159.52	1333.33	1036.38	1222.11	1049.19	1053.57	1203.76	1019.41
Mean basin elevation	m	928.54	750.10	1113.04	747.58	953.14	808.32	766.67	765.00
Max. flow distance	km	123.56	18.23	152.72	8.71	138.56	83.60	32.97	20.50
Max. flow slope	m/m	0.0038	0.0019	0.0057	0.0020	0.0048	0.0022	0.0016	0.0016
Max. stream length	km	120.09	13.46	149.49	5.99	135.43	80.43	28.48	17.74
Max. stream slope	m/m	0.0036	0.0010	0.0045	0.0014	0.0041	0.0015	0.0017	0.0013

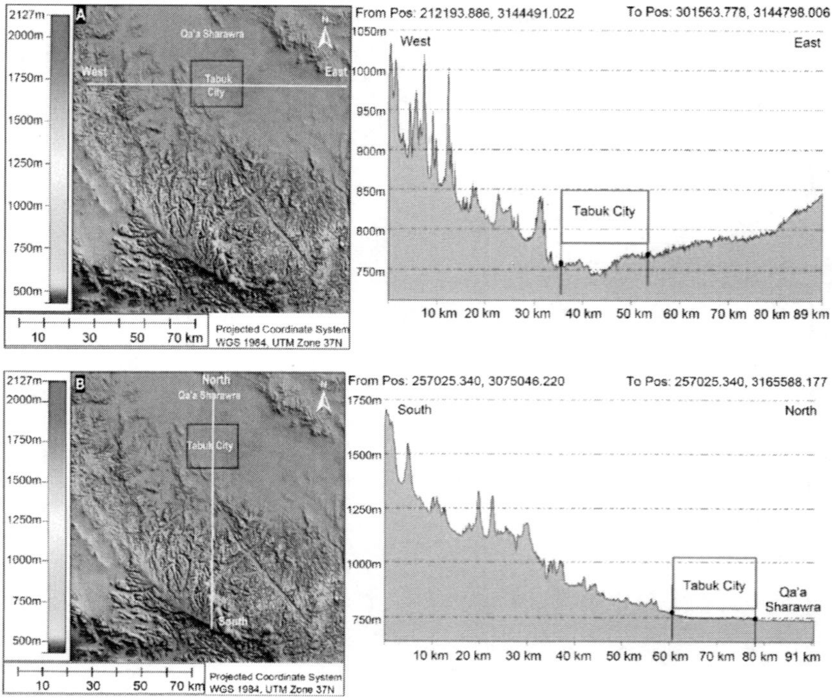

Figure 9. Cross section of Tabuk area from West to East (A), longitudinal section from South to North (B) Tabuk, KSA.

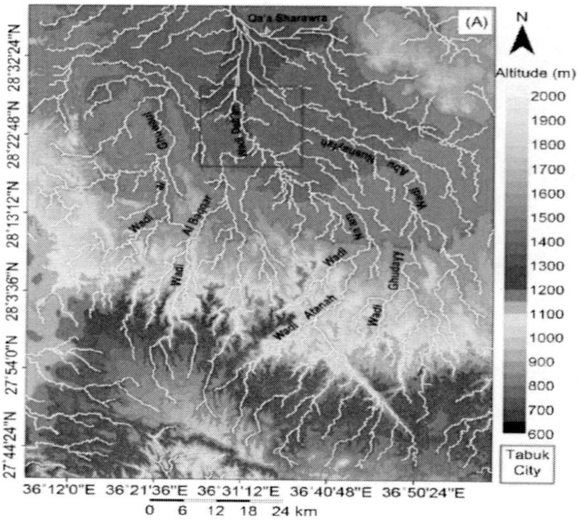

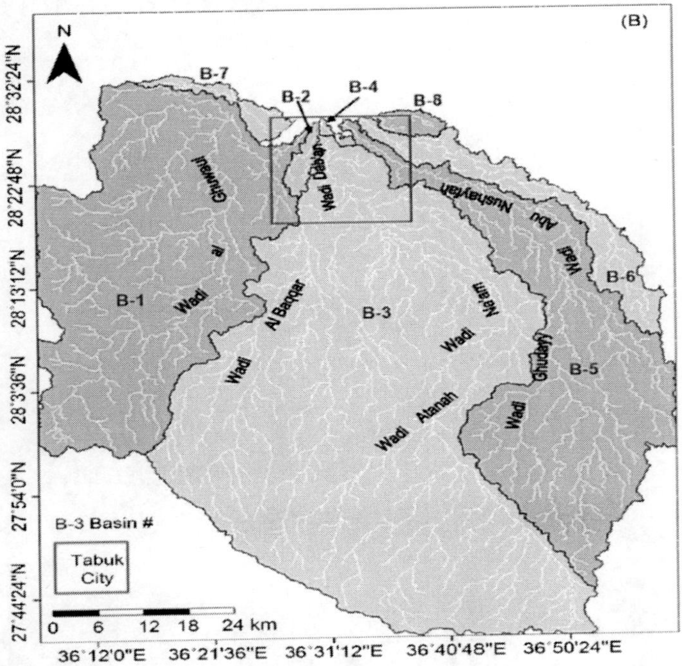

Figure 10. Drainage system combined with DEM (A), basins effect on Tabukcity (B), Tabuk area, KSA.

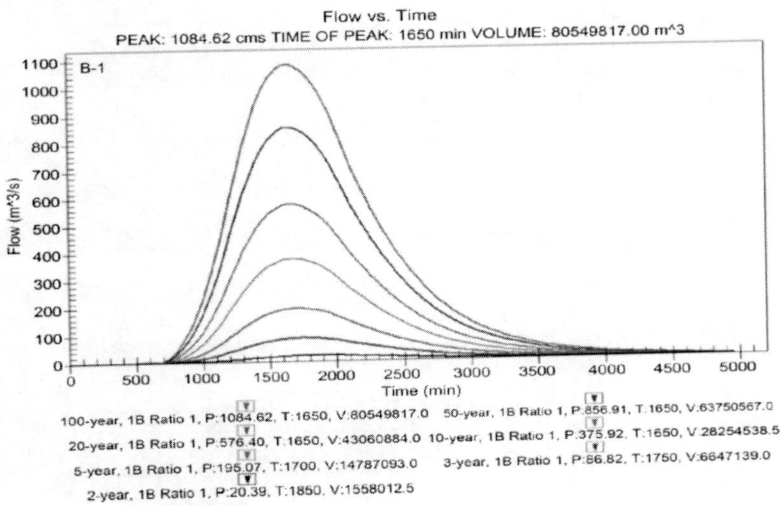

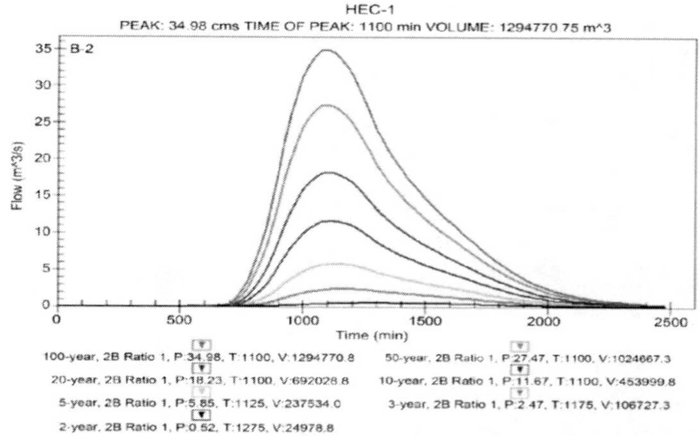
HEC-1
PEAK: 34.98 cms TIME OF PEAK: 1100 min VOLUME: 1294770.75 m^3
B-2
Flow (m^3/s)
Time (min)
100-year, 2B Ratio 1, P:34.98, T:1100, V:1294770.8
50-year, 2B Ratio 1, P:27.47, T:1100, V:1024667.3
20-year, 2B Ratio 1, P:18.23, T:1100, V:692028.8
10-year, 2B Ratio 1, P:11.67, T:1100, V:453999.8
5-year, 2B Ratio 1, P:5.85, T:1125, V:237534.0
3-year, 2B Ratio 1, P:2.47, T:1175, V:106727.3
2-year, 2B Ratio 1, P:0.52, T:1275, V:24978.8

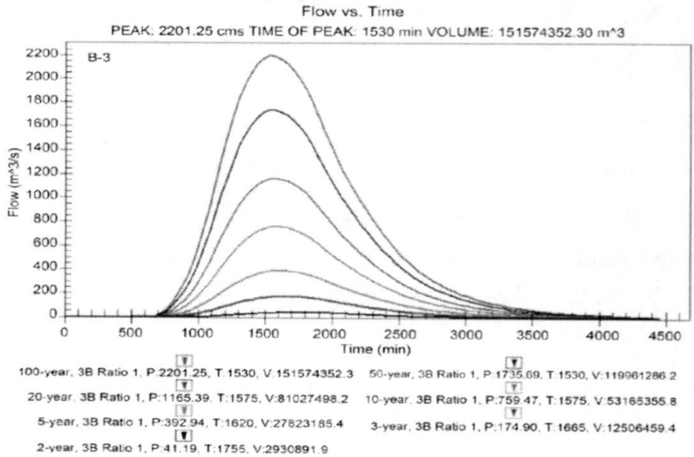
Flow vs. Time
PEAK: 2201.25 cms TIME OF PEAK: 1530 min VOLUME: 151574352.30 m^3
B-3
Flow (m^3/s)
Time (min)
100-year, 3B Ratio 1, P:2201.25, T:1530, V:151574352.3
50-year, 3B Ratio 1, P:1735.89, T:1530, V:119961286.2
20-year, 3B Ratio 1, P:1165.39, T:1575, V:81027498.2
10-year, 3B Ratio 1, P:759.47, T:1575, V:53165355.8
5-year, 3B Ratio 1, P:392.94, T:1620, V:27823185.4
3-year, 3B Ratio 1, P:174.90, T:1665, V:12506459.4
2-year, 3B Ratio 1, P:41.19, T:1755, V:2930891.9

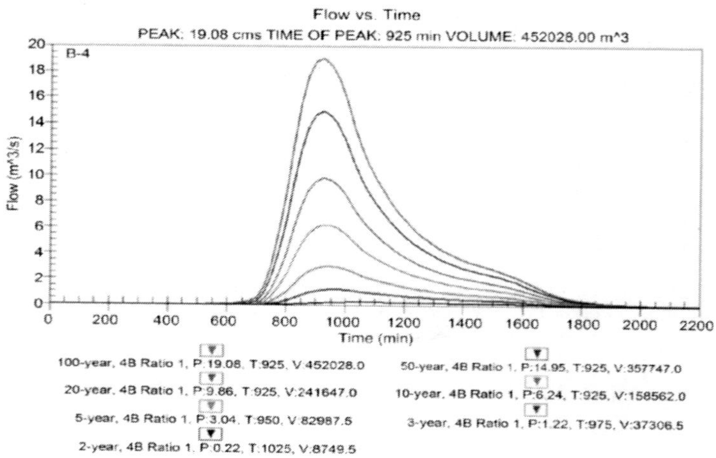
Flow vs. Time
PEAK: 19.08 cms TIME OF PEAK: 925 min VOLUME: 452028.00 m^3
B-4
Flow (m^3/s)
Time (min)
100-year, 4B Ratio 1, P:19.08, T:925, V:452028.0
50-year, 4B Ratio 1, P:14.95, T:925, V:357747.0
20-year, 4B Ratio 1, P:9.86, T:925, V:241647.0
10-year, 4B Ratio 1, P:6.24, T:925, V:158562.0
5-year, 4B Ratio 1, P:3.04, T:950, V:82987.5
3-year, 4B Ratio 1, P:1.22, T:975, V:37306.5
2-year, 4B Ratio 1, P:0.22, T:1025, V:8749.5

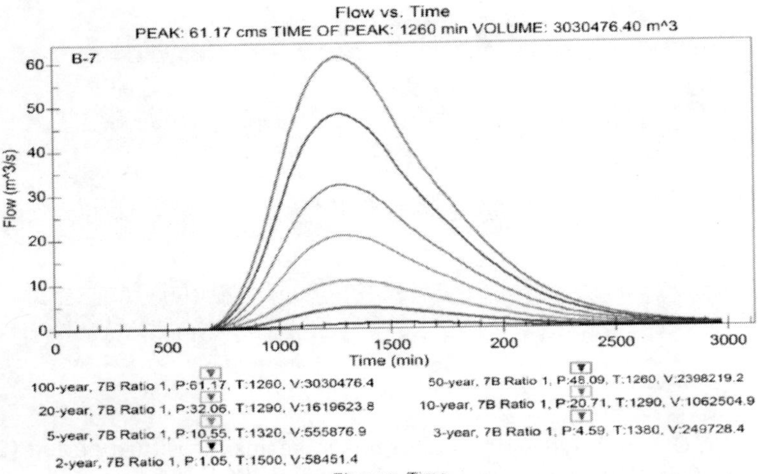
Flow vs. Time
PEAK: 61.17 cms TIME OF PEAK: 1260 min VOLUME: 3030476.40 m^3
B-7
Flow (m^3/s)
Time (min)
100-year, 7B Ratio 1, P:61.17, T:1260, V:3030476.4
50-year, 7B Ratio 1, P:48.09, T:1260, V:2398219.2
20-year, 7B Ratio 1, P:32.06, T:1290, V:1619623.8
10-year, 7B Ratio 1, P:20.71, T:1290, V:1062504.9
5-year, 7B Ratio 1, P:10.55, T:1320, V:555876.9
3-year, 7B Ratio 1, P:4.59, T:1380, V:249728.4
2-year, 7B Ratio 1, P:1.05, T:1500, V:58451.4

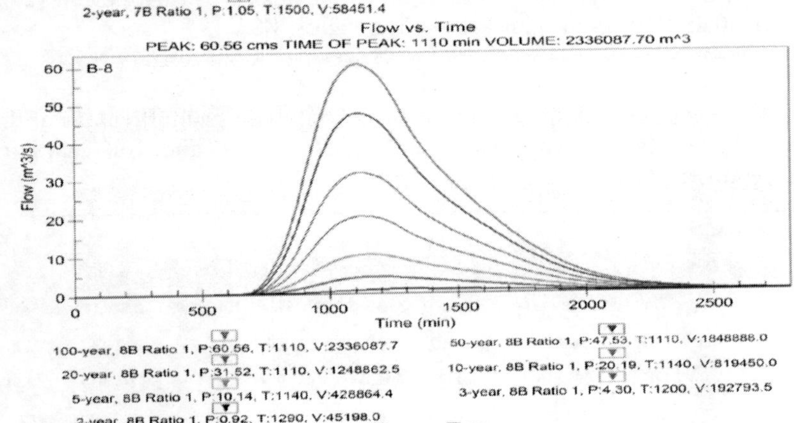
Flow vs. Time
PEAK: 60.56 cms TIME OF PEAK: 1110 min VOLUME: 2336087.70 m^3
B-8
Flow (m^3/s)
Time (min)
100-year, 8B Ratio 1, P:60.56, T:1110, V:2336087.7
50-year, 8B Ratio 1, P:47.53, T:1110, V:1848888.0
20-year, 8B Ratio 1, P:31.52, T:1110, V:1248862.5
10-year, 8B Ratio 1, P:20.19, T:1140, V:819450.0
5-year, 8B Ratio 1, P:10.14, T:1140, V:428864.4
3-year, 8B Ratio 1, P:4.30, T:1200, V:192793.5
2-year, 8B Ratio 1, P:0.92, T:1290, V:45198.0

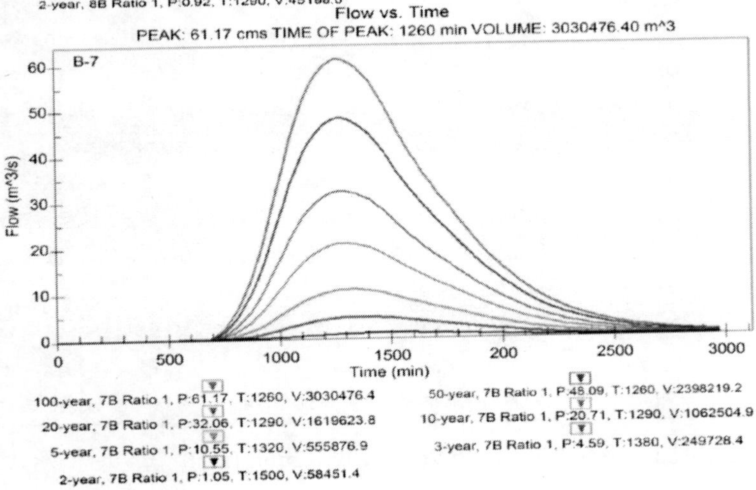
Flow vs. Time
PEAK: 61.17 cms TIME OF PEAK: 1260 min VOLUME: 3030476.40 m^3
B-7
Flow (m^3/s)
Time (min)
100-year, 7B Ratio 1, P:61.17, T:1260, V:3030476.4
50-year, 7B Ratio 1, P:48.09, T:1260, V:2398219.2
20-year, 7B Ratio 1, P:32.06, T:1290, V:1619623.8
10-year, 7B Ratio 1, P:20.71, T:1290, V:1062504.9
5-year, 7B Ratio 1, P:10.55, T:1320, V:555876.9
3-year, 7B Ratio 1, P:4.59, T:1380, V:249728.4
2-year, 7B Ratio 1, P:1.05, T:1500, V:58451.4

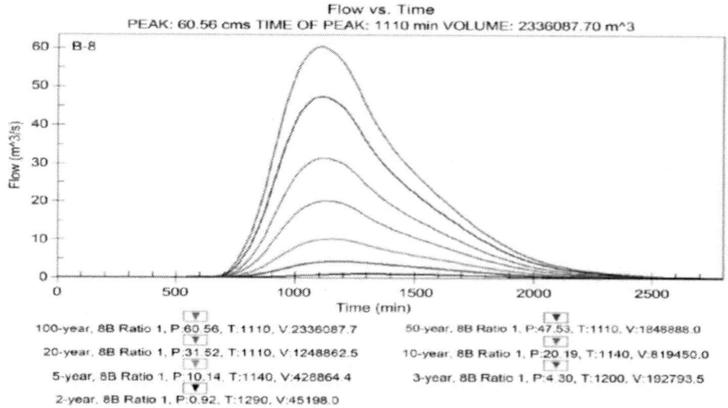

Figure 11. Outflow Hydrograph of drainage basins for return period (2, 3, 5, 10, 20, 50 and 100 years), of Tabuk City, Tabuk, KSA.

Table 4. Measured peak flood discharge and volume of runoff values of the drainage basins for return period (2, 3, 5, 10, 20, 50 and 100 years) of Tabuk region, KSA.

Return Period (Years)			100	50	20	10	5	3	2
B-1	Qp[a]	m^3/s	1084.62	856.91	576.4	375.92	195.07	86.82	20.39
	V[b]	m^3	80,549,817	63,750,567	43,060,884	28,254,538	14,787,093	6,647,139	1,558,012
B-2	Qp	m^3/s	34.77	27.25	18.04	11.57	5.79	2.45	0.52
	V	m^3	1,296,918	1,026,449	693,334	454,938	238,101	107,040	25,091
B-3	Qp	m^3/s	2201.25	1735.69	1165.39	759.47	392.94	174.9	41.19
	V	m^3	151,574,352	119,961,286	81,027,498	53,165,355	27,823,185	12,506,459	2,930,891
B-4	Qp	m^3/s	19.08	14.95	9.86	6.24	3.04	1.22	0.22
	V	m^3	452,028	357,747	241,647	158,562	82,987	37,306	8749
B-5	Qp	m^3/s	823.63	650.2	437	285.49	148.09	65.97	15.5
	V	m^3	62,802,054	49,703,878	33,572,484	22,028,350	11,528,280	5,182,032	1,214,469
B-6	Qp	m^3/s	117.52	92.85	62.62	40.99	21.35	9.56	2.25
	V	m3	10,913,010	8,636,776	5,833,464	3,827,417	2,002,903	900,207	210,917
B-7	Qp	m^3/s	61.17	48.09	32.06	20.71	10.55	4.59	1.05
	V	m^3	3,030,476	2,398,219	1,619,623	1,062,504	555,876	249,728	58,451
B-8	Qp	m^3/s	60.56	47.53	31.52	20.19	10.14	4.3	0.92
	V	m^3	2,336,087	1,848,888	1,248,862	819,450	428,864	192,793	45,198

Qp: Peak discharge (m^3/sec); [b]V: Volume of Runoff (m^3).

In Tabuk city, some parts of the watershed have runoff or flash floods on areas comprise expansive soil and hence the problem is duplicated due to first the flash flood and second the heaving of the soil. On the other hand, some locations have no flow, consequently no problems from this point of view even if the expansion soil is found. Volume of surface runoff can give singe of the potential swelling soil locations, where the expansion is linearly related to surface runoff.

The drainage pattern represented in. Delineating the drainage pattern effect on Tabuk districts is helpful in determining the directions of running water and the possible locations for accumulation (Figure 12(a)). The flow direction of surface runoff or storm water discharge should be away from the expansion soil zones. Therefore, the integration was carried out between geotechnical distribution of the expanded soil and the drainage network over Tabuk neighborhoods (Figure 12(b)). Consequently, the riskiest zones in the Tabuk city have been identified (Figure 13). It found that, the main watercourse of Wadi Dab'an and Abu Nushayfah passes through area contains expansive shale.

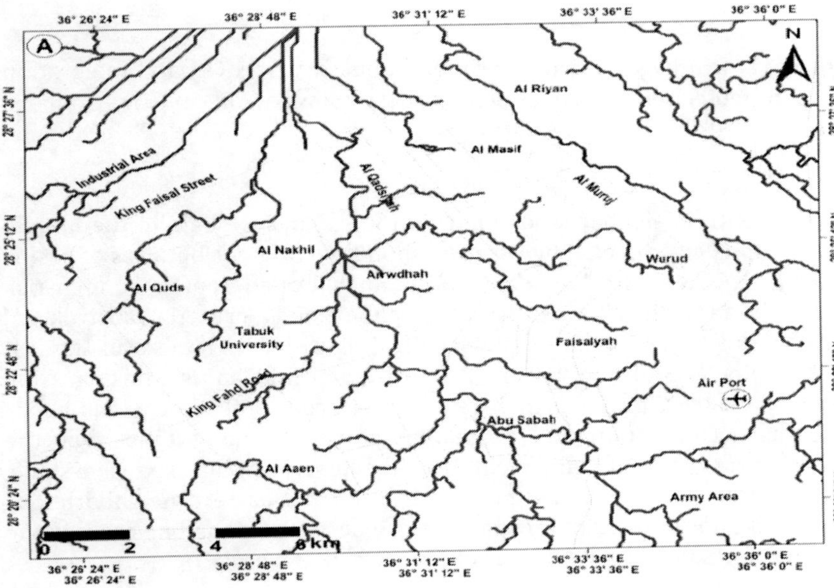

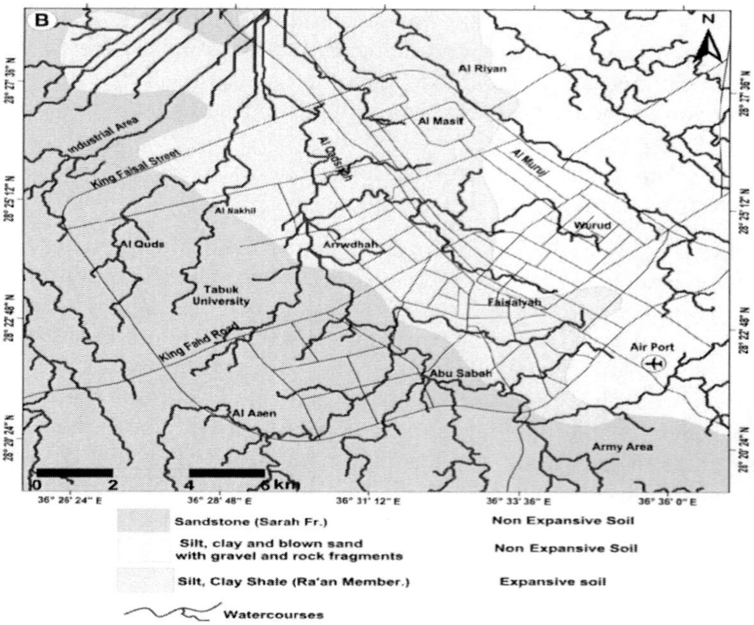

Figure 12. Drainage system overly on Tabuk districts (A) and integration between geotechnical distribution of the expanded soil and the drainage network (B) Tabuk city, KSA.

The residential neighborhoods built on the expansive shale in the middle part of Tabuk city affected by surface runoff flow of the tributaries of basins of Wadi Abu Nushayfah, Na'am and Atanah outlet, represent high-risk zones. These high-risk residential neighborhoods are Al Qadsiyah, Al Maseif, Arrwdah, Al Nakhil and Al Rajhi (Figure 13). Flatness in the last eight kilometers of the basin (B-3) of Wadi Dab'an is starting from Arrawdah district to the north. It reduces surface runoff velocity and hence increases accumulation and stagnant water for a long time over the expansive soil, consequently maximizing the heaving of the expansive shale. The low-risk zones includes the western and the eastern neighborhoods (Figure 13), it is built on non-expansion soil or/rock (sandstone rocks in the west and silt, sand with boulders cobble rocks in the east).

6. CONCLUSIONS

Tabuk city contains expanded shale, in the central part of the city. It expands when it absorbs more water and creates the potential to exert enough force on a building or other structures to cause damage. In addition, Tabuk city exposed too many flash floods, mainly from large drainage basin of Wadi Na'am, Al Baqqar and Dab'an which covered an area of about 2860 km^2 and had high peak discharge and volume of runoff 2201 m^3/s and 151,574,352 m^3, respectively and 41.19 m^3/s and 2,930,891 m^3 for returned period 100 and 2 years, respectively.

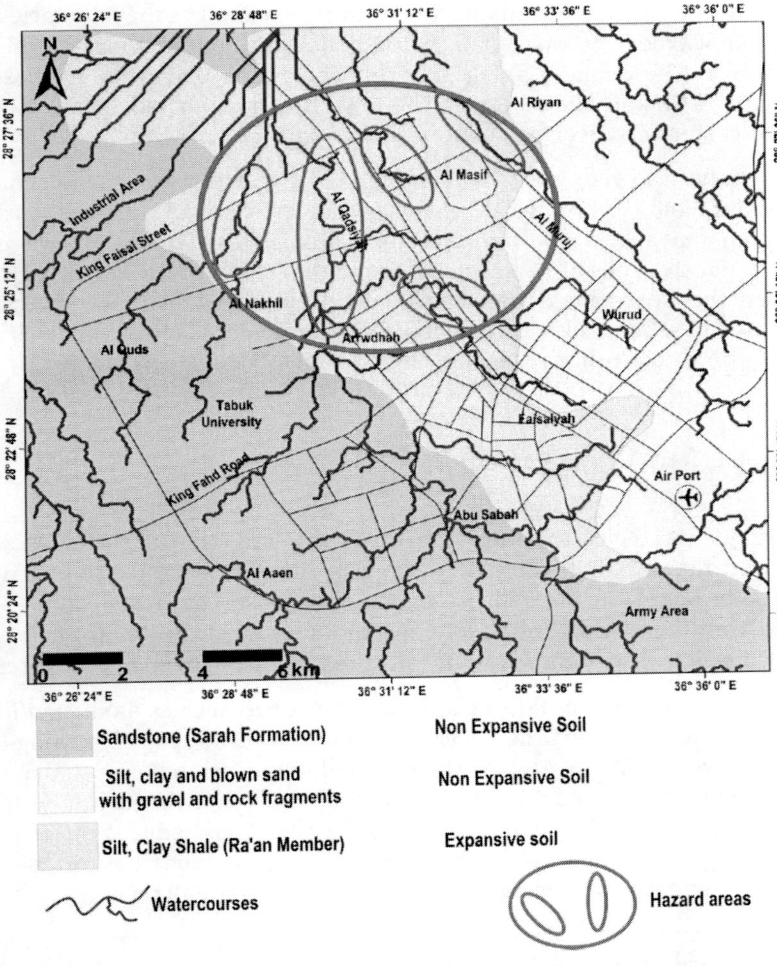

Figure 13. Hazard map of expansive shale over Tabuk districts, Tabuk, KSA.

In the event of flash floods or surface runoff, water flows and collects in some places that contain the expanded shale, causing its expansion and thus the destruction of the buildings is built upon it. The direction of flow of surface runoff or storm water discharge should be away from the expansion soil zones. The quicker the water can flow out, the better it is. The integration was carried out between geotechnical distribution maps of the expansion soil and flow of runoff, and the riskiest places had been identified in Tabuk city. The resident neighborhoods built on the expansive shale in the middle of Tabuk city and affected by surface runoff of the tributaries of the basin of Wadi Abu Nishayfah, Na'am and Atanah outlet, represent high-risk zones. These high-risk neighborhoods are Al Qadsiyah, Al Maseif, Arrwdah, Al Nakhil and Al Rajhi. The low-risk zones include the western and the eastern districts; it is non-expansion soil.

This study showed the importance of characterizing geotechnical properties and hydrological elements as part of a comprehensive environmental analysis for future urban planning. Based on the hazard map of the expansive soils of Tabuk city, it found that the western side of the city lying in the upstream of the small basin and the subsurface layers is composed of sandstone. It can suggest that, the western side of Tabuk city is suitable for future urban extension.

7. RECOMMENDATIONS

Two major strategies to mitigate hazards resulted from flash floods and runoff on the swelling soil zone (high-risk zone) in Tabuk districts. It is essentially based on prevent or reduce flow of surface runoff and flash floods from the offside Tabuk city and drainage storm water from inside Tabuk districts. The hazards can be mitigated by the following procedures:

1) Construction of water flow retardation structures such as dams, artificial ponds, and orifice divert the course of the wadi etc. in upstream along the main courses of Wadi Abu Nushayfah and Ghudayy, Na'am, Al Baqqar, Dab'an and Atanah and Ghuwaul. This would serve two goals; the first decreasing the flash floods flow to Tabuk city and thus reduces the chance of swelling soils; and the second increasing the contribution to the groundwater aquifer, which depends essentially on rainfall recharge.

2) Nature wadis and artificial open channels that pass into expansion soil areas (hazard zones) within Tabuk city should be refine and lining, to prevent infiltration into the subsurface expansion soil.

3) Benefit from the surface water discharge by matching the paved street layout within the affected districts with the direction of flow, to reduce time of concentration and increase the speed of the discharge. In addition, the

direction and slope of the main streets should be in the north direction toward Al Qa'a depression.

4) Construction of storm water network in the hazard zones of Tabuk neighborhoods.

ACKNOWLEDGEMENTS

The Research presented in this paper originated from a Master Degree in Geology from Faculty of Science, Damietta University. The authors are grateful indebted to Prof. Ahmed Basal, Geology Department, Faculty of Science, Damietta University for long-standing collaboration and fruitful scientific discussion. Also, grateful thanks to Dr. Abed El Kareem El Rosttum and Eng. Hikmat El Rosttum, owner of Soil & Foundation Co. (Safco) and Geotechnical & Environmental Co. (Geco) Companies, KSA, for offering the laboratory facility to conduct the geotechnical tests of this work.

REFERENCES

1. Seed, H.B., Woodward, R.J. and Lundgren, R. (1962) Prediction of Swelling Potential for Compacted Clays. Journal of Soil Mechanics and Foundation Engineering Division, 88, 53-87.
2. Li, J., Cameron, D.A. and Ren, G. (2014) Case Study and Back Analysis of a Residential Building Damaged by Expansive Soils. Computers and Geotechnics, 56, 89-99. https://doi.org/10.1016/j.compgeo.2013.11.005
3. Li, X., Liang, Y., Zhang, P. and Wang, Y. (2010) Research into Treatment Theory and Technique for Expansive Soil Geological Disasters. Memoirs of the Muroran Institute of Technology, 59, 165-171.
4. Kerrane, J.P. (2011) What Are Expansive Soils. Benson & Associates PC, Golden.
5. Rosenbalm, D.C. (2013) Volume Change Behavior of Expansivesoils Due to Wetting and Drying Cycles, Arizona. PhD Thesis, Arizona State University, Phoenix. [Citation Time(s):1
6. Slater, D.E. (1983) Potential Expansive Soils in Arabian Peninsula. American Society for Civil Engineering. Geothermal Engineering, 109, 744-746.
7. Abduljauwad, S.N. and Ahmed, R. (1990) Expansive Soil in Al-Qatif Area. Arabian Journal of Science and Engineering, 15, 133-144.
8. Erol, A.O. and Dhowian, A. (1990) Swell Behavior of Arid Climate Shale from Saudi Arabia. Journal of Engineering Geology, 23, 243-254. https://doi.org/10.1144/GSL.QJEG.1990.023.03.06
9. Abduljauwad, S.N. (1994) Swelling Behavior of Calcareous Clays from the Eastern Province of Saudi Arabia. Journal of Engineering Geology, 27, 333-351. https://doi.org/10.1144/GSL.QJEGH.1994.027.P4.05
10. Al-Sabtan A.A. (2005) Geotechnical Properties of Expansive Clay Shale in Tabuk, Saudi Arabia. Journal of Asian Earth Sciences, 25, 747-757. https://doi.org/10.1016/j.jseaes.2004.07.003
11. Ruwaih, I.A. (1987) Experiences with Expansive Soils in Saudi Arabia. Proceedings of 6th International Conference on Expansive Soils, New Delhi, 1-4 December 1987, 317-322.

12. Dhowian, A.W., Erol, A.O. and Youssef, A. (1990). Evaluation of Expansive Soils and Foundation Methodology in the Kingdom of Saudi Arabia. General Directorate Research Grants Programs, King Abdulaziz City for Science and Technology.
13. Dafalla, M.A. and Al-Shamrani, M.A. (2008) Performance-Based Solutions for Foundations Onexpansive Soils-Al Ghatt Region, Saudi Arabia. International Conference on Geotechnical Engineering, Chiangmai, 10-12 December 2008, 147-156.
14. Dhowian, A.W. (1984) Characteristics of Expansive Clay-Shale in the Northern Region of Saudi Arabia. Proceedings of the 5th International Conference on Expansive Soils, Adelaide, 21-23 May 1984, 316-320.
15. Dafalla M.A. and Al-Shamrani M.A. (2012) Expansive Soil Properties in a Semi-Arid Region. Research Journal of Environmental and Earth Sciences, 4, 930-938.
16. Dafalla, M.A., Al-Shamrani, M.A., Puppala, A. and Ali, H.E. (2010) Use of Rigid Foundation System on Expansive Soils. American Society for Civil Engineers, GSP (Geotechnical Special Publication), 199. https://doi.org/10.1061/41095(365)170
17. Fredlund, D.G. and Rahardjo, H. (1993) Soil Mechanics for Unsaturated Soils. Wiley, New York. https://doi.org/10.1002/9780470172759
18. Handy, R.L. (1995) The Day the House Fell: Homeowner Soil Problems from Landslides to Expansive Clays and Wet Basements. ASCE, Reston. https://doi.org/10.1061/9780784401040
19. Wray, W.K. and Meyer, K.T. (2004) Expansive Clay Soil—A Widespread and Costly GeoHazard. GeoStrata, 5, 24-28.
20. Mitchell, J.K. and Raad, L. (1973) Control of Volume Changes in Expansive Earth Materials. Proceedings of the Workshop on Expansive Clay and Shales in Highway Design and Construction, 2, 200-219.
21. Snethen, D.R. and Johnson, L.D. (1980) Evaluation of Soil Suction from Filter Paper. Miscellaneous Paper GL-80-4, US Army Engineer Waterways Experiment Station, Vicksburg.
22. El Sayed, S.T. and Rabbaa, S.A. (1986) Factors Affecting Behavior of Expansive Soils in the Laboratory and Field—A Review. Geotechnical Engineering, 17, 89-107.
23. Day, R.W. (1992) Irrigation, Drainage, and Landscaping for Expansive Soil. Journal of Irrigation and Drainage Engineering, 118, 285-290. http://webistem.com/acoustics2008/acoustics2008/cd1/data/articles/002216.pdf https://doi.org/10.1061/(ASCE)0733-9437(1992)118:2(285)
24. Federal Emergency Management Agency (FEMA) (1997) Multi Hazard Identification and Risk: Assessment. Expansive Soils.
25. Abushandi, E. and Alatawi, S. (2015) Dam Site Selection Using Remote Sensing Techniques and Geographical Information System to Control Flood Events in Tabuk City. Hydrology: Current Research, 6, 189.
26. Abushandi, E. (2016) Flash Flood Simulation for Tabuk City Catchment, Saudi Arabia. Arabian Journal of Geosciences, 9, 188. https://doi.org/10.1007/s12517-015-2192-x
27. Hussein, M.T. and Zaidi, F.K. (2012) Assessing Hydrological Elements as Key Issue for Urban Development in Arid Regions.
28. Tabuk, M. (1993) The Town Plans Report. Asia Aaero Survey Company (Personal Communication).
29. ASTM (1995) Book of Standards, Soil and Rock. Philadelphia.
30. MOWE (Ministry of Water and Electricity) (2015) Kingdom of Saudi Arabia.
31. Hyfran, M. (1998) Developed by INRS-Eau with Collaboration of Hydro-Québec Hydraulic Service (Department Hydrology) in the Framework of Hydro-Québec/CRSNG Statistical Hydrology Chair Located at INRS-Eau. http://www.wrpllc.com/books/hyfran.html
32. US Army Corps of Engineers (USACE) (1998) HEC-1 Flood Hydrograph Package User's Manual, Hydrologic Engineering Center (HEC), Davis.
33. McCuen, R.H. (2005) Hydrologic Analysis and Design. 3rd Edition, Prentice Hall, Upper Saddle River.
34. Ally, W.M., Dawdy, D.R. and Schaake, J.C. (1980) Parametric Deterministic Urban Watershed Model. Journal of the Hydraulics Division, 106, 679-690.

35. US Army Corps of Engineers (USACE) (2000) Hydrologic Modeling System HEC-HMS Technical Reference Manual. Hydrologic Engineering Center, Davis.
36. Sonbol, M.A., Mtalo, F., El-Bihery, M.A. and Abdel Motaleb, M. (2005) Watershed Modeling of WadiSudr and Wadi Al-Arbain in Sinai, Egypt. International Conference of UNESCO Flanders Fust Friend/Nile Project—Towards A Better Cooperation, The 5th Project Management Meeting & 9th Steering Committee Meeting, Sharm El-Shiekh, 12-15 November 2005, 1-12.
37. Soil Conservation Services (SCS) (1985) National Engineering Handbook, Section 4: Hydrology. US Department of Agriculture, Soil Conservation Service, Engineering Division, Washington DC.
38. Sorman, A.U., Abdulrazzak, M.J. and Elhames, A.S. (1990) Rainfall-Runoff Modeling of Amicrocatchment in the Western Region of Saudi Arabia, Hydrology in Mountainous Regions. I—Hydrological Measurements. IAHS Publ. No. 193.
39. Osterkamp, W.R., Lane, L.J. and Savard, C.S. (1994) Recharge Estimates Using a Geomorphic Distributed-Parameter Simulation Approach. Amargosa River Basin, 30, 493-507.
40. Al Dousari, S.A. and Al Ghadban, A.N. (2007) Sustainable Fresh Water Resources Management in Proceedings of World Water and Environmental Resources Congress.
41. Milewski, A., Sultan M., Yan, E.B., Becker, R., Abdeldayem, A., Soliman F. and Abdel Gelil, K. (2009) A Remote Sensing Solution for Estimating Runoff and Recharge in Arid Environments. Journal of Hydrology, 373, 1-14. https://doi.org/10.1016/ j. jhydrol.2009.04.002
42. Shadeed, S. and Al Masri, M. (2010) Application of GIS-Based SCS-CN Method in West Bank Catchments, Palestine. Water Science and Engineering, 3, 1-13.
43. Abushandi, E. and Merkel, B. (2011) Modeling Rainfall Runoff Relations Using HEC-HMS and IHACRES for a Single Rain Event in an Arid Region of Jordan. Water Resources Management, 27, 2391-2409. https://doi.org/10.1007/s11269-013-0293-4
44. Sherif, M.M., Mohamed, M.M., Shetty, A. and Almulla, M. (2011) Rainfall Runoff Modeling of Three Wadis in the Northern Area of UAE. Journal of Hydrologic Engineering, 16, 10-20. https://doi.org/10.1061/(ASCE)HE.1943-5584.0000280
45. Subyani, A.M. and Al-Ahmadi, F.S. (2011) Rainfall-Runoff Modeling in the Al-Madinah Area of Western Saudi Arabia. Journal of Environmental Hydrology, 19, 1-13.
46. Ponce, V.M. and Hawkins, R.H. (1996) Runoff Curve Number: Has It Reached Maturity? Journal of Hyrologic Engineering, 1, 9-20. https://doi.org/10.1061/(ASCE)1084-0699(1996)1:1(11)
47. Janjou, D., Halawani, M.A., Al-Muallem, M.S., Brosse, J.M., Becq-Giraudon, J.F., Dagain, J., Genna, A., Razin, P., Roobol, M.J., Shorbaji, H. and Wyns, R. (1997) Geologic Map of the Tabuk Quadrangle, Sheet 28B, Kingdom of Saudi Arabia (with Text): Saudi Arabian Deputy Ministry for Mineral Resources. Jiddah, Geoscience Map GM-137.
48. Powers, R.W., Ramirez, L.F., Redomond, C.D. and Elberg, E.L. (1966) Geology of the Arabian Peninsula, Sedimentary Geology of Saudi Arabia. US Geological Survey Professional Paper 560-D, 147.
49. Parson-Basil Consultants (1967) Agricultural and Water Resources: The Great Nafud Sedimentary Basin. Ministry of Agriculture and Water, Riyadh.
50. Lloyd, J.W. (1969) The Hydrogeology of the Southern Desert of Jordan. UNDP/FAO Publ. Tech. Rep. 1., Spec. Fund 212, Rome.
51. Al Watban, N.F. (1976) Groundwater Potentiality of Tabuk and Saq Aquifers in Qasim Region. MSc Thesis, Institute of Applied Geology, Jeddah.
52. Glintzboeckel, C. (1981) A Tentative Synopsis of the Geology of the Saudi Arabian Sedimentary Basin in Relation to Phosphate Prospecting. Report BRGM-OF-01-23, Jeddah.
53. Burdon, D.J. (1982) Hydrogeological Conditions in the Middle East. Quarterly Journal of Engineering Geology, 15, 71-82. https://doi.org/10.1144/GSL.QJEG.1982.015.02.01
54. Lozej, G.P. (1983) Geological and Geochemical Reconnaissance Exploration of the Cover Rocks in Northwestern Hijaz. Deputy Ministry for Mineral Resources Open-File Report RF-OF-03-2, Jeddah.

55. Laboun, A.A. (1986) Stratigraphy and Hydrocarbon Potential of the Paleozoic Succession of Both the Widyan and Tabuk Basins, Arabia. In: Halbouty, M., Ed., Future Petroleum Provinces of the World, American Association of Petroleum Geologists Memoir No. 50, 373-394.
56. Garfield, L.P. (1984) Mineralisation in Cover Rock Type Shelf-Sediments of the Tethys Margins. Report Riofinex, File RF-DF-01-29, Jeddah.
57. Williams, P.L., Vaslet, D., Johnson, P.R., Berthiaux, A., Le Start, P. and Fourniguet, J. (1987) Geologic Map of the Jabal Habashi Quadrangle, Sheet 26F, Kingdome of Saudi Arabia (with Text): Saudi Arabian Deputy Ministry for Mineral Resources. Jeddah, Geoscience Map GM-98A. (Scale 1: 250,000)
58. Dafalla, M.A. and Al-Shamrani, M.A. (2014) Swelling Characteristics of Saudi Tayma Shale and Consequential Impact on Light Structures. Journal of Civil Engineering and Architecture, 8, 613-623.
59. Van Der Merwe, D.H. (1964) The Prediction of Heave from the Plasticity Index and Percentage Clay Fraction of Soil. South African Institute of Civil Engineers, 6, 103-107.
60. Sridharan, A. and Prakash, K. (2000) Classification Procedures for Expansive Soils. Geotechnical Engineering, 143, 235-240.
61. https://doi.org/10.1680/geng.2000.143.4.235
62. Mitchell, J.K. and Soga, K. (2005) Fundamentals of Soil Behavior. 3rd Edition, John Wiley and Sons Inc., New York.
63. Embaby, A.A., Abu Halawa, A. and Ramadan, M. (2017) An Experimental Study to Mitigate Swelling Pressure of Expansive Tabuk Shale, Saudi Arabia. In Press.

CHAPTER 10

Influence of Crust Formation on Soil Porosity under Tillage Systems and Simulated Rainfall

Jaqueline Dalla Rosa [1,*], Miguel Cooper [2], Frédéric Darboux [3,†], João Carlos Medeiros [1], Carla Campanaro [2] and Luiz Roberto Martins Pinto [4]

[1] *Universidade Federal do Piauí, Campus Professora Cinobelina Elvas (UFPI/CPCE), CEP: 64900-000 Bom Jesus, PI, Brazil*

[2] *Departamento de Ciência do solo, Escola Superior de Agricultura "Luiz de Queiroz", Universidade de São Paulo (ESALQ/USP), CEP: 13400-000 Piracicaba, SP, Brazil*

[3] *INRA, UR0272, UR Science du sol, F-45075 Orléans, France*

[4] *Universidade Estadual de Santa Cruz, Campus Soane Nazaré de Andrade, Rodovia Jorge Amado, Km 16, Bairro Salobrinho, CEP: 45662-900 Ilhéus, Bahia, Brazil*

AUTHOR CONTRIBUTIONS

Jaqueline Dalla Rosa wrote the paper, performed the experiments, analyzed the data and coordinated the data collection. Miguel Cooper and Frédéric Darboux were Jaqueline Dalla Rosa's academic advisers and supported her in writing and proofreading the paper. João Carlos Medeiros and Carla Campanaro supported Jaqueline Dalla Rosa in performing the experiments, data collection and laboratory analysis. Luiz Roberto Martins Pinto supported Jaqueline Dalla Rosa in analyzing the data.

ABSTRACT

Surface crusts, formed by raindrop impact, degrade the soil surface structure causing changes in porosity. An experiment was conducted with the objective of evaluating the influence of the formation of a crusting layer on the porosity (percentage of area, shape and size) of a Haplic Acrisol under three tillage systems, and simulated rainfall. The tillage systems were: conventional tillage (CT), reduced tillage (RT) and no-tillage (NT). Each tillage system was submitted to different levels of simulated rainfall (0, 27, 54 and 80 mm) at an intensity of 80 $mm \cdot h^{-1}$. Undisturbed soil samples were collected and resin impregnated for image analysis in two layers: layer 1 (0–1 cm) and layer 2 (1–2 cm). Image analysis was used to obtain the pore area percentage, pore shape and size. The degradation of the soil surface and change in porosity, caused by rainfall, occurred differently in the tillage systems. In the CT and RT systems, the most pronounced pore changes caused by rainfall occurred in layer 1, but in the NT system, this change occurred in layer 2. The rainfall caused change of pore area percentage in the CT and RT systems, with reduction of complex and an increase of rounded pores. The NT system showed greater occurrence of the rounded pores (vesicles), originated by processes of wetting below the residue cover, and by alternating periods of wetting and drying. In this study, the changes in porosity were attributed to two main factors: (1) to the effect of the raindrop directly on the soil surface (for CT and RT tillage systems) and (2) water transfer processes in the soil surface (for NT systems).

Keywords: image analysis; pore shape; pore size; soil crusting; soil structure

1. INTRODUCTION

Soil functions as an environmental regulator of the hydrological and biogeochemical cycles, besides being a reservoir of biodiversity [1,2]. However, soil is also a natural resource that can be easily degraded. Erosion is one of the main forms of soil degradation, and the erosive process begins by degrading the soil surface. Surface crusts, formed by raindrop impact, degrade the soil surface structure causing changes in porosity. Soil porosity is a dynamic attribute and has been used in different ways by scientists to evaluate soil surface degradation by crusting [3]. Superficial crusts are formed under rain or irrigation through two main processes. Initially, raindrop impact disaggregates the soil surface structure into micro-aggregates, which are rearranged "in situ" without evidence of lateral movement, forming structural crusts. Secondly, depositional crusts are formed by the translocation of micro-aggregates by runoff or still depositional processes, as in puddles, which can be deposited at variable distances from the origin [4].

Studies that use soil porosity to evaluate the surface degradation by crusting are based on qualitative techniques, such as micromorphological descriptions of soil thin sections, quantitative techniques that use 2D or 3D image analysis, and also soil infiltration data to assess changes in porosity before and after the application of rainfall [3,4,5,6]. Searching for a greater detailing, Fox et al. [3] evaluated changes in soil porosity with depth in a structural crust by quantifying pore type and size distribution to distinguish between the crust layer and underlying soil.

The data on porosity obtained from image analysis or observations of the pore morphological characteristics are important to evaluate the soil structure and the processes that influence its formation or degradation [7]. Parameters such as the size and shape, distribution and the relative position of the pores are essential to evaluate changes in soil structure caused, for example, by different tillage practices [7,8,9], wetting and drying processes [10].

Macroporosity (>75 μm) is highly influenced by changes in soil structure. In a study developed in the field using conventional tillage systems, the results showed the formation of soil compaction due to successive rainfall events, and that this process impacted mainly the soil macroporosity [11]. A similar process was observed by Rousseva et al. [12] that evaluated the formation of crusts under simulated rainfall in two layers 0–3 cm and 3–6 cm. The authors observed that the occurrence of crusts caused a decrease in macroporosity in the both layers mainly due to the loss of elongated pores. The greatest reduction occurred in the 3–6 cm layer and was caused by the pressure exerted by the raindrops on the soil surface that was transferred to this deeper layer, resulting in the destruction of large pores. Similarly, in tilled soils without crop residue cover in Italy, soil crusting reduced pore connectivity [6].

The degradation of the soil surface caused by raindrop impact and, consequently, crusting may occur at different magnitudes and it depends on the soil characteristics, such as: texture; mineralogy; organic matter; initial moisture; distribution, size and aggregate stability [13-17]; and climatic factors, especially rain intensity, drop size distribution, drop impact velocity and kinetic energy, etc. [18-22].

Moreover, tillage practices, tillage intensity and crop residue cover also affect crust formation by rainfall and can influence mayor or minor degradation of the soil surface. Studies have recommended the adoption of soil conservation systems, such as reduced or no-tillage, to prevent soil structural degradation, crust formation and soil loss by erosion [23-26]. Part of this study has already been published and showed that the greater soil revolving and incorporation of crop residues, promotes greater degradation by drop impact and crust formation [27]. However, many aspects about the process dynamics of crust formation in these tillage systems still need clarification, such as the changes caused by rain events on the surface porosity and the predominant pore types.

Therefore, the objective of this study was to evaluate the changes in soil porosity in a crusting layer of a Haplic Acrisol under three tillage systems and simulated rainfall using image analysis techniques.

2. MATERIALS AND METHOD

2.1. Study Area, Soil Characteristics and Experimental Design

The experiment was carried out in the municipality of Piracicaba, São Paulo State, Brazil (coordinates: 22°42″ S, 47°36″ W) on a sandy Haplic Acrisol according to WRB 2014 [28]. The climate of the study site is tropical Cwa (Köppen classification). The landscape is gently undulating and the altitude is 542 m. The average annual precipitation for the last ten years was 1193 mm and the three maximum rainfalls in 24 h occurred in March 2002 (104 mm), May 2005 (139 mm) and January 2008 (87 mm) (Meteorological database from the University of São Paulo-ESALQ, coordinates: 22°42′0″ S–47°38′00″ W; Elevation: 546 m).

The soil chemical properties for the 0–20 cm layer are: pH in $CaCl_2$ = 5.11; OM = 32 $g \cdot dm^{-3}$; P = 5 $mg \cdot dm^{-3}$; K = 0.8 $mmolc \cdot dm^{-3}$; Ca = 23 $mmolc \cdot dm^{-3}$; Mg = 12 $mmolc \cdot dm^{-3}$; H + Al = 28 $mmolc \cdot dm^{-3}$; Al = 0.43 $mmolc \cdot dm^{-3}$; SB = 35.8 $mmolc \cdot dm^{-3}$; CEC = 63.8 $mmolc \cdot dm^{-3}$; V = 56%. Where: pH in $CaCl_2$: pH extracted with $CaCl_2$ (0.01 M); OM: Organic Matter; P: Phosphorous; K: Potassium; Ca: Calcium; Mg: Magnesium; H + AL: Hydrogen + Aluminum (potential acidity); SB: Sum of bases; CEC: effective cation exchange capacity; V: base saturation. The chemical analyses were performed according to Raij and Quaggio (1997) [29]. The particle size distribution of the studied soil, with sand particles separation for the 0–5 cm layer, and the general particle size distribution for the other layers are described in Table 1. The particles size distribution was analyzed according to Gee and Bauder (1986) [30].

Table 1. Particle size distribution for the soil studied.

Layer	Clay	Silt	Sand					
(cm)			VFS	FS	MS	CS	VCS	Total Sand
					$g \cdot kg^{-1}$			
0–5	164	20	164	471	155	20	6	816
5–20	179	38	-	-	-	-	-	783
40–60	257	52	-	-	-	-	-	691
80–100	262	37	-	-	-	-	-	701

VFS: very fine sand (0.105–0.053 mm); FS: fine sand (0.210–0.105 mm); MS: medium sand (0.50–0.210 mm); CS: coarse sand (1.00–0.50 mm); VCS: very coarse sand (2–1.00 mm).

In November 2008, the soil was tilled in all the experimental area (one plowing + three harrowing procedures) to standardize the conditions of the soil surface and a cover crop was sown (Brachiaria decumbens Stapf., without seed incorporation and sowing density of 13 $kg \cdot ha^{-1}$).

The experimental design was completely randomized with micro-plots and three repetitions. The treatments consisted of three tillage systems (CT, RT and NT) and four simulated rainfall volumes (0, 27, 54 and 80 mm) applied on the micro-plots. The effects of the treatments were evaluated in two soil layers (layer 1: 0–1 cm and layer 2: 1–2 cm) and in two consecutive years (2009 and 2010). In 2010 the experiment was installed in the same plots of the previous year.

2.2. First Year of Experiment (2009)

In early April 2009, the fully developed brachiaria was mechanically mowed and the regrowth was controlled with a systemic herbicide (Glyphosate), at a dosage of 4.0 $L \cdot ha^{-1}$. The average dry matter produced by brachiaria amounted to 8.400 $kg \cdot ha^{-1}$. The dry matter was determined with the aid of a metal square (0.5 m × 0.5 m). All plant material retained in the square was cut near the soil surface, dried in stove and determined the dry matter in $kg \cdot ha^{-1}$.

To evaluate the effect of soil tillage on the pore soil system, in June 2009, we established three tillage systems, with different soil mobilization intensities: conventional tillage (CT) performed with one plowing and two disking procedures; reduced tillage (RT) with heavy disking; and no-tillage (NT), without mobilization of the soil surface. The soil tillage depths were 25–30 cm in CT and 15–20 cm in RT. After each preparation, the remaining dry matter cover on the soil surface was 0% for CT, 70%–80% for RT and 100% for NT. Each tillage system was performed in 7 m × 30 m plots, in which 12 micro-plots of 1 m × 1 m, were installed randomly.

To assess the effect of rainfall on the soil surface porosity, we applied different simulated rainfall amounts (0, 27, 54 and 80 mm). Each rainfall amount was applied only once in the different micro-plots, with three repetitions, totaling 12 micro-plots for each soil tillage system. The simulated rainfall intensity was 80 $mm \cdot h^{-1}$ (±5 $mm \cdot h^{-1}$), applied shortly after the soil tillage operations with a rainfall simulator operated at 2.4 m above the ground. The simulator consisted of an oscillating nozzle type VeeJet Spray Nozzle (H/U-80100, Spraying Systems Co., City, State abbrev. if USA, Country). The mean weighted diameter (MWD) of the raindrops ranged from 1.5 to 2.5 mm. The rainfall was applied to an area corresponding to 1 m^2. Studies using rainfall simulators are frequently used to evaluate surface degradation and crust formation [27], aggregate breakdown [22,31] and soil erosion [32].

2.3. Soil Sampling, Impregnation and Porosity Analysis

For the porosity analysis, the samples were collected from the experimental area 24 h after the rainfall simulation. In each micro-plot, undisturbed samples were collected, which consisted of a block of soil (7 cm × 5 cm × 12 cm), surface-oriented, and placed in special paper boxes. In the laboratory, the samples were air-dried for 15 days and then in a ventilated oven at 40 °C for 48 h. Next, the samples were impregnated with polyester resin [33-35]. After full impregnation, hardening and drying, the blocks were cut into vertical surface-oriented slices (approximately 0.5 cm thick). A representative slice was chosen and sanded and polished with a lapping machine.

To evaluate the effects of the rainfall amount on soil porosity in depth, the image analysis was sequenced in two layers, (i) "layer 1", directly affected by the raindrops, corresponding to the first 1 cm of soil (0–1 cm) and (ii) "layer 2", right below the crust, with a thickness also of 1 cm (1–2 cm), which can be indirectly affected by the pressure exerted by the crust. Each layer (layer 1 and 2) was analyzed using three micro-photographs of 1 cm × 1 cm.

Knowing that pore distribution and morphology is important to assess soil structure, these attributes were analyzed by image analysis techniques, as proposed by Murphy et al. [34] and Ringrose-Voase [8]. Two-dimensional images (digital microphotographs, 1 cm × 1 cm), with 1024 × 768 pixels resolution and 10× magnification, were obtained with a CCD Sony®camera attached to a Zeiss® stereomicroscope. An ultraviolet light source was used to separate the pores from the soil matrix at image acquisition. The images were processed and analyzed using the Noesis® Visilog 5.4 software, Manufacturer, City, State abbrev. if USA, Country).

The soil porosity determined by the image analysis was expressed in percentage (%) of the image area occupied by pores. The pores were classified as: rounded (Round.), elongated (Elong.) and complex (Comp.), according to two shape indices (I1 and I2), as described in Cooper et al. [36]; and according to size: small (S) (156–15,600 μm^2), medium (M) (15,600–156,000 μm^2) and large (L) (>156,000 μm^2), as described in Pagliai et al. [37,38].

To help understand the distribution of pores after rainfall application, thin sections were prepared for each tillage system at the highest rainfall amount (80 mm) for analysis in a scanning electron microscope at 35× magnification. The thin sections were vertically oriented prepared and carbon-coated [35].

2.4. Second Year of Experiment (2010)

After soil sampling in June 2009, the experimental area remained fallow until November 2009, when the brachiaria was sown again (13 kg·ha^{-1}) and nitrogen (200 kg·ha^{-1} of N) was applied superficially without incorporation. All the steps until the second sample collection, in June 2010, followed the same protocol of the previous year (2009), with the tillage and rainfall applied in the same micro-plots. The dry matter produced by brachiaria in 2010 was 10.400 kg·ha^{-1}. In the second year of the experiment, the percentage of remaining dry matter cover after tillage on the soil surface was 0% for CT, 80%–85% for RT and 100% for NT.

During the fallow period, that preceded soil tillage (December 2008 to May 2009, and July 2009 until May 2010), the experimental area was hit by natural rains. The natural rainfall accumulated for the study period amounted to 554 mm (December 2008 to May 2009) and 1.348 mm (July 2009 to May 2010) (Meteorological database from the University of São Paulo—ESALQ).

2.5. Statistical Analysis

We analyzed the area percentage of the image occupied by each type of pore according to shape and size: rounded small (Round_S), rounded medium (Round_M), rounded large (Round_L), elongated small (Elong_S), elongated medium (Elong_M), elongated large (Elong_L), complex small (Comp_S), complex medium (Comp_M), complex large (Comp_L) and total pore area (TPA).

We performed ANOVA and F tests to evaluate the effects of the variation factors (soil tillage, simulated rainfall, soil layers) by years using the GLM Procedure in SAS (2002–2004). Soil tillage type was compared by Tukey's test ($p < 0.05$) and regression analyses were done to analyze the impact of rainfall on soil structure. To understand the relationships between the types of pores and the effects of other factors in the experiment we performed multivariate analyses using the PCA technique and biplot graphs [39,40]. Biplot graphs were created considering the first two principal components with greater variance and eigenvalues greater than 1.0. [41,42,43]. The PCA was calculated by means of a data matrix to test the relationships between (1) soil tillage; (2) rainfall; (3) soil tillage and rainfall interaction, and to select the most important pores. The relevance of variables was judged according to the direction, orientation and magnitude of their vectors. The PCA and the biplot graphs were prepared with 'Biplot and Singular Value Decomposition Macros for Excel ©' [42].

Due to the large number of variables evaluated in this work (soil tillage, rain, soil layer and year), we decided to conduct some exploratory PCA's to better understand the effects of soil tillage and simulated rainfall amount on the area percentage occupied with pores. Based on these exploratory PCA's, which showed differences for the years (and soil layers (), we proceeded with the PCA analysis for each layer and each year (for layer 1 and layer 2).

3. RESULTS

The analysis of variance demonstrated significant effect for soil tillage, rainfall and layer for 2009 and 2010 (Table 2). The effect of the tillage systems is observed in Table 3, where the different types of pores showed similar behavior in the three tillage systems evaluated for both years. The % TPA and Comp_L showed higher values in CT and RT, decreasing in NT. Already the % area of rounded pores was higher in NT and lower in the CT and RT.

Significant effects were observed for rainfall applied to CT and RT tillage systems in 2009 (Figure 1). In both tillage systems, the increase in the amount of rain caused a decrease in the % area of Comp_L pores and TPA. For both tillage systems (CT and RT) there is no difference in the % pore area between 27 and 54 mm rain amounts, this means that the % of Comp_L pore area and TPA decrease between 0 and 27 mm rain depth, do not change between 27 and 54 mm and decrease again with 80 mm of rain amount (Figure 1).

Table 2. Analysis of variance for percentage of pore area in the two years of the experiment and for the studied soil.

Variation Factors/Porosity	Round_S	Round_M	Elong_M	Elong_L	Comp_M	Comp_L	TPA
			2009				
Soil Tilllage	0.016	0.0271	0.032	0.034	ns	0.0002	<0.0001
Layer	ns	ns	ns	ns	0.0343	ns	ns
Soil tillage*Layer	ns	ns	ns	ns	ns	ns	ns
Rainfall	0.0262	ns	0.0082	ns	0.0044	0.0034	0.0017
			2010				
	Round_S	Round_M	Round_L	Elong_L	Comp_M	Comp_L	TPA
Soil Tilllage	ns	ns	0.001	0.0073	0.0007	ns	ns
Layer	0.014	0.0065	ns	0.032	ns	0.0005	0.0003
Soil tillage*Layer	ns	ns	ns	ns	ns	ns	ns
Rainfall	0.015	0.0005	ns	0.0002	ns	0.005	0.011

Round_S: rounded small; Round_M: rounded medium; Elong_M: elongated medium; Elong_L: elongated large; Comp_M: complex medium; Comp_L: complex large; TPA: total pore area; Round_L: rounded large; ns: no significance.

Table 3. Effect of the tillage systems on pore area percentage in the two years of the experiment and for the studied soil.

Pores/Soil Tillage	Round_S	Round_M	Round_L	Elong_M	Elong_L	Comp_M	Comp_L	TPA
				2009				
CT	1.31 AB	2.42 AB	0.21 ns	0.47 AB	0.35 AB	0.13 ns	49.01 B	53.89 B
RT	1.22 B	1.98 B	0.14	0.42 B	0.25 B	0.10	50.19 B	54.31 B
NT	1.50 A	2.60 A	0.25	0.60 A	0.40 A	0.15	43.56 A	49.05 A
				2010				
CT	2.02 ns	3.72 ns	0.57 A	0.76 ns	0.82 AB	0.12 A	35.50 ns	43.52 ns
RT	1.99	3.72	0.35 B	0.83	0.66 B	0.17 B	33.97	41.69
NT	2.09	4.02	0.67 A	0.72	0.94 A	0.11 A	31.06	39.61

CT: Conventional Tillage; RT: Reduced Tillage; NT: No-Tillage; Round_S: rounded small; Round_M: rounded medium; Round_L: rounded large; Elong_M: elongated medium; Elong_L: elongated large; Comp_M: complex medium; Comp_L: complex large; TPA: total pore area; Different letters indicate significant differences between tillage systems by Tukey test ($p < 0.05$); ns: no significance.

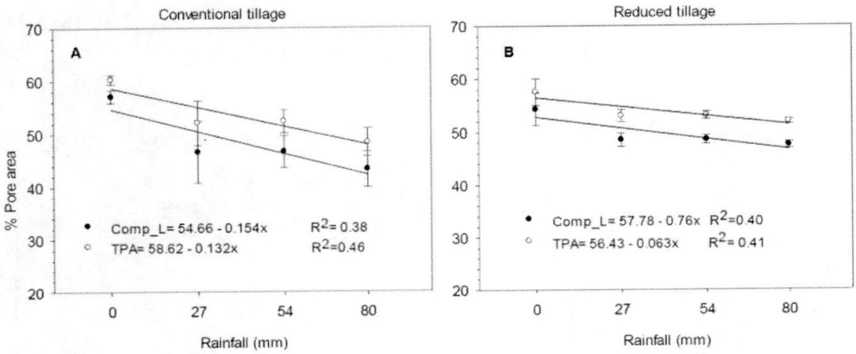

Figure 1. Effect of applied rainfall on percentage of pore area, in conventional (**A**) and reduced tillage (**B**) for layer 1 in 2009 year. Comp_L: complex large; TPA: total pore area. Error bars are standard errors.

For the year 2010, similarly to what was observed in 2009, the significant effects of the applied rainfall occurred only in CT and RT. The increase of the applied rain volume in CT caused a reduction in the % area of Comp_L pores and TPA, and an increase in the area occupied by rounded pores. In RT a quadratic effect was observed for Comp_L, TPA, Round_S and Elong_L (Figure 2). Unexpectedly, for the 80 mm rain amount there was an increase in Comp_L pore area percentage and TPA (Figure 2A), and a reduction of Round_S and Elong_L. Still in this tillage system, the % area of Round_M pores increased with the increase in rain amount. For NT, rainfall amount did not have a significant effect on pore shape and size.

3.1. Effects of Years (2009 and 2010) and Layers on the Percentage of the Image Area Occupied by Pores

The percentage of pore area in 2009 differed significantly from that in 2010 (Figure 3A), suggesting to evaluate the years separately. The year 2009 explained 42% of the variability of the pore area, while 2010 explained 58%. In 2009, a greater association with large complex pores (Comp-L) was observed (Figure 3B), while 2010 showed a better relationship with rounded and elongated pores (Figure 3C). The greatest relation with the pores Comp_L in 2009 is because the area had an initial tillage (in December 2008) and after the development of the brachiaria the tillage was applied again, for each treatment (CT and RT). Thus, in a short period the area was intensively revolved. From 2009 to 2010, the area remained without tillage (fallow) for one year, hence the greater relationship with rounded and elongated pores. The Comp_M pores correlated positively with Elong_M, orthogonally with Comp_L, Round_M and Round_S, and negatively with Elong_L and Round_L.

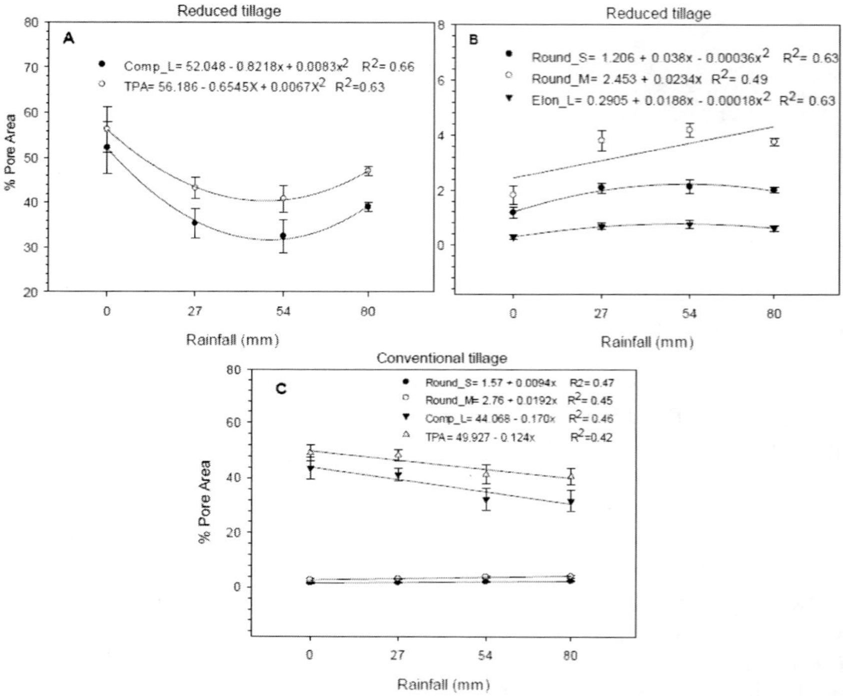

Figure 2. Effect of applied rainfall on percentage of pore area, for layer 1 in 2010 year, in reduced (**A,B**) and conventional tillage (**C**). Round_S: rounded small; Round_M: rounded medium; Elong_L: elongated large; Comp_L: complex large; TPA: total pore area. Error bars are standard errors.

3.2. Effect of Soil Tillage and Rainfall Amount on the Percentage of the Area Occupied by Pores in Layer 1 (2009 and 2010)

For layer 1 in 2009 (Figure 4A), the total variability of the pore area percentage, due to the different tillage systems and rain amounts, was 89% (74% explained by PC1 and 15% PC2). In addition, for layer 1 in 2010 (Figure 4B), the PC1 explained 63% and PC2 24% of pore area percentage variability, which represents 87% of the total variation. In both years, we observed the formation of some groups. These groups show clusters in the graph due to soil tillage and the distribution of these clusters are a result of the applied rainfall amounts.

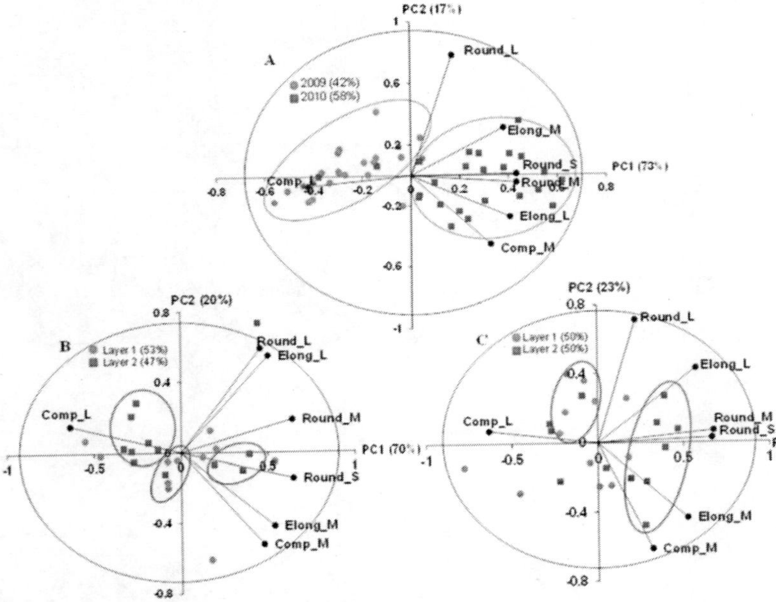

Figure 3. (**A**) PCA of the percentage of pore area in both years of study (2009 and 2010); (**B**) PCA of the percentage of pore area in the layers for 2009 and (**C**) PCA of the percentage of pore area in the layers for 2010 for a Haplic Acrisol, municipality of Piracicaba, São Paulo State, Brazil. Pore shapes (Round: rounded; Elong: elongated; Comp: complex) and sizes (S: 156–15,600 μm^2; M: 15,600–156,000 μm^2 and L: >156,000 μm^2). PC1: Principal Component 1; PC2: Principal Component 2.

The change in pore shapes due to the rainfall applied in 2009 and 2010 was observed with the reduction of Comp_L favoring the formation of smaller pores (Figure 4A,B). This effect was also observed in the regression analysis (Figure 1 and Figure 2).

The change of pore area percentage due to the applied rainfall was similar in the two years of the study for each tillage system. The CT system showed a change clearly associated with the simulated rainfall applied, which is due to the raindrop impact. Immediately after tillage (no rainfall = CT0) there is a higher percentage of Comp_L pore area. With the increase of the rainfall amount (54 and 80 mm) we observed an increase in the percentage of rounded pore area (Round_M and Round_S) (Figure 1A, Figure 2C and Figure 4A,B), and a reduction of the Comp_L and in the percentage of the total pore area (TPA).

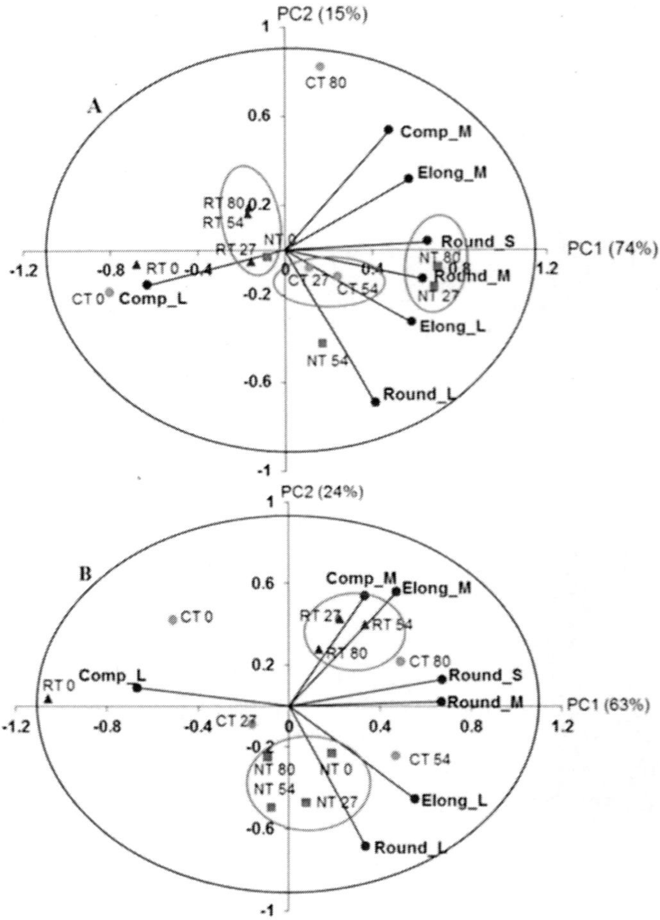

Figure 4. (**A**) PCA for the percentage of pore area in layer 1 for 2009, and (**B**) PCA for the percentage of pore area in layer 1 for 2010, in a Haplic Acrisol, municipality of Piracicaba, São Paulo State, Brazil. Pore shapes (Round: rounded; Elong: elongated; Comp: complex) and sizes (S: 156–15,600 μm^2; M: 15,600–156,000 μm^2 and L: >156,000 μm^2). PC1: Principal Component 1; PC2: Principal Component 2; CT: Conventional Tillage; RT: Reduced Tillage; NT: No-Tillage. 0, 27, 54 and 80 mm: simulated rainfall amount.

The pore changes related to the rainfall amount observed in the CT system also occurred in the RT system. CT and RT systems showed reduction of Comp_L pores and TPA as rainfall increased in 2009 and 2010. In 2010 there was a larger increase of small pores and reduction of large pores with the applied rainfalls.

The NT system showed a high association with Round_L and Elong_L in both years, demonstrating that the pores did not undergo major changes in the study period. The NT system did not provide a clear change of the pore area percentage with the increase of applied rainfall amount as observed in the CT and RT systems.

3.3. Effect of Soil Tillage and Rainfall Volume on the Percentage of the Area Occupied by Pores in Layer 2 (2009 and 2010)

The total variability of the pore area percentage in layer 2 in 2009, due to the different tillage systems and rain amounts, was 92% (71% explained by PC1 and 21% by PC2) (Figure 5A). In 2010 year, the total variability was 85% (58% explained by PC1 and 27% explained by PC2) (Figure 5B).

In 2009, the CT and RT systems showed greater association with Comp_L pore area percentage than with rounded and elongated pores. Changes in the pore shapes of these systems were not observed with the increase of simulated rainfall amount, as observed in layer 1 (Figure 5A). In 2010 this behavior continued in the CT system, but the RT system showed an increase of Round_L and Elong_L pore areas and, consequently, a decrease of Comp_L pore area (Figure 5B). The NT system, for both years, showed greater association with elongated and rounded pores (Figure 5A,B), except where rainfall was not applied (0 mm), which showed greater association with Comp_L pores. This behavior is different to the one observed in layer 1.

4. DISCUSSION

4.1. Effect of Tillage on Soil Porosity

Soil tillage affected the structure and pore arrangement of the studied soil. The higher percentage of large complex pore area in the CT and RT systems compared to NT, occurred due to the tillage system employed (Table 3). This effect was more pronounced in the first year, where the % Comp_L pores was greater in CT and RT. Similarly, other authors have reported an increase in the percentage of large pores (>500 μm) in CT and RT systems due to soil tillage, and most of the total porosity was represented by few large-sized pores [9,44,45].

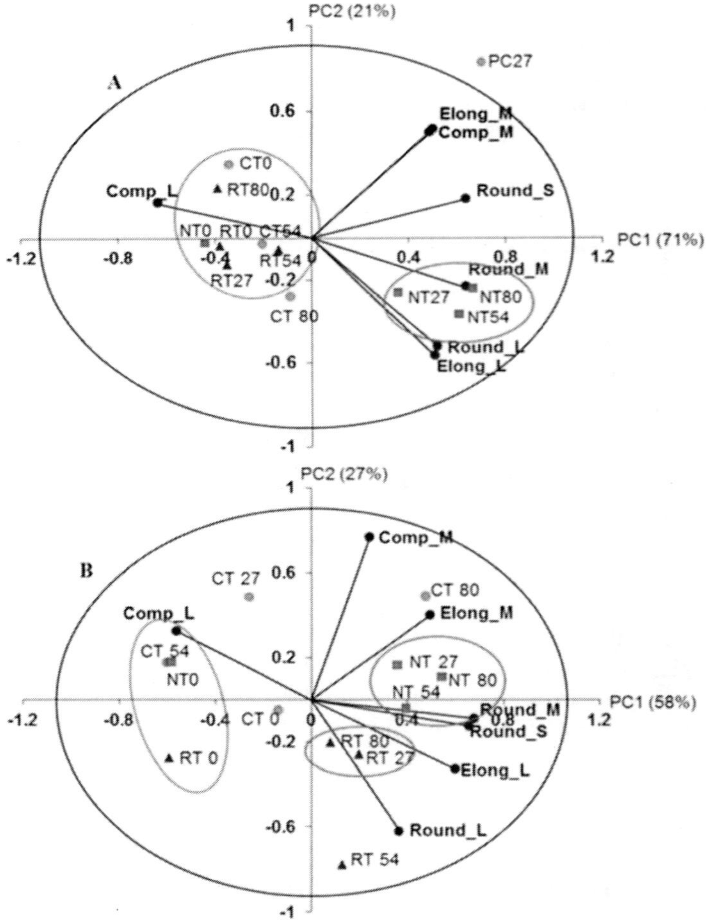

Figure 5. (**A**) PCA for the percentage of pore area in layer 2 for 2009 and (**B**) PCA for the percentage of pore area in layer 2 for 2010, in a Haplic Acrisol, municipality of Piracicaba, São Paulo State, Brazil. Pore shapes (Round: rounded; Elong: elongated; Comp: complex) and sizes (S: 156–15,600 μm^2; M: 15,600–156,000 μm^2 and L: >156,000 μm^2). PC1: Principal Component 1; PC2: Principal Component 2; CT: Conventional Tillage; RT: Reduced Tillage; NT: No-Tillage: 0, 27, 54 and 80 mm: simulated rainfall amount.

The NT system demonstrated greater relationship with rounded and elongated pores in both years (Table 3). A greater presence of rounded pores called "vesicles" was observed (Figure 6 and Figure 7C) in relation to other tillage systems that can be explained by the lack of ploughing and seedbed preparation. These "vesicles" were possibly originated by air entrapment due to particle transportation and deposition at short distances that were detached by possible rapid wetting events under the soil cover. This process occurred with the natural rain during alternating periods of wetting and drying. Chong-Feng et al. [15] also observed that the dominant process in the formation of crusts in covered soil was the eluviation process.

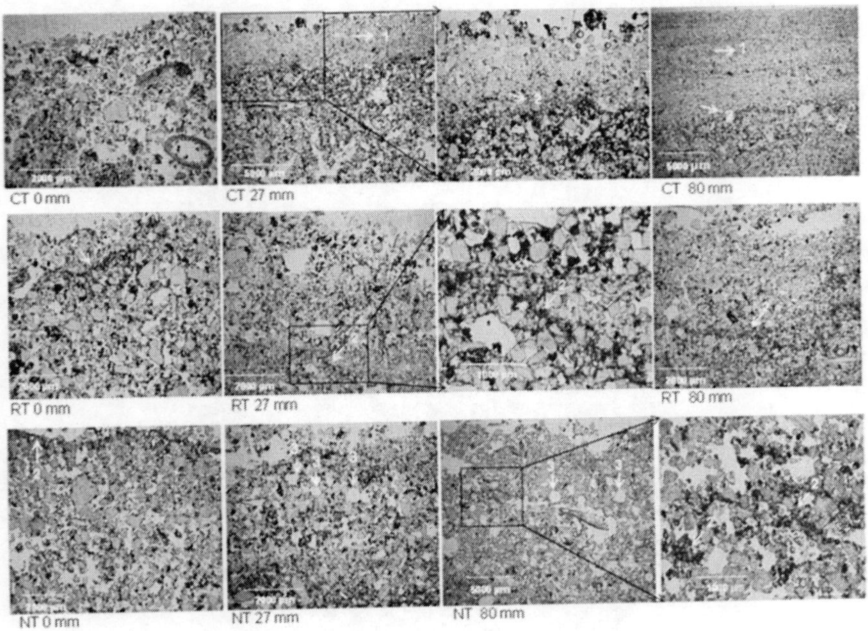

Figure 6. Images of thin sections vertically oriented for the different tillage systems and rainfall amount applied. 1: crust formed, quartz grains arranged after simulated rainfall; 2: lamination of accumulated fine material; 3: rounded pores (vesicles); CT: Conventional Tillage; RT: Reduced Tillage; NT: No-Tillage; 0, 27, 54 and 80 mm: simulated rainfall amount.

Pore characteristics depend on the structure and the structure evolution processes that occur in the soil. Therefore, the occurrence of certain pore shapes and sizes reflect the soil structural evolution processes that occurred in the soil.

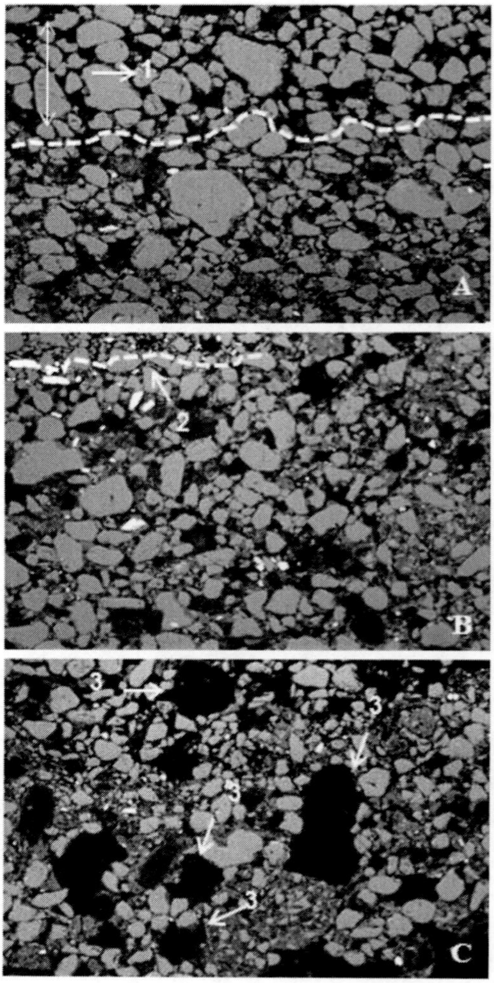

Figure 7. Thin sections vertically oriented and carbon-coated. Scanning electron microscope images with 35× magnification. (**A**): CT—Conventional Tillage; (**B**): RT—Reduced Tillage; (**C**): NT—No-Tillage. Images (**A**–**C**) refer to maximum rainfall amount applied (80 mm). 1: crust formed, quartz grains arranged after rain applied; 2: lamination of accumulated fine material; 3: rounded pores (vesicles).

For example, vesicles are more numerous and larger as the number of wetting and drying cycles increases [46]. These pores are characterized by having smoothed walls with rounded and regular shape [35], furthermore, they show little or no connection to other pores, and, thus, hinders water distribution in the soil. Elongated pores (channels and chambers) are usually caused by biological activity and plant roots, while the drying-wetting processes can lead to planar pores [47,48].

The reduction in soil tillage or the adoption of alternative systems to conventional tillage practices can preserve soil structure with a balanced pore size distribution [45,49] and, consequently, confer proper water distribution in the soil profile.

4.2. Effect of Simulated Rainfall on Crust Formation and Porosity

The simulated rainfall degraded the soil surface and caused crust formation. Crust formation was not only caused by rainfall but was also a consequence of the soil characteristics and tillage as observed in the micromorphological descriptions done by Dalla Rosa et al. [27] studying the same soil. In CT, the formation of crusts was observed after 27 mm of rain. A higher rain amount (80 mm) increased the degradation of the surface structure and the thickness of the crust layer (Figure 6; CT 80 mm). Greater degradation occurred in the CT system, because the soil surface, after tillage, remained fully exposed to rainfall. The raindrop impact on the bare soil surface promotes mechanical breakdown of aggregates [12], splash and crusting. The crusts formed in CT were classified as sieving crusts [27]. These are primarily developed by the separation of soil particles by the direct impact of the raindrops after which, with water infiltration, the dispersed material is reorganized into layers [4]. The differentiation of the quartz grain sizes and the accumulation of fine material inside the crusts forming distinct layers (Figure 6 and Figure 7A) result mainly from mechanical sieving, caused by the rainfall infiltration, which deposits the fine particles deeper in the soil [4,27,44]. Clay translocation and accumulation form fine internal laminae (link capping) caused by particle dispersion and deposition as water infiltrates in the soil (Figure 6, CT 27 and CT 80 mm; and Figure 7A). The dominance of quartz grains in the crust layer is due to the sandy particle size class of the surface horizon of this soil.

Under RT and NT, the changes in the surface conditions and the formation of crusts did not evolve with the increase of the rain amount (volumes), differently to what was observed in CT. The detachment by single-raindrop impact decreases with the increase in soil strength [21]. In this case, these tillage systems keep mulch on the soil surface, covering the whole surface or partially by semi-incorporation of the residues, which promotes greater resistance to breakdown by raindrop impact.

The soil surface was more heterogeneous in RT. In this tillage system the crusts were classified as Sieving crusts for the uncovered areas of the soil surface directly affected by rain, and as Slaking crusts, characterized as been more irregular and with incipient orientation, for the covered areas [27]. The fine material accumulated in the surface layer (Figure 7 RT and NT) can result from physical and chemical dispersion of clay particles, which are redistributed in the surface layer or migrate to deeper layers in the soil with water infiltration, concentrating in the form of lamina or simply accumulating in specific areas [17].

These changes in the soil surface (rearrangement of quartz grains and fine material accumulation in the first few centimeters of the soil surface) caused by the raindrop impact also modified the pore system. The changes caused a reduction of large complex pores (originated by soil tillage) and an increase in rounded and elongated pores (originated and/or changed primarily by the direct rainfall on the soil surface). In 2009 there was a reduction of the percentage area of Comp_L pores and TPA observed for CT and RT systems (Figure 1). In 2010 for CT, a reduction was also observed in the % area of Comp_L pores and TPA that resulted in an increase in the percentage area of rounded pores, especially Round_S and Round_M, (Figure 2C). The volume of rainfall applied influenced this pore shape and size transformation, the highest amount of rainfall (80 mm) resulted in the highest reduction of Comp_L pores and increase of rounded pores (Figure 2C). These results agree with Fox et al. [3], where the pore frequency was inversely proportional to pore area, as elongate and irregular pores were broken into smaller rounded pores under the compacting, detachment and splash effect of the raindrop impact. Rousseva et al. [12] also observed a strong decrease in the frequency of the largest pores, especially elongated porosity, while the frequency of the smaller pores remain almost constant and sometimes even increased with the accumulated simulated rainfall. However, for the same soil used in this study, Castilho et al. [50] found that crust formation caused by natural rainfall did not affect significantly the pore morphology.

The quadratic effect observed for RT in 2010 (Figure 2A,B) showed an increase of Comp_L and TPA, and a decrease of Round_S and Elong_L between the 54 and 80 mm rainfalls. This behavior could be explained by changes in the soil structure caused by the high volume rainfalls that result in the entrapment of air bubbles between aggregates and soil particles, and/or air entrapment in the interior of the aggregates that can cause aggregate breakdown and in sandy materials increase the amount of complex pores and, consequently, TPA.

Other authors have reported a reduction in porosity by crust formation resulting from rainfall incidence [4,8,26]. Chong-Feng et al. [15] observed that the pore space of the soil surface declined across time in both covered and uncovered treatments, but the changes were greatest in the uncovered treatments. This reduction was attributed to the impact of the raindrops on the soil surface [1]. Therefore, the complex (continuous) and large pores were destroyed and/or transformed into elongated and rounded (non continuous) pores. The pores in the crust layer tend to be rounded and disconnected from one another, and become inefficient to conduct water [6]. The reduction of irregular and continuous pores, which are responsible for water conduction, usually affects the soil-water-plant relations and degrades soil structure [44].

In layer 2, changes in the area of pores occurred in the RT system (2010) and NT system (2009 and 2010) resulting in the reduction of Comp_L pores and an increase of rounded pores. Possibly this change in porosity was caused by a transfer of the rainfall effect from the surface layer 1 to the sub-layer (layer 2), because in these tillage systems layer 1 remained protected by plant residue cover. This effect, as previously discussed, was probably induced by wetting and by the effect of alternating periods of wetting and drying caused by natural rains during the fallow period. This process can cause dispersion and eluviation of fine particles internally in the soil altering the soil porosity [15,17]. On the other hand, Rousseva et al. [12] also evaluating the porosity after rainfall in two layers, 0–3 cm and 3–6 cm, observed that the greatest reduction occurred in the 3–6 cm layer, and the reason given by the authors was the transfer to the sublayer of the pressure exerted by the raindrops on the soil surface.

The degradation of the soil surface and crust formation by the rainfall effect was more pronounced in the CT system, followed by RT and NT systems (Figure 6). The greatest pore changes caused by the rainfall effect also occurred in the CT system (changes of complex to rounded pores). On the other hand, even without direct effect of rainfall on the soil surface, the NT system also demonstrated presence of rounded pores, this could be associated with soil particles detached during the soil wetting process, transported and deposited at short distances favoring subsurface air entrapment that form the vesicles (Figure 6 and Figure 7) also described by Valentin (1991) [51].

5. CONCLUSIONS

Soil porosity assessed by image analysis showed differences between the studied years (2009 and 2010).

In conventional tillage and reduced tillage systems, the greatest pore changes caused by the effect of rainfall amount occurred in soil layer 1, due to the direct raindrop impact. In the no-tillage system, the major changes occurred in layer 2, by the process of water transfer.

Simulated rainfall caused surface degradation and reduction of pore area percentage in conventional tillage and reduced tillage systems, with the reduction of complex and increase of rounded pores.

The no-tillage system showed greater occurrence of the rounded pores (vesicles), originated by processes of wetting below the residue cover, and by alternating periods of wetting and drying.

Image analysis techniques and PCA explained satisfactorily the changes caused in the soil porosity by the use of different tillage systems and simulated rainfall amount.

ACKNOWLEDGMENTS

The authors are thankful to the São Paulo Research Foundation (FAPESP) for the scholarship granted to the first author and financial support (Process 2008/53591-08). To CAPES for the sandwich Ph.D. scholarship (Process 1278/10-0). The authors also thank CNPq for the research fellowship.

REFERENCES

1. Keesstra, S.D.; Geissen, V.; Mosse, K.; Piiranen, S.; Scudiero, E.; Leistra, M.; van Schaik, L. Soil as a filter for groundwater quality. *Curr. Opin. Environ. Sustain.* **2012**, *4*, 507–516.
2. Brevik, E.C.; Cerdà, A.; Mataix-Solera, J.; Pereg, L.; Quinton, J.N.; Six, J.; van Oost, K. The interdisciplinary nature of SOIL. *Soil* **2015**, *1*, 117–129.
3. Fox, D.M.; Bryan, R.B.; Fox, C.A. Changes in pore characteristics with depth for structural crusts. *Geoderma* **2004**, *120*, 109–120.
4. Valentin, C.; Bresson, L.M. Morphology, genesis and classification of surface crusts in loamy and sandy soils. *Geoderma* **1992**, *55*, 225–245.
5. Usón, A.; Poch, R.M. Effects of tillage and management practices on soil crust morphology under a Mediterranean environment. *Soil Till. Res.* **2000**, *54*, 191–196.
6. Panini, T.; Torri, D.; Pellegrini, S.; Pagliai, M.; Sanchis, M.P.S. A theoretical approach to soil porosity and sealing development using simulated rainstorms. *Catena* **1997**, *31*, 199–218.
7. Rasa, K.; Eickhorst, T.; Tippkötter, R.; Yli-halla, M. Structure and pore system in differently managed clayey surface soil as described by micromorphology and image analysis. *Geoderma* **2012**, *173*, 10–18.
8. Ringrose-Voase, A.J. Micromorphology of soil structure: description, quantification, application. *Aust. J. Soil Res.* **1991**, *29*, 777–813.

9. Pagliai, M.; Pezzarossa, B.; Mazzoncini, M.; Bonari, E. Effects of tillage on porosity and microstructure of a loam soil. *Soil Technol.* **1989**, *2*, 345–358.
10. Pires, L.F.; Cooper, M.; Cássaro, F.A.M.; Reichardt, K.; Bacchi, O.O.S.; Dias, N.M.P. Micromorphological analysis to characterize structure modifications of soil samples submitted to wetting and drying cycles. *Catena* **2008**, *72*, 297–304.
11. Schiettecatte, W.; Jin, K.; Yao, Y.; Cornelis, W.M.; Lu, J.; Wu, H.; Verbist, K.; Cai, D.; Gabriels, D.; Hartmann, R. Influence of simulated rainfall on physical properties of a conventionally tilled loess soil. *Catena* **2005**, *64*, 209–221.
12. Rousseva, S.; Torri, D.; Pagliai, M. Effect of rain in the macroporosity at the soil surface. *Eur. J. Soil Sci.* **2002**, *53*, 83–94.
13. Hu, F.N.; Xu, C.Y.; Li, H.; Li, S.; Yu, Z.H.; Li, Y.; He, X. Particles interaction forces and their effects on soil aggregates breakdown. *Soil Till. Res.* **2015**, *147*, 1–9.
14. Ma, R.M.; Li, Z.X.; Cai, C.F.; Wang, J.G. The dynamic response of splash erosion to aggregate mechanical breakdown through rainfall simulation events in Ultisols (subtropical China). *Catena* **2014**, *121*, 279–287.
15. Chong-Feng, B.U.; Gale, W.J.; Qiang-Gu, C.I.; Shu-Fang, W.U. Process and mechanism for the development of physical crusts in three typical Chinese Soils. *Pedosphere* **2013**, *23*, 321–332.
16. Fox, D.M.; Le Bissonnais, Y. A process-based analysis of the influence of aggregate stability on surface crusting, infiltration, and interrill erosion. *Soil Sci. Soc. Am. J.* **1998**, *62*, 717–724.
17. Agassi, M.; Shainberg, I.; Morin, J. Effect of electrolyte concentration and soil sodicity on infiltration rate and crust formation. *Soil Sci. Soc. Am. J.* **1981**, *45*, 848–851.
18. Salles, C.; Poesen, J. Rain properties controlling soil splash detachment. *Hydrol. Process.* **2000**, *14*, 271–282.
19. Sharma, P.P.; Gupta, S.C. Sand detachment by single rain drops of varying kinetic energy and momentum. *Soil Sci. Soc. Am. J.* **1989**, *53*, 1005–1010.
20. Torri, D.; Sfalanga, M.; Del Sette, M. Splash detachment: runoff depht and soil cohesion. *Catena* **1987**, *14*, 149–155.
21. Al-Durrah, M.M.; Bradford, J.M. The mechanism of raindrop splash on soil surfaces. *Soil Sci. Soc. Am. J.* **1982**, *46*, 1086–1090.
22. Darboux, F.; Dalla Rosa, J.; Cooper, M. Conditions for the occurrence of slaking and other disaggregation processes under rainfall. *Hydrology* **2016**, *27*, 1–9.
23. Sun, Y.; Zeng, Y.; Shi, Q.; Pan, X.; Huang, S. No-tillage controls on runoff: A meta-analysis. *Soil Till. Res.* **2015**, *153*, 1–6.
24. Armand, R.; Bockstaller, C.; Auzat, A.V.; van Dijk, P. Runoff generation related to intra-field soil surface characteristics variability. Application to conservation tillage context. *Soil Till. Res.* **2009**, *102*, 27–37.
25. Castro, L.G.; Cogo, N.P.; Volk, L.B.D. Alterations in soil surface roughness by tillage and rainfall in relation to water erosion. *Rev. Bras. Cienc. Solo* **2006**, *30*, 339–352.
26. Pagliai, M.; Vignozzi, N.; Pellegrini, S. Soil structure and the effect of management practices. *Soil Till. Res.* **2004**, *79*, 131–143.
27. Dalla Rosa, J.; Cooper, M.; Darboux, F.; Medeiros, J.C. Processo de formação de crostas superficiais em razão de sistemas de preparo do solo e chuva simulada. *Rev. Bras. Cienc. Solo* **2013**, *37*, 400–410. (In Portuguese)
28. Food and Agriculture Organization of the United Nations (FAO). *World Reference Base for Soil Resources*; Food and Agriculture Organization of the United Nations: Rome, Italy, 2014.
29. Raij, B.V.; Quaggio, J.A.; Cantarella, H.; Ferreira, M.E.; Lopes, A.S.; Bataglia, O.A. *Análise de Solo para Fins de Fertilidade*; Fundação Cargill: Campinas, Brazil, 1987; p. 170. (In Portuguese)
30. Gee, G.W.; Bauder, J.W. Particle size analysis. In *Methods of Soil Analysis*, 2nd ed.; Klute, A., Ed.; American Society of Agronomy; Soil Science Society of America: Madison, WI, USA, 1986; pp. 383–411.
31. Legout, C.; Leguédois, S.; Le Bissonnais, Y. Aggregate breakdown dynamics under rainfall compared with aggregate stability measurements. *Eur. J. Soil Sci.* **2005**, *56*, 225–238.

32. Lassu, T.; Seeger, M.; Peters, P.; Keesstra, S.D. The Wageningen rainfall simulator: Set-up and calibration of an indoor nozzle-type rainfall simulator for soil erosion studies. *Land Degrad. Dev.* **2015**, *26*, 604–612.
33. Murphy, C.P. *Thin Section Preparation of Soils and Sediments*; A.B. Academic Publication: Berkhamsted, UK, 1986; p. 149.
34. Murphy, C.P.; Bullock, P.; Turner, R.H. The measurement and characterization of voids in soil thin sections by image analysis. Part I. Principles and techniques. *J. Soil Sci.* **1977**, *98*, 498–508.
35. Castro, S.S.; Cooper, M.; Santos, M.C.; Vidal-Torrado, P. Micromorfologia do solo: Bases e aplicações. *Tópicos Ciência Solo* **2003**, *3*, 107–164. (In Portuguese)
36. Cooper, M.; Vidal-Torrado, P.; Chaplot, V. Origin of microaggregates in soils with ferralic horizons. *Sci. Agricola***2005**, *62*, 256–263.
37. Pagliai, M.; La Marca, M.; Lucamante, G.; Genovese, L. Effects of zero and conventional tillage on the lengh and irregularity of elongated pores in a clay loam soil under viticulture. *Soil Till. Res.* **1984**, *4*, 433–444.
38. Pagliai, M.; La Marca, M.; Lucamante, G.; Genovese, L. Micromorphometric and micromorphological investigations of a clay loam soil in viticulture under zero and conventional tillage. *Soil Sci. Soc. Am. J.* **1983**, *34*, 391–403.
39. Gabriel, K.R. The Biplots graphic display of matrices with application to principal component analysis. *Biometrika***1971**, *58*, 453–467.
40. Kroonenberg, P.M. Introduction to Biplots for GxE Tables. 1997. Available online: http://www.fsw.leidenuniv.nl/~kroonenb/document/biplot.zip (accessed on 12 February 2012).
41. Anderson, W.F.; Beute, M.K.; Wynne, J.C.; Wongkaew, S. Statistical procedures for assessment of resistance in a multiple foliar disease complex of peanut. *Phytopathology* **1990**, *80*, 1451–1459.
42. Moraaguilera, G.; Nietoangel, D.; Teliz, D.; Campbell, C.L. Development of a prediction model for papaya ringspot in Veracruz, Mexico. *Plant Dis.* **1993**, *77*, 1205–1211.
43. Lipkovich, I.; Smith, E.P. Biplot and Singular Value Decomposition Macros for Excel©. *J. Stat. Softw.* **2002**, *7*, 1–15. Available online: http://www.jstatsoft.org/v07/i05 (accessed on 12 February 2012).
44. Pagliai, M.; Raglione, M.; Panini, T.; Maletta, M.; La Marcac, M. The structure of two alluvial soils in Italy after 10 years of conventional and minimum tillage. *Soil Till. Res.* **1995**, *34*, 209–223.
45. Mermut, A.R.; Grevers, M.C.J.; Jong, E. Evaluation of pores under different management systems by image analysis of clay soils in Saskatchewan, Canada. *Geoderma* **1992**, *53*, 357–372.
46. Bresson, L.M.; Valentin, C. Soil surface crust formation: contribution of micromorphology. In *Developments in Soil Science*; Ringrose-Voase, A.J., Humphreys, G.S., Eds.; Elsevier: Amsterdam, The Netherlands, 1993; pp. 737–762.
47. Thompson, M.L.; Singh, P.; Corak, S.; Straszheim, W.E. Cautionary notes for the automated analysis of soil pore-space images. *Geoderma* **1992**, *53*, 399–415.
48. Vandenbygaard, A.J.; Protz, R.; Tomlin, A.D. Changes in pore structure in a no-till chronosequense of silt loam soils, Southern Ontario. *Can. J. Soil Sci.* **1999**, *79*, 149–160.
49. Pagliai, M. Micromorphometric and micromorphological investigations on the effect of compaction by pressures and deformations resulting from tillage and wheel traffic. In *Soil Compaction and Regeneration*; Monnier, G., Goss, M.J., Eds.; AA Balkema: Rotterdam, The Netherlands, 1987; pp. 31–38.
50. Castilho, S.C.P.; Cooper, M.; Juhász, C.E.P. Influence of crust formation under natural rain on physical attributes of soils with different textures. *Rev. Bras. Cienc. Solo* **2011**, *35*, 1893–1905.
51. Valentin, C. Surface crusting in 2 alluvial soils of northern Niger. *Geoderma* **1991**, *48*, 201–222.

CHAPTER 11

Three-Dimensional Geologic Modeling and Groundwater Flow Modeling above a CO_2 Sequestration Test Site

Erin Carlock, Eric W. Peterson*, David H. Malone

Department Geography-Geology, Illinois State University, Normal, IL, USA

ABSTRACT

As temperatures rise and climate change becomes an increasingly important issue, geologic carbon dioxide (CO_2) sequestration is a viable solution for reducing greenhouse gas emissions. Subsurface 3-D modeling and groundwater flow modeling were completed as a component of a CO_2 sequestration feasibility study in the city of Decatur, Illinois. The Decatur Archer Daniels Midland Company Ethanol Plant (ADM) serves as the injection site for a CO_2 sequestration project within a deep saline reservoir. Petrel was successfully used to model the glacial deposits in the area. The 3-D geologic model shows the Peoria Silt, Wedron Formation, and Cahokia Formation at the surface with the Wedron Formation holding up the steep slopes along the east and west banks of Lake Decatur. The groundwater flow model outlined the location of a local groundwater divide and showed flow from the injection site would flow towards Lake Decatur, reaching the lake in 80 days.

Keywords: Carbon Sequestration, Petrel, Hydrogeology, MODFLOW, Glacial Sediments

1. INTRODUCTION

Many feasibility and sustainability studies have been conducted to test the concept of carbon dioxide (CO_2) sequestration. The Weyburn Project in Canada [1] [2] , the Sleipner Field in the North Sea [3] [4] , the deep saline storage partnership among the Midwest Geological Sequestration Consortium (MGSC), Archer Daniels Midland Company (ADM) and Schlumberger Carbon Services in Decatur, Illinois [5] are examples of large-scale sequestration projects. The Weyburn Project transports CO_2 from the Dakota Gasification Company in North Dakota along a 330 kilometers (km) [206 miles (mi)] pipeline to an oil and gas field in southeast Saskatchewan, Canada [6]. After transfer within pipeline, the CO_2 is injected in an oil field in Canada for enhanced oil recovery purposes and to be sequestered in the subsurface. The Weyburn Project started injecting 5000 tons of CO_2 per day in 2000 resulting in a total of 7-million tons of CO_2 successfully injected and sequestered by 2006 [2].

Another large-scale project is active in the Sleipner Field in the North Sea [3]. Statoil captures CO_2used in the production of natural gas and injects it 1000 m deep beneath the seabed into a saline reservoir. The project began sequestering CO_2 in 1996 and proved to be viable by 2000, at which point 1 million tons per year were being injected into a reservoir that can hold up to 600 billion tons [7].

The MGSC project is designed to inject 1 million metric tons of anthropogenic CO_2 into a deep saline reservoir housed in the Mt. Simon Sandstone. The project is on the Archer Daniels Midland Company (ADM) property in Decatur, Illinois (Figure 1). ADM is an agricultural processing company that produces food-grade ingredients, animal feed, and renewable fuels. The ADM plant is an essential component to this feasibility study because they will be providing not only the land to install the injection well but also the pressurized CO_2 gas that will be injected into the subsurface [8] [9]. With a 500 m thickness, the Mt. Simon Sandstone has storage capacity and a permeability that allows for injection of fluids [5] [10]. Injection began in November 2011, with a total 999,215 tons injected in November 2014.Within the Illinois Basin, Cambrian strata are used to store commercial natural gas, and vertical migration of the natural gas to the surface has been observed. Given the similar geology and physical properties of the fluids, CO_2 leakage is a possibility. Lake Decatur is about 1.60 km east of the injection site and serves as a major water source for the city of Decatur. Therefore, some uncertainty remains as to the dynamics of the injected CO_2 that could pose a risk to water quality in Lake Decatur.

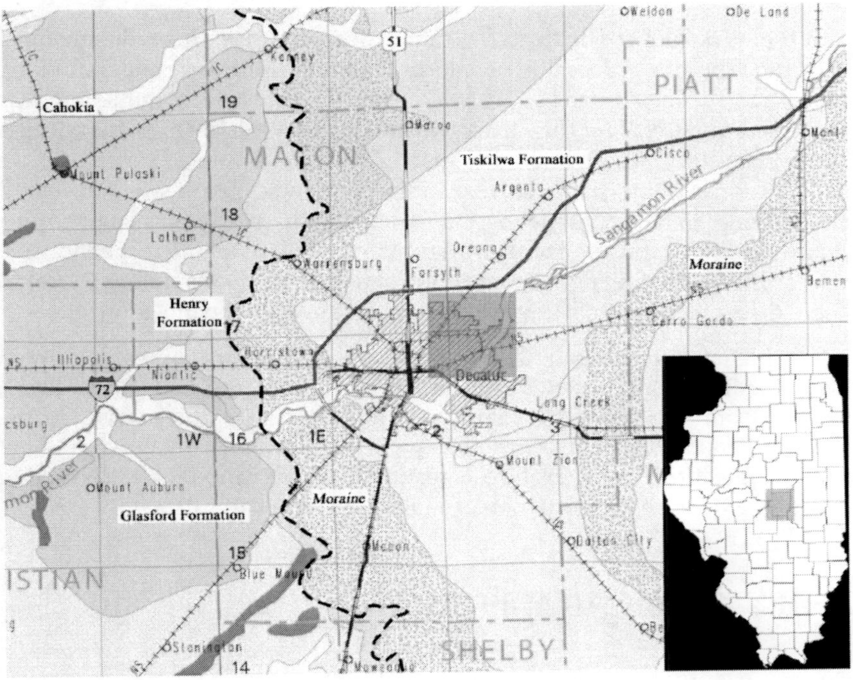

Figure 1. Surficial Geologic Map of the Macon County area, including Decatur (Modified from [11]). Green is Wisconsin till of the Tiskilwa Formation (moraines are indicated by stippling). The pink is various members of the Glasford Formation. Yellow is the Cahokia Formation. Light orange is the Henry Formation. Dark orange is the Illinois Pearl Formation. The dashed line is the Wisconsin terminal moraine.

Potential risks associated with geologic sequestration include leakage, migration, transport of metals or other contaminants in the subsurface, and chemical reactions that may occur in the subsurface [12] - [14]. In the case that the gas migrates vertically, there is the potential for interaction with surface or near surface waters and release into the atmosphere in a hazardous state and concentration.

To better develop this area as a CO_2 sequestration site, information about the near-surface geology and hydrogeology is required. This geologic and hydrogeologic information is crucial to understanding possible migration pathways in the shallow subsurface. With this work, we addressed two main questions: 1) what is the structure and stratigraphy of the shallow geologic units in Decatur area surrounding the ADM site? and 2) what are the characteristics of groundwater flow locally in the Decatur area?

With the advent of sophisticated visualization software, three-dimensional (3D) modeling is becoming a more common tool for understanding subsurface geology (e.g. [15] - [19]. The 3D subsurface modeling was completed to determine the structure and distribution of Quaternary units that overlie the sequestration site and to test the ability of Petrel [20] to model shallow Quaternary (i.e. discontinuous) units. Petrel was developed for bedrock geology modeling, fault modeling, reservoir engineering, geophysical interpretation and stratigraphic analysis making it the industry standard for modeling petroleum reservoirs. This project employs Petrel to model discontinuous Quaternary deposits that overlie the injection site.

In the unlikely event that CO_2 migrates upward, the groundwater flow in the immediate area needs to be understood. Both the Sangamon River, flowing from northeast to southwest, and Lake Decatur serve as the municipal and industrial water sources and lie hydrologically downgradient from the injection site. Therefore, a groundwater flow model was created to investigate the rate and direction of shallow groundwater flow.

2. METHODS AND MATERIALS

2.1. Geologic Conditions

The Cambrian to Pennsylvanian bedrock strata [21] are overlain by up to 91 m of Quaternary-age glacial deposits. From exploratory drilling, glacial deposits from three separate glaciations were observed (Figure 2).

The Cahokia Formation is recent gravel, sand, silt, and clay that is adjacent to the Sangamon River. It is variable in color and is typically less than 5 m thick in the study area.

The Peoria Silt is a wind-blown silt, clay, and fine sand deposit. Within the exploratory holes, the Peoria Silt was seen as a thin 0.5 - 1.0 m surficial layer. It is light olive brown to light yellowish brown in color, laminated, and leached.

The Tiskilwa Formation is a very poorly-sorted deposit composed of very fine silts up to cobbles and boulders [22]. Classified as a diamicton, the Tiskilwa is grayish brown to dark gray in color with a thin, pale yellow oxidized layer at the top. It is calcareous throughout, very stiff, and massive.

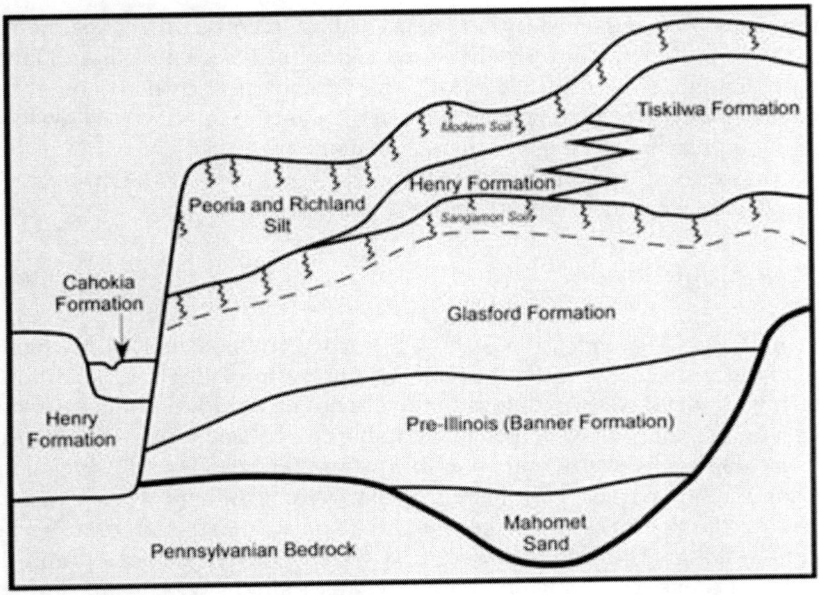

Figure 2. Schematic cross section of the Quaternary units in the study area.

Beneath the Tiskilwa Formation are the Sangamon Geosol, Roby Silt and Glasford Formation, which collectively make up the Illinois glacial deposits seen in the study area [23]. The Sangamon Geosol, an organic-rich, green to gray diamicton, formed on top of Illinois sediments. As an ancient soil horizon, the geosol is oxidized and leached.

The Glasford Formation is a till unit of the Illinois succession. Like the Tiskilwa, it is very poorly sorted, containing very fine silt to cobbles and boulders [24]. Reddish gray in color, the Glasford Formation is calcareous with some beds of poorly sorted sand and gravel.

Pre-Illinois units occur below the Glasford. The Banner Formation is the till member of the Pre-Illinois succession. Present only in the subsurface, the Banner Formation is described in core data as a diamicton loam unit that was dark grayish brown in color, calcareous and very hard.

2.2. Data Sources

Several different types of well data were incorporated into the project (Figure 3). Water well data for Macon County were acquired from the Illinois State Geological Survey (ISGS) ILWATER database [25]. Some water well data are only available in paper form, having never been digitized.

These data were retrieved from files at the ISGS records office. Coal boring data were obtained from a website hosting all of the coal wells in Macon County maintained by the ISGS [26]. The Geological Records Union at the ISGS provided all original paper copies of oil and gas wells and coal wells in Macon County that were not in the online databases. Lastly, the ISGS drilled eight (8) exploratory holes in the study area as part of the MGSC project [5]; the core data were incorporated into the project.

2.3. Geologic Model

For the Petrel 3D model, the well data needed to be sorted by township, range, and section to identify the wells within the modeling area. Additional filtering removed wells with logs that were not descriptive, did not provide accurate locations, or were not translatable to geologic units. The data for the useable wells were then put into a spreadsheet. As the Sangamon Geosol is the marker bed between the Wisconsin and Illinois glacial episodes, the presence of the geosol served as a lithostratigraphic "picks" on the well logs. The well logs were first sorted into five sections (Table 1). Anything above the Sangamon Geosol was called "Wisconsin" for the Wisconsin Episode glacial deposits, anything representing the Sangamon Geosol was called "Sangamon", anything below the Sangamon Geosol and above any second soil horizon was called "Illinois" for the Illinois Episode deposits, any well that hit bedrock and had any indication of a second ancient soil horizon was called "Pre-Illinois" for the Pre-Illinois Episode deposits. From wells that hit bedrock, the shale or limestone, a classification of "Bedrock" was used. Upon review, the "Wisconsin" was subdivided into two additional layers: The "Cahokia" and the "Peoria". Additionally, the "Sangamon" was combined with the "Illinois", and the "Pre-Illinois" and "Bedrock" were combined.

The next step was identifying the surface elevations for all of the wells. Using ESRI's ArcGIS, a Digital Elevation Map (DEM) acquired from the ISGS Geospatial Data Clearinghouse, four quadrangle maps, and a layer created with the well locations were incorporated into a GIS database. The well layer was created by making a spreadsheet of all of the wells that were be used for the model. API and X, Y data (in UTM format) were included for each well. A surface elevation for each well was determined and incorporated into the well database.

The well data were sorted into separate spreadsheets such that each of the five lithostratigraphic sections had their own spreadsheet. These sections only included the top of each lithostratigraphic unit from each well to create well top data. These well top spreadsheets included the API, X, Y locations, elevation, and lithology.

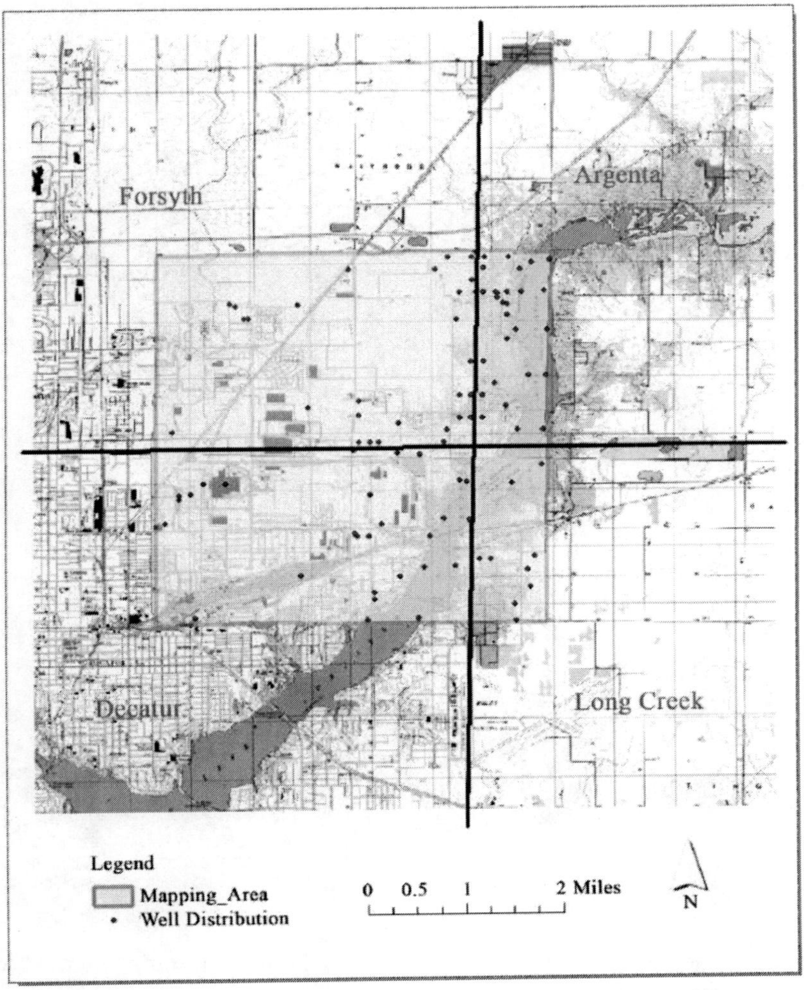

Figure 3. Petrel 3D model domain area indicated in blue. Water well and exploratory boring areas are indicated by red dots. The black lines delineate the four 7.5 minute quadrangles in the area.

Using Petrel, a new project was created with five sets of "well tops" folders. Each layer (Cahokia, Peoria, Wisconsin, Illinois, and Bedrock) was imported into their corresponding well top folder, which in Petrel are represented as a series of dots that make up the well layers (Figure 4). Each unit was created as a separate layer to ease subsequent editing/adjustment. Next, surfaces were created out of the well top data to generate the model (Figure 5) for the Cahokia, Peoria, Wisconsin, Illinois, and Bedrock (Table 1).

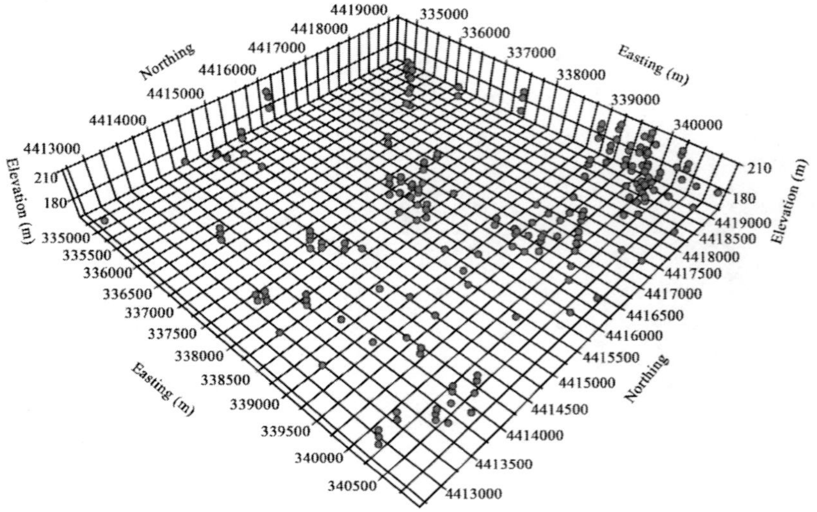

Figure 4. Petrel well top data for initial 3D model.

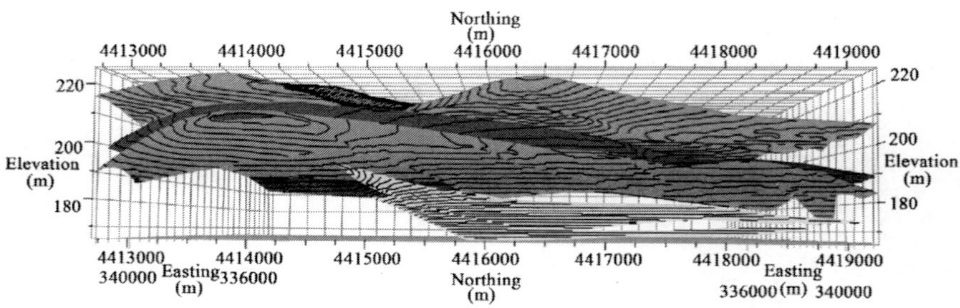

Figure 5. Geologic horizons interpreted by Petrel for initial model from well top data.

2.4. Groundwater Flow Model

To establish and confirm hydrogeologic boundary conditions for the local (shallow) groundwater flow model, GFLOW [27], an analytical element method was used to simulate regional flow through the surficial deposits comprised of diamicton tills and sand and gravel layers that total 15 m in thickness. On the regional scale the hydraulic conductivity (K) was assumed to be homogeneous and isotropic. The Sangamon River, to the east, Spring Creek, to the west, and Lake Decatur, along the east and southeast edge of the domain were near-field elements represented as constant head (Dirichlet) conditions. Model parameters were measured on site, calibrated by the model, or estimated using data acquired from the ISGS exploratory holes (Table 2). An initial simulation was conducted with GFLOW to establish regional flow and to aid in the development of local flow conditions (Figure 6).

Table 1. Key to geology and axes for all Petrel screenshots.

Key	
X-Axis	UTM Zone 16 horizontal grid value.
Y-Axis	UTM Zone 16 vertical grid value.
Z-Axis	Elevation above MSL in feet.
Vertical Exaggeration	5
Cahokia Alluvium	Yellow
Peoria Silt	Purple
Wisconsin Till	Green
Illinois Till	Orange
Bedrock	Blue

The local groundwater flow model was developed using MODFLOW [28]. Using a uniform 0.15 km x 0.15 km cell (0.23 km^2), a 40 row × 40 column grid represented the domain (Figure 7). Aquifer parameters and conditions used in the GFLOW model, with the exception of K, were used for the MODFLOW development. Rather than the 12 m/day, the local model calibrated to a K value of 0.12 m/day. The boundary conditions for the model domain were represented as constant heads, with values established from the GFLOW simulation (Figure 6). Spring Creek, Sangamon River, and Lake Decatur were also incorporated into the domain as constant head cells. Head values for Lake Decatur were obtained from a gauging station about 2 km downstream of the study area. (Haring, Pers. Com.). Recharge for the simulation was set at 10% (0.101 m/yr) of the average annual precipitation rate for Decatur, Illinois of 1.01 m/yr [29]. The use of 10% is consistent with other groundwater flow simulations in central Illinois.

Particle tracking, using MODPATH [30] , was used to further delineate the location of a groundwater divide identified with the GFLOW simulation (Figure 6). The horizontal migration was simulated by centering particles on the injection site and forward modeling. The time for the first particle to reach Lake Decatur signified the travel time.

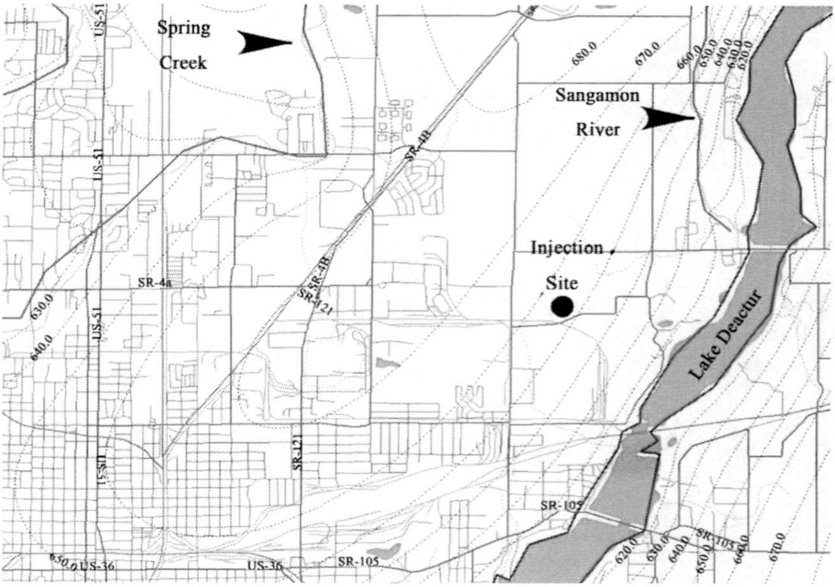

Figure 6. GFLOW model of the area with flow lines in gray. Also shows Lake Decatur and the sequestration injection site.

3. RESULTS

3.1. Geologic Model

The Petrel model simulates Quaternary deposits overlying Pennsylvanian bedrock in the Decatur area, indicating glacial sediment thicknesses up to 50 m (Figure 8 & Figure 9). The Peoria Silt thins to the northwest and to the east where the lake is present. The resistive Wisconsin till is responsible for creating higher slopes along the east and west edges of the Lake (Figure 8). The Peoria Silt terminates at the top of these slopes along the western boundary of Lake Decatur and picks up along the flat lying areas west of the lake. Waters of the Sangamon River completely eroded the Wisconsin till unit beneath the lake and exposed the lower Illinois till. Fluvial processes associated with the Sangamon River have deposited a thin layer of Cahokia Alluvium within the floodplain and along the bottom of Lake Decatur where the Peoria silt and Wisconsin till units had previously been deposited.

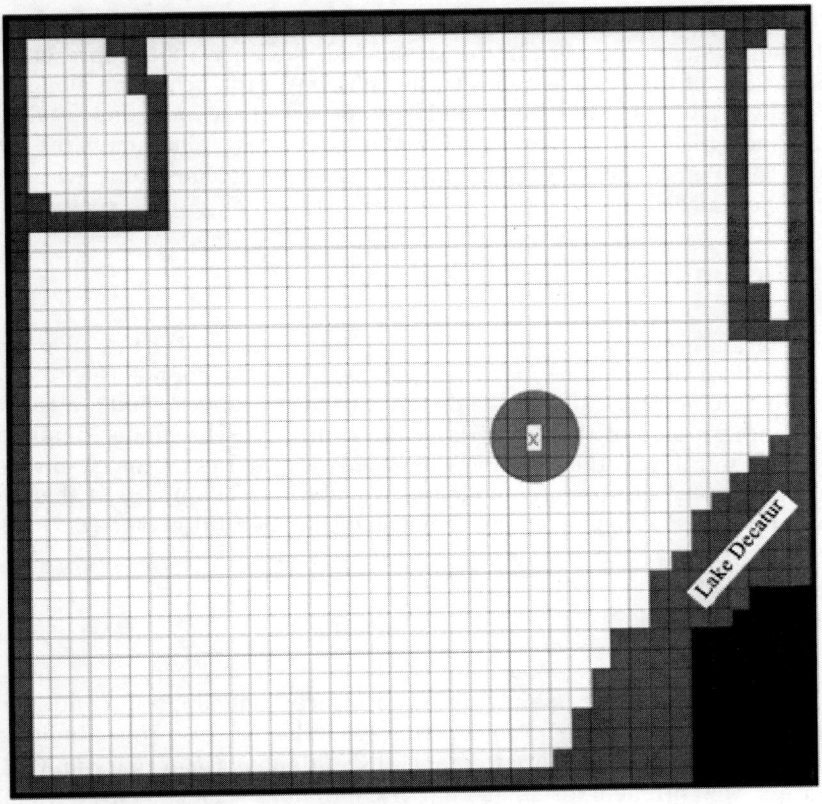

Figure 7. The 40 × 40 MODLFOW grid over study area with boundary conditions included. Blue cells represent constant head conditions. Black cells are inactive cells. Gray circle represents the location of the sequestration site.

Table 2. Parameters used for GFLOW model.

Aquifer parameters	Value	Source
Base elevation	134 m	Field data
Thickness	15 m	Field data
Porosity	0.2	Estimate
Hydraulic conductivity	12 m/day	Model calibration
Recharge	0.24 m/year	Model calibration

The surface layer of the geologic model (Figure 6) includes the Cahokia alluvium, which is present in the streambed, the Peoria Silt, representing about 85% of the surface deposits, the Wisconsin till, occurring along the lake edges and in the northwest and east corners of the modeling area, and the Illinois till, which is present below the thin Cahokia alluvium in the lake bed. The Petrel model of the surficial geology is consistent with the published surficial geology map of the area [31].

3.2. Groundwater Flow Model

The local groundwater flow model identified a NE-SW trending groundwater divide in the northwestern portion of the domain (Figure 10). Flow on the northern side of the divide is to the northwest, and flow along the southern side of the divide is to the southeast. The model simulated groundwater flow at the injection site would move to the southeast towards Lake Decatur. The forward modeling with MODPATH indicated that the particles placed at the injection site reached Lake Decatur in 80 days.

4. DISCUSSION

While Petrel is used mainly for subsurface modeling of continuous stratigraphic units in the petroleum and gas industry, this project indicates that Petrel has the capability to model discontinuous Quaternary units. To address the problem of modeling discontinuous units, a stratigraphic column consistent across all wells is needed; thus, an absent unit in a well is assigned a thickness of zero (0) meters. Where this occurs, the model result is a pinch out of the unit, excluding it accordingly. The modeling process in Petrel also required simplification (i.e. upscaling) of the Quaternary units because they are heterogeneous, with discontinuous lenses of sediments of various size, i.e. sand and gravel lenses. While Petrel is capable of distinguishing facies within units given high resolution log data [17] [19], the discontinuous nature of the lenses may be too detailed to model accurately. We did not address this issue. Rather, we modeled the materials on a unit/formation level (low resolution), homogenizing the units, instead of at the facies level. This simplified the model and produced fewer layers. The standard input procedures were also altered in order to model the Quaternary units. An additional method deviation was importing surfaces of well top data separately rather than collectively. This method allowed alterations of surfaces to be more efficient; if a surface needed to be altered, which it did frequently, only one layer was modified rather than the entire model.

A concern for broader project is that there are thick sand and gravel bodies within the glacial deposits or fractures in the bedrock surface that may act as significant migration pathways for gas if there is a shallow leak or if gas migrates vertically from deeper in the bedrock units. It is assumed that the overlying till can act as a cap layer given the combined thickness of the glacial units and the overall composition as being fine-grained, poorly sorted and believed to possess a low permeability and porosity. However, the unweathered till can have significant vertical and horizontal fracture permeability that will allow migration [32]. Additionally, all glacial till units have extensive lenses of sand and gravel that can act as lateral and/or vertical migration pathways.

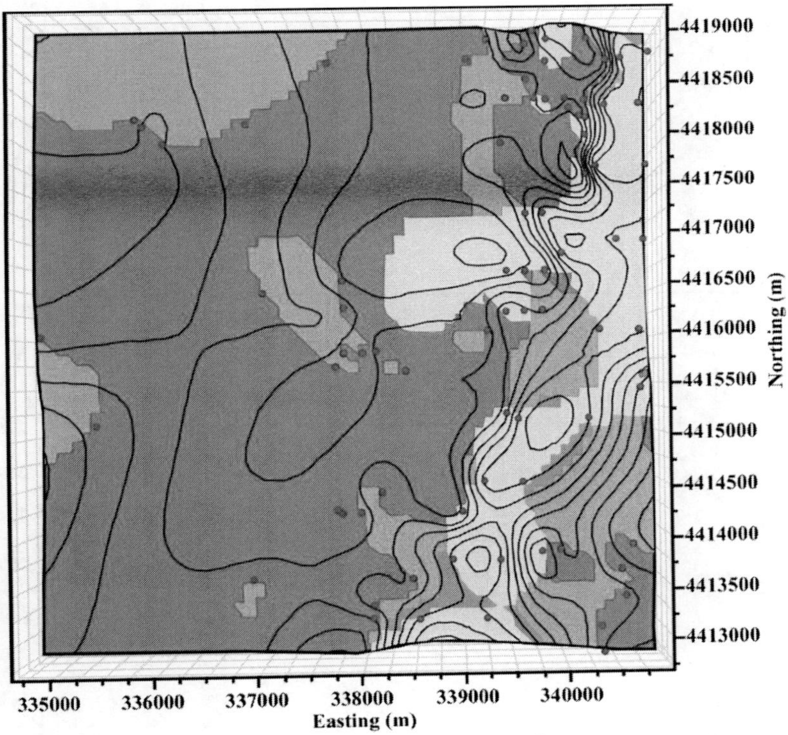

Figure 8. Plan view of the final Petrel model illustrating the surficial geology. The black circle is the estimated extent of the gas plume once it is injected and red dots show well distribution. Contour interval is 5 meters.

Fluid flow in the shallow subsurface will be transported towards Lake Decatur. Any vertical leakage from the deep saline reservoir would take years, if not 100s of years, based upon vertical fluid migration from an oil and gas field [33]. Additionally, the successful injection of nearly 1,000,000 tons of CO_2 [6] reveals the feasibility of the reservoir. However, once in the glacial sediment, migration to the lake would be on the order of 10s to 100s of days. Since glacial units vary in composition such as the presence of discontinuous sand and gravel lenses, flow rates through the units will vary as compared to the model. However, the general direction of groundwater flow will remain the same, towards Lake Decatur.

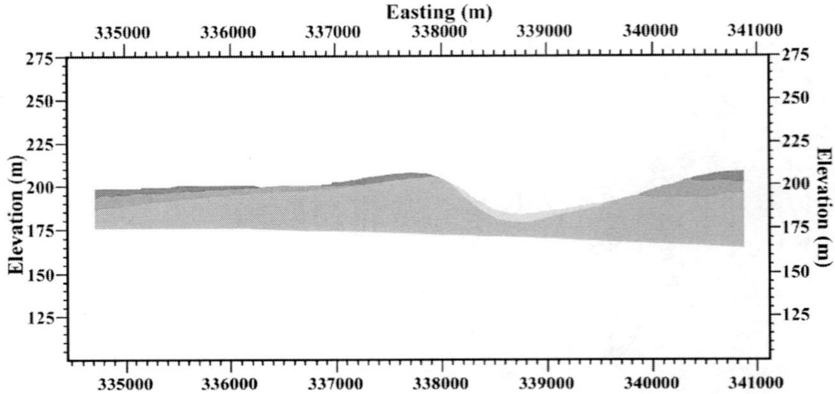

(a)

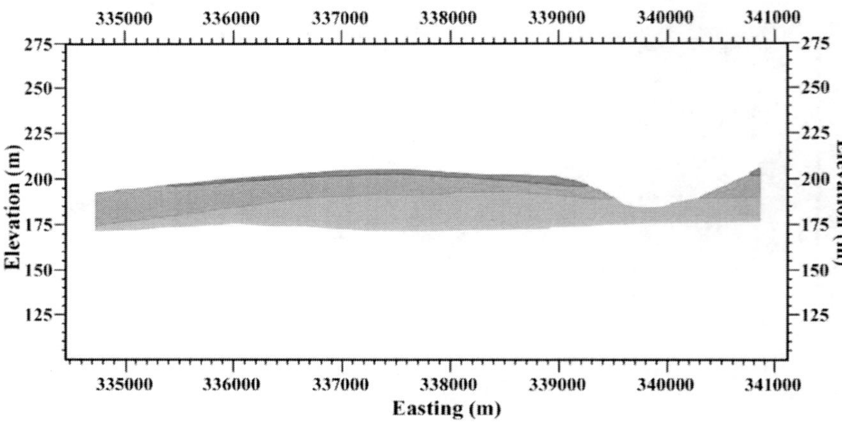

(b)

Figure 9. Petrel generated (a) east-west cross section along the southern margin of the geologic model (4412800 line); (b) east-west cross section along the 4415200 line of the geologic model.

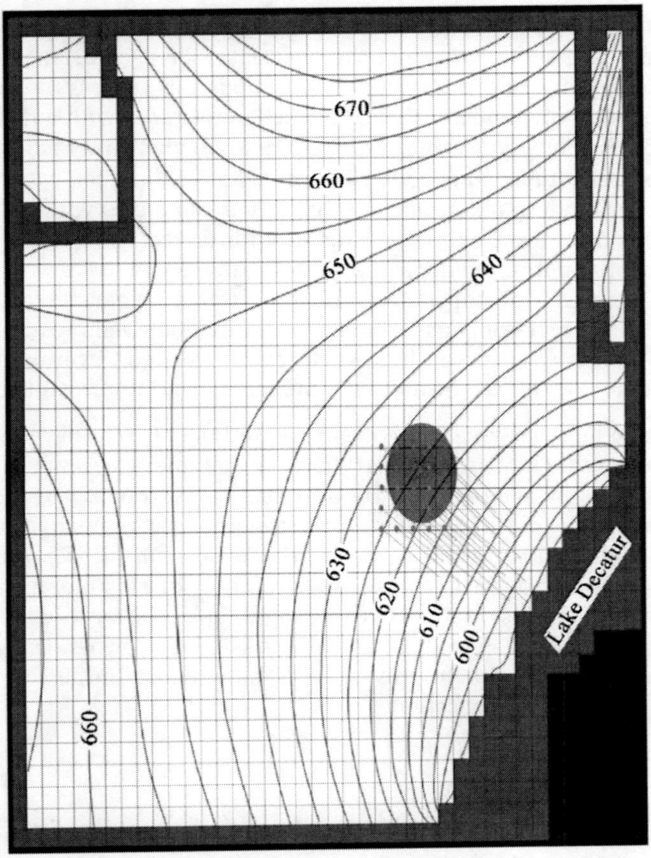

Figure 10. Average head values are in red (contoured in feet). Projected gas plume at depth is the gray circle. The red X marks the position of the injection well. Results of the average flow MODPATH simulation with particles and flow paths in purple. This represents the position of the particles after 80 days.

5. CONCLUSIONS

Petrel is used traditionally for subsurface modeling of deep, continuous bedrock systems by the petroleum and gas industry. This project successfully tested the program's ability to model discontinuous Quaternary units. To combat this problem of modeling discontinuous units, a consistent stratigraphic column across all wells was used to so where a unit was absent in a well, it was included with a thickness of 0'. Using this method, a successful geologic model was created with stratigraphic units pinching out as needed honoring the well data. The Quaternary units also had to be simplified in order for the Petrel modeling process to work. Since there are mostly till deposits above bedrock in this area with a thin layer of loess on the surface, modeling was from a somewhat low resolution. Any sand deposit encountered in a well log less than 0.67 m thick was combined with the surrounding material. This simplified the facies within the major till units resulting in a simplified model with fewer layers. The standard Petrel model input procedures were also altered in order to map the Quaternary units. Well top data were imported as separate stratigraphic surfaces rather than being associated with individual wells resulting in the seamless ability to alter or edit surfaces as needed. Overall it can be concluded that Petrel has the ability to model discontinuous geologic units.

The resulting model is an accurate representation of Quaternary deposits in the Decatur area. They are up to 48 m thick in some places and thinner or eroded completely where Lake Decatur has cut through. The Peoria Silt covers most of the surface except on the topographic high slopes along the edges of Lake Decatur. In these places, the Wisconsin till is present along with the flat, higher areas in the northwest corner of the mapping area. The Cahokia alluvium fills in the streambed of Lake Decatur where it has completely eroded the Peoria Silt and Wisconsin till deposits. The geometry of the Quaternary units in this area is somewhat simple, though the composition is mostly consistent.

REFERENCES

1. White, D., et al. (2004) Greenhouse Gas Sequestration in Abandoned Oil Reservoirs: The International Energy Agency Weyburn Pilot Project. GSA Today, 14, 4-10. http://dx.doi.org/10.1130/1052-5173(2004)014<004:GGSIAO>2.0.CO;2
2. Petroleum Technology Research Centre (2015) Weyburn-Midale: The IEAGHG Weyburn-Midale CO2 Monitoring and Storage Project. http://ptrc.ca/projects/weyburn-midale
3. Statoil (2015) Technology & Innovation. http://www.statoil.com/ en/ technologyinnovation/pages/default.aspx
4. Arts, R., et al. (2004) Monitoring of CO2 Injected at Sleipner Using Time-Lapse Seismic Data. Energy, 29, 1383-1392. http://dx.doi.org/10.1016/j.energy.2004.03.072
5. Midwest Geological Sequestration Consortium (2016) Deep Saline Storage. http://www.sequestration.org/mgscprojects/deepsalinestorage.html
6. Preston, C., et al. (2005) IEA GHG Weyburn CO2 Monitoring and Storage Project. Fuel Processing Technology, 86, 1547-1568. http://dx.doi.org/10.1016/j.fuproc.2005.01.019
7. Bartlett, K., Pinsker, L.M. and Reed, C. (2003) Demonstrating Carbon Sequestration. Geotimes, 48, 22.
8. Ball, D. Gupta, N. and Metzger, B. (2007) Energy Industry Examining CO2 Sequestration Options. Oil & Gas Journal, 105, 20.
9. Wells, A.W., et al. (2007) The Use of Tracers to Assess Leakage from the Sequestration of CO2 in a Depleted Oil Reservoir, New Mexico, USA. Applied Geochemistry, 22, 996-1016. http://dx.doi.org/10.1016/j.apgeochem.2007.01.002

10. Peterson, E.W., Martin, L.I. and Malone, D.H. (2015) Identification of Potential Vertical Gas Migration Pathways above Gas Storage Reservoirs. World Journal of Environmental Engineering, 3, 23-31. http://dx.doi.org/10.12691/wjee-3-2-1
11. Stiff, B.J. (2000) Surficial Deposits of Illinois. Illinois State Geological Survey.
12. Lovelace, B. (2008) Can We Protect Our Ground Water If We Deploy Large-Scale Carbon Sequestration Technologies? Ground Water, 46, 805-806. http://dx.doi.org/10.1111/j.1745-6584.2008.00496_3.x
13. Pruess, K. (2008) On CO2 Fluid Flow and Heat Transfer Behavior in the Subsurface, Following Leakage from a Geologic Storage Reservoir. Environmental Geology, 54, 1677-1686. http://dx.doi.org/10.1007/s00254-007-0945-x
14. Vasco, D.W., et al. (2010) Satellite-Based Measurements of Surface Deformation Reveal Fluid Flow Associated with the Geological Storage of Carbon Dioxide. Geophysical Research Letters, 37, 1-5.
15. Blendinger, W., Brack, P., Norborg, A.K. and Wulff-Pedersen, E. (2004) Three-Dimensional Modelling of an Isolated Carbonate Buildup (Triassic, Dolomites, Italy). Sedimentology, 51, 297-314. http://dx.doi.org/10.1046/j.1365-3091.2003.00622.x
16. Soller, D.R., Price, S.D., Kempton, J.P. and Berg, R.C. (1999) Three-Dimensional Geologic Maps of Quaternary Sediments in East-Central Illinois. United States Geological Survey 0607937041.
17. Wagle, J., Malone, D.H., Peterson, E. and Tranel, W.L. (2016) Porosity Controls on Secondary Recovery at the Loudon Field, South-Central Illinois. Interpretation, 4, T1-T13. http://dx.doi.org/10.1190/int-2015-0078.1
18. Carlock, D.C., Thomason, J.F., Malone, D.H. and Peterson, E.W. (2016) Stratigraphy and Extent of the Pearl-Ashmore Aquifer, McHenry County, IL, USA. World Journal of Environmental Engineering, 4, 6-18.
19. Kimple, D., Peterson, E.W. and Malone, D.H. (2015) Stratigraphy and Porosity Modeling of Southern Central Illinois Chester (Upper Mississippian) Series Sandstones Using Petrel. World Journal of Environmental Engineering, 3, 82-86.
20. Schlumberger (2010) Petrel Introduction G & G Manual. Schlumberger, Houston.
21. Bergstrom, R.E., Piskin, K. and Follmer, L.R. (1976) Geology for Planning in the Springfield-Decatur Region, Illinois. Illinois State Geological Survey, Urbana.
22. Hansel, A.K. and Johnson, W.H. (1996) Wedron and Mason Groups: Lithostratigraphic Reclassification of Deposits of the Wisconsin Episode, Lake Michigan Lobe Area. Illinois State Geological Survey, Champaign.
23. Kempton, J.P., Morse, W.J. and Visocky, A.P. (1982) Hydrogeologic Evaluation of Sand and Gravel Aquifers for Municipal Groundwater Supplies in East-Central Illinois. Illinois State Water Survey and the Illinois State Geological Survey.
24. Follmer, L.R., McKay, E.D., Lineback, J.A. and Gross, D.L. (1979) Wisconsinan, Sangamonian, and Illinoian Stratigraphy in Central IllInois. Illinois State Geological Survey.
25. Illinois State Geological Survey (2015) Illinois Water Well (ILWATER) Interactive Map. http://isgs.illinois.edu/ilwater
26. Illinois State Geological Survey (2015) Coal Mines in Illinois Viewer (ILMINES). http://isgs.illinois.edu/ilmines
27. Haitjema, H.M. (1995) Analytic Element Modeling of Groundwater Flow. Academic Press, Cambridge.
28. McDonald, M.G. and Harbaugh, A.W. (1988) A Modular Three-Dimensional Finite-Difference Ground-Water Flow Model. Techniques of Water-Resources Investigations of the United States Geological Survey, United States Geological Survey, Washington DC.
29. National Oceanic and Atmospheric Administration (2002) 11: Illinois. National Oceanic and Atmospheric Administration, Asheville.
30. Pollock, D.W. (1994) User's Guide for MODPATH/MODPATH-PLOT. Version 3: A Particle Tracking Post-Processing Package for MODFLOW, the US Geological Survey Finite-Difference Ground-Water Flow Model.
31. Roche, E. and Malone, D.H. (2008) Surficial Geology of the Forsyth Quadrangle, Macon County, Illinois, in 2008 EDMAP 2008. Illinois State Geological Survey, Champaign.

32. Keller, C.K., Kamp, G.V.D. and Cherry, J.A. (1986) Fracture Permeability and Groundwater Flow in Clayey Till Near Saskatoon, Saskatchewan. Canadian Geotechnical Journal, 23, 229-240. http://dx.doi.org/10.1139/t86-032
33. Weaver, T.R., Frape, S.K. and Cherry, J.A. (1995) Recent Cross-Formational Fluid Flow and Mixing in the Shallow Michigan Basin. Geological Society of America Bulletin, 107, 697-707. http://dx.doi.org/10.1130/0016-7606(1995)107<0697:RCFFFA>2.3.CO;2

CHAPTER 12

Assessment of the Sensitivity of Streamflow Simulations to Changes in Patch Resolution Using GIS Based Hydro-Ecologic Model

Samson G. Mengistu[1], Melkamu A. Ali[2*], Fuad A. Yassin[2]

[1]*National Hydrology Research Center, Environment Canada, Saskatoon, Canada*

[2]*Global Institute for Water Security, University of Saskatchewan, Saskatoon, Canada*

ABSTRACT

Eight different patch configurations were investigated to analyze the effect of patch characterization/formation in streamflow simulation, using the Regional Hydro-Ecologic Simulation Systems (RHESSys) model. It is investigated for eight different patch configurations of a subcatchment of the Turkey Lakes Watershed, Ontario. The model's hydrological parameters are calibrated for each of these patch configurations and the performance of the simulations is evaluated. Results indicate that both the nature of the flow simulation and the calibrated parameter values are sensitive to patch configuration. The best simulation results were obtained for the patch configuration with the highest spatial variation of climate, stream network and hillslope conditions across the subcatchment. Different patch configurations also lead to markedly different calibrations of the model's hydrological parameters ($54.26 < k < 119.13$; and $1.02 < m < 2.28$), which has implications for the physical interpretation and transferability of the calibrated parameter values.

Keywords: RHESSys, Patch, Model Calibration, Flow Simulation

1. INTRODUCTION

Fundamental physical processes in hydrology such as rainfall, evaporation, transpiration, infiltration, surface and ground water flow and connectivity can be precisely described by physical laws at small spatial and temporal scales [1]. However, one of the unresolved key challenges in hydrology is how the laws can be applied at scales by far larger and temporally and spatially more heterogeneous than the small scale they have been proved valid [1] [2].

Over the last few decades, considerable studies have been made in using large scale numerical hydrologic models developed to understand environmental processes and predict environmental changes at various spatial and temporal scales [3] - [5] . Most of these studies have employed GIS-coupled physically based distributed hydrologic models. In these models, the spatial variability is taken into account by dividing the study area into several smaller elements or regular mesh pixels over which point scale laws or measurement data at point or local scales are upscaled to predict water dynamics at larger scales [1]. There is a general assumption that change in scales can be accommodated by using effective parameters that should be determined during calibration of these models. Examples are MIKE-SHE [6] [7] and TOPKAPI [8].

This grid parameterization approach, however, is considered as inadequate by some authors. [9] describes the use of effective parameter values as inadequate to the scaling problem. In addition, effective parameter based distributed models are heavily over parameterized and computationally not efficient, as they result in the application of a set of model parameters to each grid cell where the small scale physical laws are employed [10] . Hence, various researchers have proposed the concept of using homogeneous modeling entities, which are aggregated area of similar hydrologic behavior on the basis of hydrology, land use, soil, vegetation, slope and so on. Examples include the simple scaling and multiscaling framework [11] ; the Representing Element Area concept [12] ; hydrologically similar units or "patches" [13] ; the Hydrological Response Units concept of [14] ; and basin-scale model equations (e.g. [15]). The hydrologically similar units approach as compared to the grid approach is believed to be more efficient computationally as specific set of models parameters is applicable to each type of entities rather than individual grids.

Despite these various attempts, a number of recent reviews have identified that scaling issues are at the heart of most hydrologic problems which are not yet resolved [1] [5] [16] - [21] . [17] noted that many of the better known scaling procedures in hydrologic modeling did neglect the important components of the complex hydrologic processes that we observe in the field. For example, the amount of information contained in hydrologically similar units like patches is highly dependent on the various assumptions considered during their formation. The way that these modeling entities have been formed or characterized may have impact on simulation results of the hydrologic regime.

This study, therefore, aims to investigate how the characterization of hydrologically similar units affects simulation output in hydrological models. The simulations are performed using the Regional Hydro-Ecologic Simulation Systems (RHESSys) [22], in which "patches" represent the finest scale of the spatial hierarchy, i.e. the basic modeling entities on which energy and water components are simulated. These patches are formed by the intersection of each coarser level of the hierarchy and other attributes of the watershed [22] It is important to understand the sensitivity of the model to patch characterization before applying the model for the purpose of investigating the effect of potential management practices in catchment's hydrologic regimes, both water quantity and quality, investigated at the outlet of a catchment. Particularly, this study investigates the impact of patch characterization on simulated flow regime, and on the calibration of RHESSys's main hydrological parameters: saturated hydraulic conductivity, k, and, decay of saturated hydraulic conductivity with depth, *m*.

2. DESCRIPTION OF THE HYDROLOGICAL MODEL

RHESSys is a hydro-ecological, GIS-based model that is designed to simulate fluxes of water (evaporation, transpiration, stream flow, and soil water), nutrients (net photosynthesis, plant growth), and carbon in the landscape. The model uses the Soil-Plant-Atmosphere Continuum concept [23] within a distributed watershed context to represent the spatial distribution of hydrologic and ecophysiological processes and characteristics of a catchment [22]. RHESSys is composed of three process submodels: a meteorological model (based on MT- CLIM; [24]), an eco-physiological model (based on BIOME-BGC; [25]), and two distributed hydrologic models. TOPMODEL [26] is a "quasi"-distributed hydrologic model that distributes soil water according to the statistical distribution of the topographically defined soil wetness index of the patches. An adapted version of DHVSM

[27] is a distributed hydrologic model, where saturated subsurface through flow and overland flow are explicitly routed between contiguous patches. The explicit routing, whilst computationally not as efficient as TOPMODEL, produces hydrologic patterns that more closely matched empirical data [28]. Therefore, for this study, the explicit hydrologic routing approach is used.

The RHESSys model is structured as a spatially nested hierarchical representation of the landscape (Figure 1), in which different hydrological, climatological, and ecosystem processes are partitioned. Different processes are modeled at each of these levels within the hierarchy. At the base of the hierarchy is the drainage network. Within the drainage network, levels in the hierarchy, range from climate zones (representing orographic effects on precipitation), hillslopes (representing horizontal water, carbon, and nitrogen transfers), and patches (representing vertical water, C, and N cycling). A detailed description of the parameters and processes at each level of the hierarchy is provided in [22]. The patches, which represent the finest scale of the spatial hierarchy, are formed by the intersection of each coarser level of the hierarchy and other attributes of the watershed [22]. Therefore the patches are homogeneous units in terms of climate, hillslope, soils, riparian areas, vegetation, and land use.

The RHESSys model requires a large number data inputs, including both spatial data, in the form of GIS- generated maps, and nonspatial data, in the form of tables (Figure 2). The maps, which are usually GIS generated, represent spatially variable properties, such as topography, vegetation, soil, and land use.

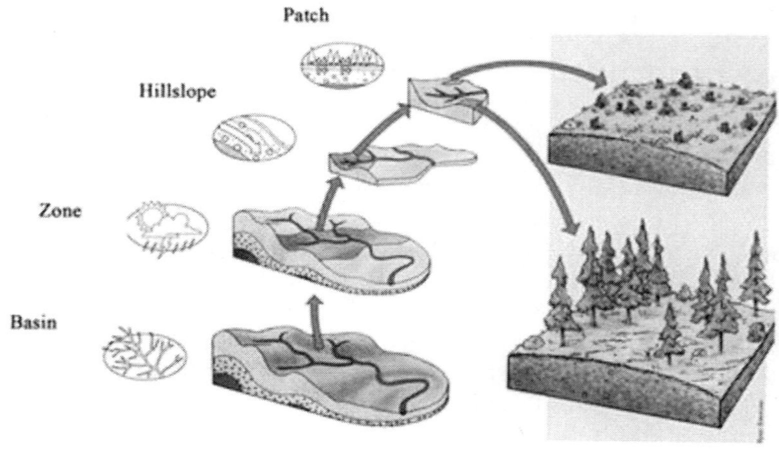

Figure 1. The spatial hierarchy within RHESSys [29]

The tables include temporally variable climate data (daily precipitation, maximum and minimum temperatures) and default values for physiological characteristics of the vegetation and physical characteristics of the soils.

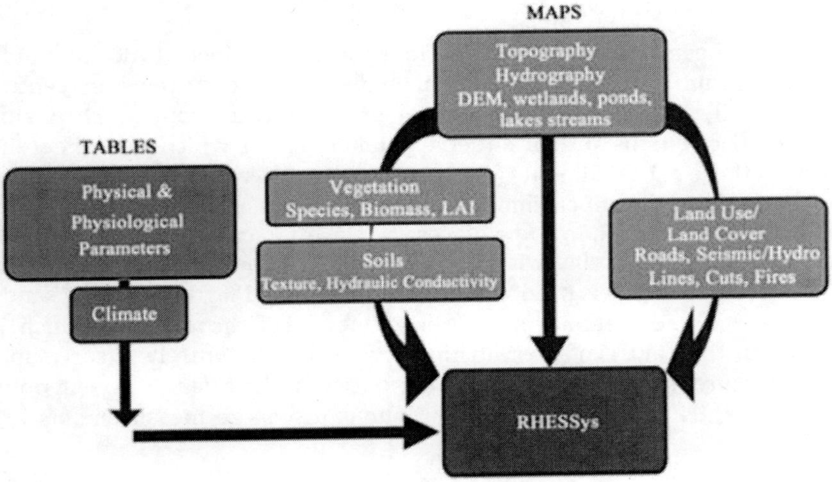

Figure 2. Required inputs for the RHESSys model. Inputs to the model include spatial data (maps) generated within a GIS and non-spatial data (tables) [29]

3. STUDY AREA DESCRIPTION

This research was conducted on a subcatchment of the Turkey Lake Watershed (TLW). The TLW has an area of 10.5 km^2 and located within sugar maple dominated forest of the Algoma Highlands of Central Ontario (47°80'30''N, 84°82'50''W; Figure 3), on the Precambrian shield approximately 60 km north of Sault Ste. Marie [30] . [31] noted that the climate within the TLW is continental and is strongly influenced by the proximity of Lake Superior, with mean annual precipitation and temperature of 1200 mm and 5.0°C respectively, for the period 1981 to 2006. A snow pack typically persists from late-November, early December through to late-March, early April. Peak stream discharge occurs during snowmelt and again in October to November during autumn storms. Almost 35% of the average precipitation is in the form of snowfall and spring snowmelt accounts for 30% - 60% of annual runoff [32]

Vegetation in TLW is relatively undisturbed and is predominantly mature to overmature tolerant hardwood forest (approximately 90% sugar maple, Acer saccharum Marsh.) that underwent a light selective harvesting ("high grading") in the 1950s [33] . Details on the physical and biological characteristics of the watershed are given by [34]

TLW has been used as an experimental basin since 1980 to study anthropogenic perturbations in Canadian Shield ecosystems in central Ontario [30] It is divided into a number of subcatchments each having hydrometric stations at their outlet. Catchment (C38), which is used herein, is one of these subcatchments (Figure 3), with an area of 64,625 m^2. Most of the area within the subcatchment has an elevation of nearly 405 m above sea level (Figure 4(a)). In some places the elevation can go as high as 450m that makes the total relief within the watershed to be approximately 45 m. The soil (Figure 4(b)) is predominantly silty loam (71%), followed by sandy loam (15%), and wetland area (14%). Outside of the wetland, which is located in the centre of the catchment, land use is entirely sugar maple forest (Figure 4(c)). The subcatchment outlet is identified as a weir point (47°2'52"N, 84°24'29"W) where continuous discharge measurements are available.

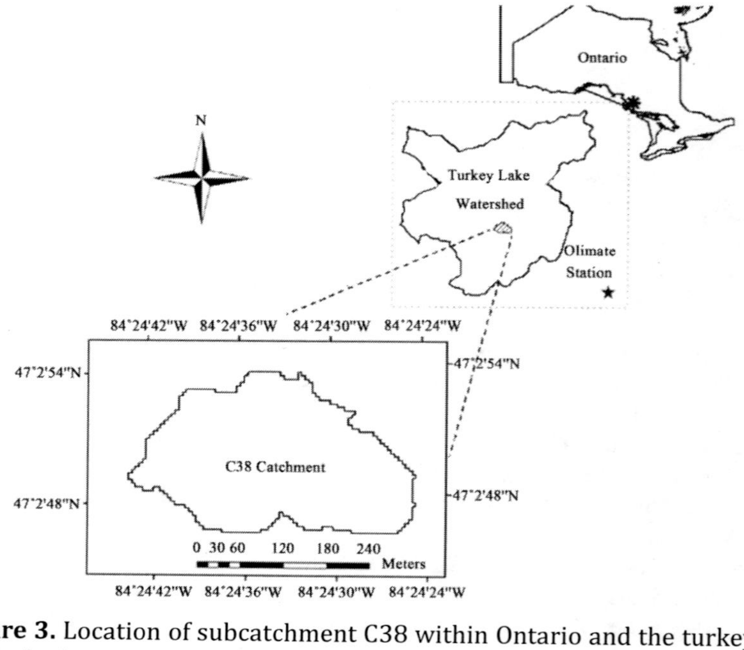

Figure 3. Location of subcatchment C38 within Ontario and the turkey lake watershed.

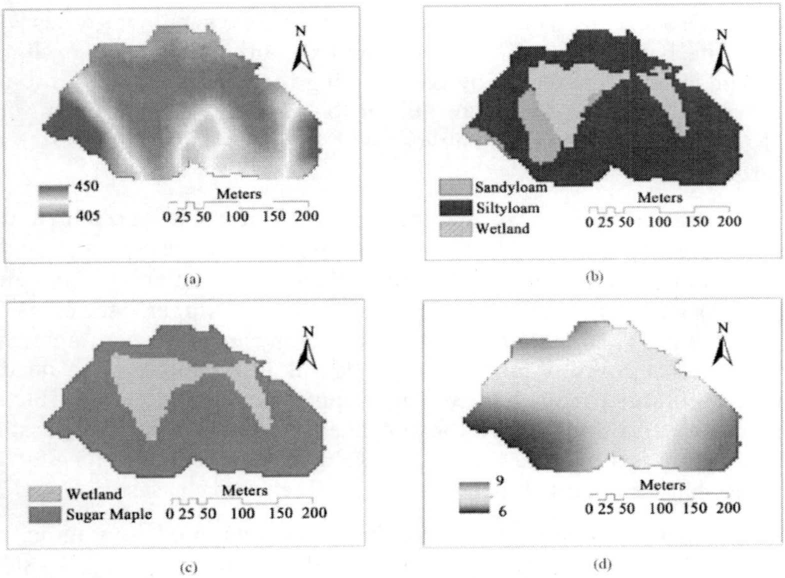

Figure 4. Spatial properties of subcatchment C38. (a) DEM; (b) Soil distribution; (c) Land use and (d) Leaf-area index.

4. MODEL SETUP

The stream flow at the outlet of C38 was modelled for eight patch configuration scenarios. The simulation results were compared to see the effect of patch resolution on the flow regime at the outlet of the catchment.

Spatial data were derived from terrain analysis of a 5m resolution Digital Elevation Model (DEM) obtained from the Ontario Base Map of the Ontario Ministry of Natural Resources (OMNR), and preprocessed using the Terrain Analysis System [34] Topographic attributes, including both primary (slope, aspect, specific contributing area) and secondary (wetness index) attributes, were derived from established terrain analysis techniques [26] [36] -[38]

Two sets of stream networks were derived from the DEM, using specific contributing area thresholds of 2000 m^2 and 5000 m^2. The stream networks were buffered to consider the heterogeneity in topography of riparian area near the streams during the creation of patches. Similarly, two sets of hillslopes were derived from the DEM, also with thresholds of 2000 m^2 and 5000 m^2. Finally, two sets of climate zones were created. Since there is no statistical correlation denoting orographic/elevation effect on precipitation, climate zones were determined arbitrarily. As such, two climate zones were created, one having one climate zone and the other having two climate zones. The latter was created by reclassifying the processed DEM into two zones: areas with elevation from 405 m to 427.4 m classified in the first zone and areas with elevation between 427.4 m and 450 m classified in the second zone.

A digital vegetation map was adapted from the Forest Resource Inventory (FRI) surveyed by OMNS in 1995. Vegetation classes were amalgamated to include conifer, hardwood, poplar, and black spruce. Canopy Leaf Area Index (LAI) was derived from LANDSAT Thematic Mapper data using the methods of [39]. A detailed description of the physiological parameters for the vegetation classes is provided by [29] . Soil attributes were derived from data obtained from Forest Landscape Productivity Survey (FLaPS) maps created and updated by the OMNR. In RHESSys soils are classified based on their textural classes. The model contains default values for the hydraulic characteristics of most common soil textural classes.

Patches, which represent a unique homogeneous modeling units, are created by overlaying spatial layers of the climate zone, hill slope, vegetation, soils and buffered stream networks [22] . Since there is only one soil map, and one vegetation map, eight different patch configurations (PC1 to PC8) can be obtained by overlying these with different combinations of the climatic zones, hill slopes and stream networks (Table 1; Figure 5). The number of patches in each configuration depends on the overlay and varies between 93 and 168 (Table 1).

Temporal climate data (daily precipitation, daily maximum temperature, and daily minimum temperature) for a 5 year calibration period (June 1986 to July 1991) were obtained from a station close to the southeastern corner of TLW (Figure 3) where daily meteorological data around the watershed have continuously been measured since 1980.

For each of the eight patch configurations, the model was calibrated for optimal flow regime simulation. All calibrations were based on continuous flow measurement at the outlet of subcatchment, covering the entire simulation period (June 1986 to July 1991). Before calibration, the catchment hydrologic and nutrient condition was stabilized by spinning up of the model since the year 1600. The calibrations optimize the soil parameters that define soil transmissivity: the saturated hydraulic conductivity, k, and the decay of saturated hydraulic conductivity with depth, m. Model performance is measured with the Nash-Sutcliffe efficiency coefficient, E, which indicates the agreement between the observed and simulated discharges ($E = 0$ represents poor agreement and $E = 1$ represents complete agreement; [40]). For each patch configuration, a Monte-Carlo calibration was applied, in which 75 different combinations of m and k were simulated.

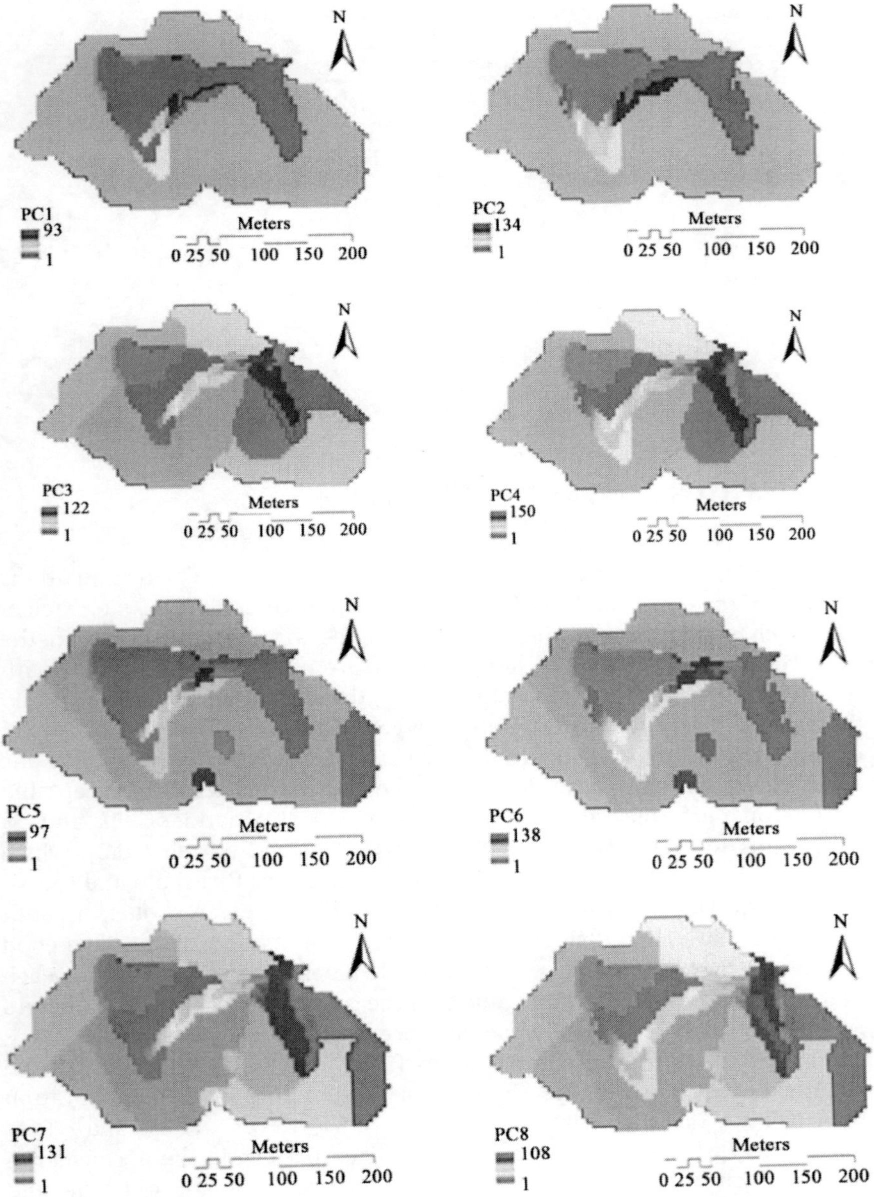

Figure 5. Maps of the eight patch characterization scenarios used in this study project.

Table 1. Range of values of input files for patch creation

Patch name	Hillslope and stream threshold values		Number of zones	Number of patches
	Hillslope	Stream		
Patch 1	5000	5000	1	93
Patch 2	5000	2000	1	134
Patch 3	2000	5000	1	122
Patch 4	2000	2000	1	159
Patch 5	5000	5000	2	97
Patch 6	5000	2000	2	138
Patch 7	2000	5000	2	131
Patch 8	2000	2000	2	168

5. RESULTS AND DISCUSSION

The Nash-Sutcliffe efficiencies, resulting from the Monte-Carlo simulations for each of the eight patch configurations, are shown in Figure 6. For each of the patch configurations there appears to be a systematic trend in the relation between E and k, although the nature of that relation varies with the different patch configurations. On the other hand, there is no discernable trend in the relation between E and m. Thus, the hydraulic conductivity, k, has a stronger influence over E, than the decay in hydraulic conductivity with depth, m. Most patch configurations result in acceptable simulation performances ($E > 0.5$), for most combinations of k and m. Exceptions to this are PC2 ($-1.4 < E < 0$), PC6 ($-1.0 < E < 0.2$), which indicates that the combination of low stream network threshold and coarse hillslope thresholds is not suitable. In PC8, with low thresholds for both stream network and hillslope, most k and m combinations resulted in good simulation performance ($E > 0.6$). Optimal values for k and m vary for each patch category (Table 2), indicating that the calibration of these parameters is sensitive to the way the patches are defined. In general, the patch configurations with two climate zones (i.e. PC5, PC6, PC7 and PC8) have slightly better performance than the corresponding patch configuration with only one climate zone (i.e. PC1, PC2, PC3 and PC4, respectively).However, there is no generalized pattern in the performance of the patch configurations with different thresholds for hillslope and stream network definition, other than that the two configurations with a high thresholds for hillslope definition and a low threshold for stream network definition (PC2 and PC6) both perform poorly ($E = -0.02$ and $E = 0.20$, respectively). The total simulated flow and the relative difference to the total observed flow during the simulation period also reflect these observations (Table 2).

These quantitative results are a first indication that model simulation is sensitive to the way patches are defined. For each patch configuration, the optimal m and k values (Table 2) were selected for further analysis and flow simulation at the outlet of the subcatchment. Qualitative comparison of the observed and simulated daily specific discharge (Figure 7) shows that the model captures the observed discharge variations reasonably well, again with the exception of PC2 and PC6. However, the model seems to be overly responsive in 1986 and 1988, simulating higher peak flows than were observed during the summer periods. The model also under predicts the flow at the beginning of the simulation period, especially in scenarios with river network threshold (PC1, PC3,PC5 and PC7).Additionally, the spring runoff in the year 1988 was consistently underpredicted, which may be due to a delay in the melting of snow cover. Overall, however, the timing of flow events is well captured in all cases, with the exception of PC2 and PC6. Qualitative comparison of the yearly simulated and observed runoff, based on water years (from June to May in the following year), shows that the total annual flow is consistently underpredicted in most patch configurations, except PC2, PC6 and PC8 (Figure 8). Configurations PC2 and PC6 resulted in overprediction of the total yearly runoff for all the simulation years. Again, it is clear that patch scenario 8 has the best simulation result, with the least difference between the simulated and the observed annual average runoff values.

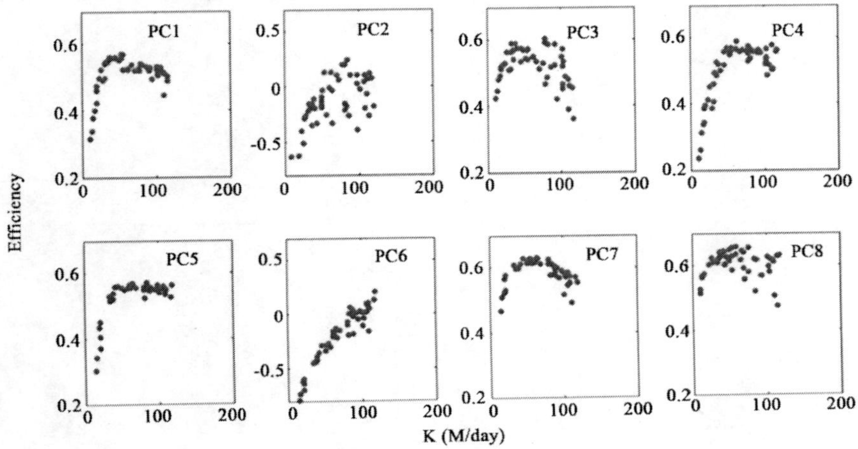

(a)

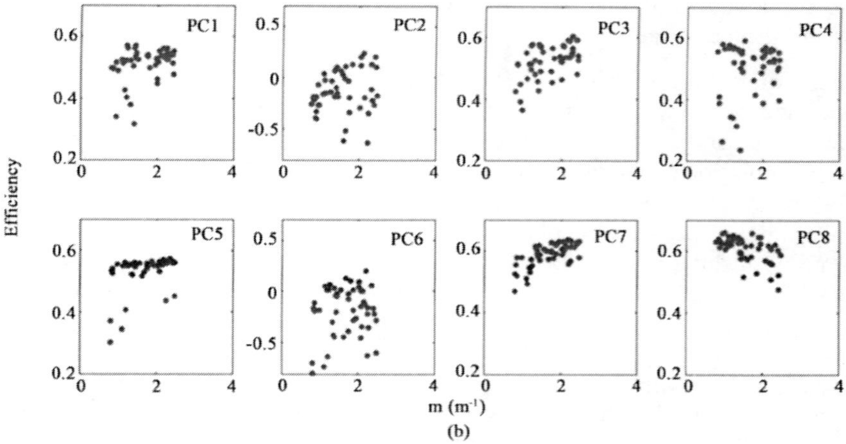

Figure 6. Maps showing distribution of the Nash Sutcliffe efficiency values that correspond to selected 75 k (a) and m (b) values generated during model calibration.

Table 2. Table that shows the best m and k combinations and the resulting total simulated flow simulated for the eight patch scenario projects.

	m	k	Efficiency(E)	Simulated	Observed	% Difference
Patch 1	1..43	54.26	0.57	2827.74	3294.52	14.17
Patch 2	2.28	101.35	-0.02	5142.64	3294.52	56.10
Patch 3	2.28	79.30	0.60	2954.98	3294.52	10.34
Patch 4	1.51	63.63	0.59	3760.48	3294.52	14.14
Patch 5	2.21	119.13	0.57	2864.30	3294.52	13.06
Patch 6	2.21	119.13	0.20	4727.80	3294.52	43.51
Patch 7	2.18	65.18	0.63	2938.58	3294.52	10.80
Patch 8	1.02	59.55	0.66	3285.43	3294.52	0.28

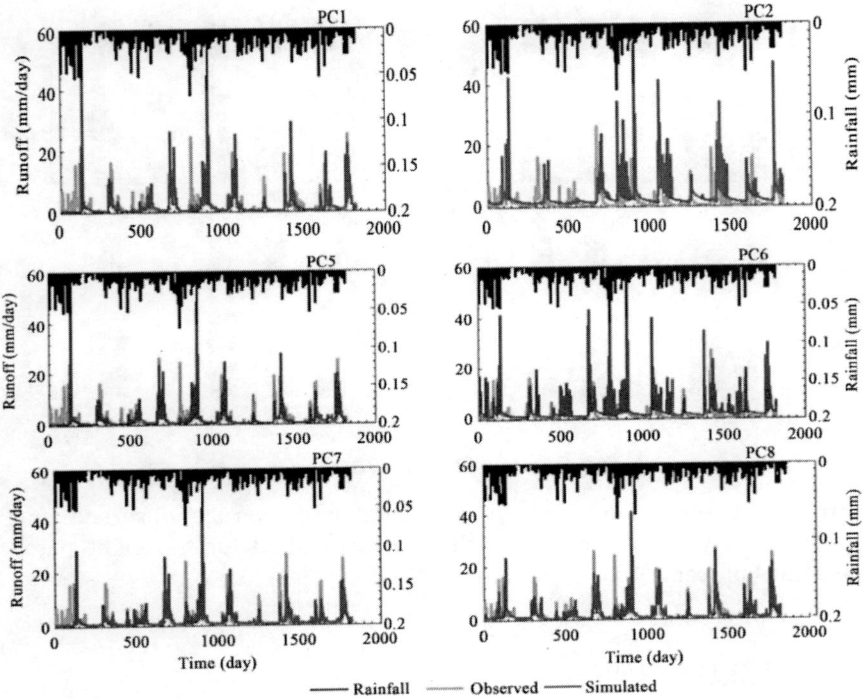

Figure 7. Daily time series curves of the observed and simulated flows at the outlet of the c38 watershed for the eight patch classification scenarios.

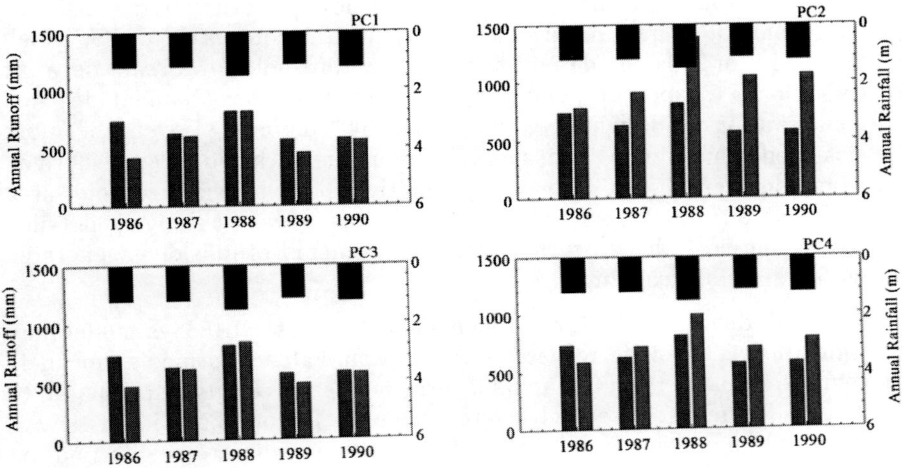

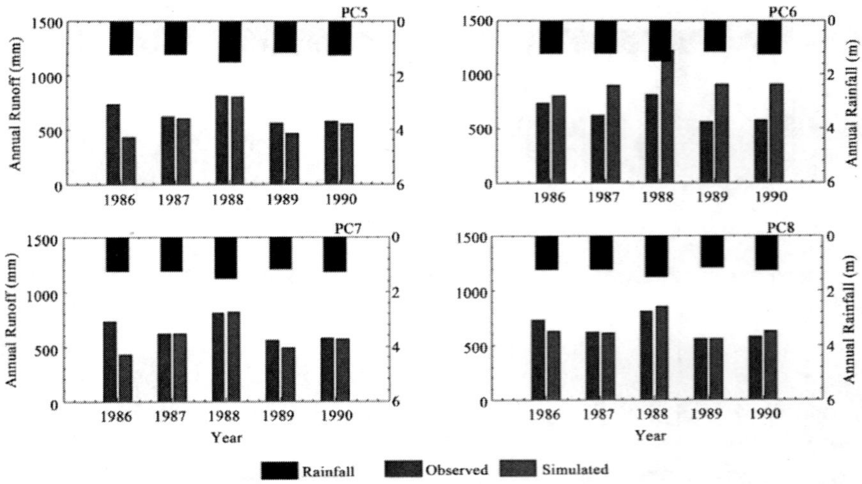

Figure 8. Comparison of the yearly total and average observed and simulated flows at the outlet of the c38 watershed for the eight patch classification scenarios.

6. CONCLUSIONS

Eight patch configuration scenarios for subcatchment 38 of the Turkey Lake Watershed were calibrated for hydrological flow simulation with the RHESSys model. Most patch configurations could, after calibration, be made to give reasonable estimation of observed flow patterns for the five-year water simulation period, with the exception of the patches defined by a high threshold for hillslope definition and a low threshold for stream network definition (PC2 and PC6). The best simulation result was obtained for patch configuration PC8, which was formed by combining layers of more classified climate zones and layers of stream network and hillslope derived using smaller threshold value. Similarly, the smaller threshold area specified during stream network delineation gave a relatively detailed stream network that in turn influenced the detail of hillslopes generated within the subcatchment.

Based on these results, it can be concluded that the RHESSys model flow simulation is sensitive to patch classification. Patches formed from finely aggregated patch forming inputs layers give better results as compared to patches formed from coarsely aggregated inputs. From the point of view of computation efficiency and memory, it may be very interesting to find the threshold level at which patch forming inputs should be aggregated so that flow simulation is not affected under this threshold.

The most notable result, however, is that the model's k and m parameters are dependent on the patch configuration and vary notably between the different patch configurations ($54.26 < k < 119.13$; and $1.02 < m < 2.28$). This dependency poses problems for the physical interpretation of these parameter values, and for their transferability to neighboring catchments.

Patch configuration in RHESSys modeling applications is often arbitrary or undocumented. This study, however, has shown that the result and the accuracy of flow simulations could be affected by the assumptions made during patch formation process.

ACKNOWLEDGEMENTS

The authors acknowledge Catchment Research Facility (CRF) of Western University, Canadian Forest Services of Natural Resources Canada, and National Water Research Institute and Meteorological Services of Environment Canada for providing the TLW datasets used in this study. We also acknowledge Dr. Marco Van De Wiel for his assistance with editing the manuscript.

REFERENCES

1. Dehotin, J. and Braud, I. (2007) Which Spatial Discretization for Which Distributed Hydrological Model? Hydrology and Earth System Sciences Discussions, 4, 777-829.
2. Refsgaard, J.C., Thorsen, M., Jensen, J.B., Kleeschulte, S. and Hansen, S. (1999) Large Scale Modelling of Groundwater Contamination from Nitrate Leaching. Journal of Hydrology, 221, 117-140.
3. Jung, G., Wagner, S. and Kunstmann, H. (2012) Joint Climate-Hydrology Modeling: An Impact Study for the Data-Sparse Environment of the Volta Basin in West Africa. Hydrology Research, 43, 231.
4. Li, L., Ngongondo, C.S., Xu, C.-Y. and Gong, L. (2013) Comparison of the Global TRMM and WFD Precipitation Datasets in Driving a Large-Scale Hydrological Model in Southern Africa. Hydrology Research, 44, 770.
5. Zhang, X., Drake, N.A., and Wainwright, J. (2013) Spatial Modelling and Scaling Issues. In: Wainwright, J. and Mulligan, M. Eds., Environmental Modelling, John Wiley & Sons, Ltd., Chichester, UK, 69-90. http://dx.doi.org/10.1002/9781118351475.ch5
6. Abbott, M.B., Bathurst, J.C., Cunge, J.A., O'Connell, P.E. and Rasmussen, J. (1986) An Introduction to the European Hydrological System—Systeme Hydrologique Europeen, "SHE", 1: History and Philosophy of a Physically-Based, Distributed Modelling System. Journal of Hydrology, 87, 45-59.
7. Abbott, M.B., Bathurst, J.C., Cunge, J.A., O'Connell, P.E. and Rasmussen, J. (1986) An Introduction to the European Hydrological System—Systeme Hydrologique Europeen, "SHE", 2: Structure of a Physically-Based, Distributed Modelling System. Journal of Hydrology, 87, 61-77.

8. Ciarapica, L. and Todini, E. (2002) TOPKAPI: A Model for the Representation of the Rainfall-Runoff Process at Different Scales. Hydrological Processes, 16, 207-229. http://dx.doi.org/10.1002/hyp.342
9. Beven, K. (1995) Linking Parameters across Scales: Subgrid Parameterizations and Scale Dependent Hydrological Models. Hydrological Processes, 9, 507-525.
10. Beven, K. (2000) On the Future of Distributed Modelling in Hydrology. Hydrological Processes, 14, 3183-3184. http://dx.doi.org/10.1002/1099-1085(200011/12)14:16/17<3183::AID-HYP404>3.0.CO;2-K
11. Gupta, V.K., Mesa, O.J. and Dawdy, D.R. (1994) Multiscaling Theory of Flood Peaks: Regional Quantile Analysis. Water Resources Research, 30, 3405-3421.
12. 12. Wood, E.F., Sivapalan, M., Beven, K. and Band, L. (1988) Effects of Spatial Variability and Scale with Implications to Hydrologic Modeling. Journal of Hydrology, 102, 29-47.
13. 13. Becker, A. (1992) Criteria for a Hydrologically Sound Structuring of Large Scale Land Surface Process Models. In: O'Kane, J.P., Ed., Advances in Theoretical Hydrology, Elsevier, Amsterdam, 97-111. http://dx.doi.org/10.1016/b978-0-444-89831-9.50014-4
14. Flügel, W.-A. (1995) Delineating Hydrological Response Units by Geographical Information System Analyses for Regional Hydrological Modelling Using PRMS/MMS in the Drainage Basin of the River Bröl, Germany. Hydrological Processes, 9, 423-436.
15. Kavvas, M.L., Chen, Z.Q., Tan, L., Soong, S.T., Terakawa, A., Yoshitani, J., et al. (1998) A Regional-Scale Land Surface Parameterization Based on Areally-Averaged Hydrological Conservation Equations. Hydrological Sciences Journal, 43, 611-631.
16. Arrigo, J.A.S. and Salvucci, G.D. (2005) Investigation Hydrologic Scaling: Observed Effects of Heterogeneity and Nonlocal Processes across Hillslope, Watershed, and Regional Scales. Water Resources Research, 41, W11417.http://dx.doi.org/10. 1029/2005wr004032
17. Blöschl, G. (2001) Scaling in Hydrology. Hydrological Processes, 15, 709-711. http://dx.doi.org/10.1002/hyp.432
18. Cerdan, O., Le Bissonnais, Y., Govers, G., Lecomte, V., van Oost, K., Couturier, A., et al. (2004) Scale Effect on Runoff from Experimental Plots to Catchments in Agricultural Areas in Normandy. Journal of Hydrology, 299, 4-14.http://dx.doi.org/10.1016/j.jhydrol.2004.02.017
19. Corwin, D.L., Hopmans, J. and de Rooij, G.H. (2006) From Field- to Landscape-Scale Vadose Zone Processes: Scale Issues, Modeling, and Monitoring. Vadose Zone Journal, 5, 129-139. http://dx.doi.org/10.2136/vzj2006.0004
20. Schulze, R. (2000) Transcending Scales of Space and Time in Impact Studies of Climate and Climate Change on Agrohydrological Responses. Agriculture, Ecosystems & Environment, 82, 185-212. http://dx.doi.org/10.1016/S0167-8809(00)00226-7
21. Sivapalan, M., Grayson, R. and Woods, R. (2004) Scale and Scaling in Hydrology. Hydrological Processes, 18, 1369-1371. http://dx.doi.org/10.1002/hyp.1417
22. Tague, C.L. and Band, L.E. (2004) RHESSys: Regional Hydro-Ecologic Simulation System—An Object-Oriented Approach to Spatially Distributed Modeling of Carbon, Water, and Nutrient Cycling. Earth Interactions, 8, 1-42. http://dx.doi.org/10.1175/10873562(2004)8<1:RRHSSO>2.0.CO;2
23. Philip, J.R. (1966) Plant Water Relations: Some Physical Aspects. Annual Review of Plant Physiology, 17, 245-268. http://dx.doi.org/10.1146/annurev.pp.17.060166.001333
24. Running, S.W., Nemani, R.R. and Hungerford, R.D. (1987) Extrapolation of Synoptic Meteorological Data in Mountainous Terrain and Its Use for Simulating Forest Evapotranspiration and Photosynthesis. Canadian Journal of Forest Research, 17, 472-483. http://dx.doi.org/10.1139/x87-081
25. Running, S.W. and Coughlan, J.C. (1988) A General Model of Forest Ecosystem Processes for Regional Applications I. Hydrologic Balance, Canopy Gas Exchange and Primary Production Processes. Ecological Modelling, 42, 125-154. http://dx.doi.org/10.1016/0304-3800(88)90112-3
26. Beven, K.J. and Kirkby, M.J. (1979) A Physically Based, Variable Contributing Area Model of Basin Hydrology. Un modèle à base physique de zone d'appel variable de l'hydrologie

du bassin versant. Hydrological Sciences Bulletin, 24, 43-69. http://dx.doi.org/10.1080/02626667909491834
27. Wigmosta, M.S., Vail, L.W. and Lettenmaier, D.P. (1994) A Distributed Hydrology-Vegetation Model for Complex Terrain. Water Resources Research, 30, 1665-1679. http://dx.doi.org/10.1029/94WR00436
28. Tague, C.L. and Band, L.E. (2001) Evaluating Explicit and Implicit Routing for Watershed Hydro-Ecological Models of Forest Hydrology at the Small Catchment Scale. Hydrological Processes, 15, 1415-1439. http://dx.doi.org/10.1002/hyp.171
29. Sanford, S.E. (2005) Scale-Dependence of Natural Variability of Stream Flow Parameters in a Forested Drainage Basin on the Boreal Shield. M.S. Thesis, University of Western Ontario, London, Ontario.
30. Monteith, S.S., Buttle, J.M., Hazlett, P.W., Beall, F.D., Semkin, R.G. and Jeffries, D.S. (2006) Paired-Basin Comparison of Hydrological Response in Harvested and Undisturbed Hardwood Forests during Snowmelt in Central Ontario: I. Streamflow, Groundwater and Flowpath Behaviour. Hydrological Processes, 20, 1095-1116. http://dx.doi.org/10.1002/hyp.5956
31. Webster, K.L. (2007) Topographic Controls on Soil Carbon Dioxide Efflux. PhD Thesis, University of Western Ontario, London, Ontario.
32. Murray, C.D. and Buttle, J.M. (2005) Infiltration and Soil Water Mixing on Forested and Harvested Slopes during Spring Snowmelt, Turkey Lakes Watershed, Central Ontario. Journal of Hydrology, 306, 1-20. http://dx.doi.org/10.1016/j.jhydrol.2004.08.032
33. Laporte, M.F., Duchesne, L.C. and Morrison, I.K. (2003) Effect of Clearcutting, Selection Cutting, Shelterwood Cutting and Microsites on Soil Surface CO2 Efflux in a Tolerant Hardwood Ecosystem of Northern Ontario. Forest Ecology and Management, 174, 565-575. http://dx.doi.org/10.1016/S0378-1127(02)00072-5
34. Jeffries, D.S. and Semkin, R.G. (1982) Basin Description and Information Pertinent to Mass Balance Studies of the Turkey Lakes Watershed. Rep. 82-01, Turkey Lakes Watershed, Algoma, Ontario, Canada.
35. Lindsay, J.B. (2005) The Terrain Analysis System: A Tool for Hydro-Geomorphic Applications. Hydrological Processes, 19, 1123-1130.http://dx.doi.org/ 10.1002/hyp.5818
36. Band, L.E. (1986) Topographic Partition of Watersheds with Digital Elevation Models. Water Resources Research, 22, 15-24. http://dx.doi.org/10.1029/WR022i001p00015
37. Quinn, P., Beven, K., Chevallier, P. and Planchon, O. (1991) The Prediction of Hillslope Flow Paths for Distributed Hydrological Modelling Using Digital Terrain Models. Hydrological Processes, 5, 59-79.http://dx.doi.org/10.1002/hyp.3360050106
38. Zevenbergen, L.W. and Thorne, C.R. (1987) Quantitative Analysis of Land Surface Topography. Earth Surf. Process. Landforms, 12, 47-56. http://dx.doi.org/10.1002/esp.3290120107
39. Nemani, R., Pierce, L., Running, S. and Band, L. (1993) Forest Ecosystem Processes at the Watershed Scale: Sensitivity to Remotely-Sensed Leaf Area Index Estimates. International Journal of Remote Sensing, 14, 2519-2534. http://dx.doi.org/10.1080/01431169308904290
40. Nash, J.E. and Sutcliffe, J.V. (1970) River Flow Forecasting through Conceptual Models Part I—A Discussion of Principles. Journal of Hydrology, 10, 282-290. http://dx.doi.org/10.1016/0022-1694(70)90255-6

CHAPTER 13

Spatiotemporal Variations in Snow and Soil Frost—A Review of Measurement Techniques

Angela Lundberg [1,*], David Gustafsson [2], Christine Stumpp [3], Bjørn Kløve [4,†] and James Feiccabrino [5,†]

[1]*Civil, Environmental and Natural Resources Eng., Luleå University of Technology (LTU), SE-971 87 Luleå, Sweden*

[2]*Research and Development, Swedish Meteorological and Hydrological Institute (SMHI), SE-601 76 Norrköping, Sweden*

[3]*Helmholtz Zentrum München, German Research Center for Environmental Health—Institute of Groundwater Ecology, Ingolstädter Landstraße 1, D-857 64 Neuherberg, Germany*

[4]*Water Resources and Environmental Eng. Research Group, University of Oulu, FI-900 14 Oulu, Finland*

[5]*Department of Water Resources Eng., Lund University, SE-221 00 Lund Sweden*

AUTHOR CONTRIBUTIONS

Angela Lundberg wrote most of the text but all co-authors have read and commented on the entire text. David Gustafsson was main responsible for the sections about the Radar and the Low Frequency Impedance Band Techniques and Christine Stumpp for the section about Tracer Techniques. Björn Klöve made minor contributions under the headings Introduction and Conclusions and James Feiccabrino under the heading Precipitation.

ABSTRACT

Large parts of the northern hemisphere are covered by snow and seasonal frost. Climate warming is affecting spatiotemporal variations of snow and frost, hence influencing snowmelt infiltration, aquifer recharge and river runoff patterns. Measurement difficulties have hampered progress in properly assessing how variations in snow and frost impact snowmelt infiltration. This has led to contradicting findings. Some studies indicate that groundwater recharge response is scale dependent. It is thus important to measure snow and soil frost properties with temporal and spatial scales appropriate to improve infiltration process knowledge. The main aim with this paper is therefore to review ground based methods to measure snow properties (depth, density, water equivalent, wetness, and layering) and soil frost properties (depth, water and ice content, permeability, and distance to groundwater) and to make recommendations for process studies aiming to improve knowledge regarding infiltration in regions with seasonal frost. Ground-based radar (GBR) comes in many different combinations and can, depending on design, be used to assess both spatial and temporal variations in snow and frost so combinations of GBR and tracer techniques can be recommended and new promising methods (auocostics and self potential) are evolving, but the study design must be adapted to the scales, the aims and the resources of the study.

Keywords: snow; seasonal frost; spatiotemporal variation; measurement techniques

1. INTRODUCTION

Increased air temperature will change snow accumulation and melt patterns, influence seasonal frost thickness and duration, snowmelt infiltration and recharge patterns Barnett, et al. [1,2,3,4]. Climate change will also impact snow cover and soil frost processes causing more dynamic freeze/thaw periods. Changes in climate and the related changes in snow hydrology influence water resources in many ways, affecting water quantity and quality. Snowmelt water dominates river runoff and aquifer recharge providing the main source of water in many regions. Snowmelt contains nutrients and pollutants [2,5] and pollutant flux processes are central elements of environmental risk assessment. Better understanding of snowmelt infiltration and recharge processes are thus needed to understand and adapt to climate change and recognize impacts of future land management schemes on water resources. This is essential for understanding consequences and for effective implementation of the protective actions required by the European Commission Water Framework Directive [6], Floods Directive [7] and Groundwater Directive [8] to meet good water quantity and quality. Quantifying future evolution of infiltration and recharge in northern climates thus requires both, trustworthy forecasting of future climatic variables, and knowledge about spatiotemporal variations in present recharge processes.

Important snow resources in high latitudes and mountains with seasonal snow and frost are relatively unknown parts of the hydrological cycle with quite few high quality measurements. This is alarming as mountain regions receive high amounts of precipitation and thus also greatly contribute to runoff and groundwater recharge. In these environments, changes in spatiotemporal patterns of snow accumulation and melt are principal controls on hydrologic responses to climate change [9]. How valleys and mountaintops are influenced by climate change seems to vary. Some studies of mountain climate show faster warming of high elevations than of lower elevations (western European Alps [10]) while the high elevations have been cooling lately in the Rocky Mountains, Colorado, US [11]. Most studies conclude that less snow will fall in a warmer climate [12,13,14,15]. However, effects of warmer climate on snowfall will depend on regional factors. Warmer climate generally means shorter snow season and thus decreasing snowfall. On the other hand, snowfall is more frequent and more intense at air temperatures just below zero than at colder temperatures, (maximum air humidity increases with air temperature), increasing snow accumulation for such climatic conditions [16]. Depending on which process becomes predominant, snow accumulation may increase or decrease as an effect of warmer climate [17]. Based on simulations of 21st century climate and using 20 global climate models, Räisinen [16] showed, that even if peak snow water equivalent (SWE) generally decreased, it increased in the coldest areas. Snow depth trends over northern Eurasia are increasing in the northeast and decreasing in the west [18]. An overview of the complicated pattern of observed changes in snow cover duration and maximum snow depths in different climatic regions is provided by Brown and Mote [19], indicating that snow cover duration was the snow variable with the strongest climate sensitivity varying with altitude and climate regime. Long-term records of annual snow accumulation in Alpine regions advise that substantial, regional scale shifts in snowpack features have occurred with associated changes in the frequency and timing of freeze–thaw cycles [5].

Changes in snow covers will influence soil frost and recharge which is important for groundwater renewal and storm flow quantity and quality.

There is no direct relation between frost depth and infiltration or recharge amount since infiltration depends on both snowmelt mass and soil ice content. Winters with shallow frost do not typically produce more infiltration than winters with deep frost since winters with shallow frost usually have thicker snow cover [20,21,22] When frost indicators and snowmelt runoff were compared in small catchments no proof for larger or faster runoff for winters with deep frost were found [23]. Even if studies on seasonally frozen ground show that frost can limit or totally prevent infiltration, this is far from always the case, and to only focus on frost depth is not enough. Frozen soil is seldom impermeable, and many studies show that infiltration takes place also through thick frost layers if the soil surface contains air filled pores [24,25,26]. However, if snowmelt or rainwater freezes, forming an ice layer at top of the soil surface, this blocks infiltration even if the ice layer is thin [22,27,28,29]. Frozen soil infiltration capacity decreases drastically with the ice content of the soil [30,31]. The ice content of the frozen soil and especially of the topsoil is thus important to access when studying the impact of seasonal frozen ground on infiltration and recharge. As the water content of soil also influence ice content, wet soils form more impermeable soils. The behavior of groundwater in seasonal frozen ground regions is thus dominated by the spatiotemporal extent of hydraulic isolation between the water on top of and below the near-surface frozen zone [32]. The close relationship between snow and frost depths to control the snowmelt separation into infiltration (and recharge) or runoff should be better understood [33]. To increase process understanding a combination of well-designed field experiments and modeling studies is of importance [32].

Topography, soil properties and vegetation cover influence snowmelt and infiltration in complex, unexpected ways. Sandy south-facing hill slopes were for example found to generate greater surface runoff than north-facing slopes and finer-textured soils [34]. Snowmelt input to groundwater recharge can thus display systematic and significant spatial variability due to the influences of elevation, slope, aspect and forest cover on both snow accumulation and snowmelt processes [35]. Turbulent flux dominated sites melt faster than sheltered sites [35]. And snow accumulation and melt was shown to differ significantly between clear-cut, juvenile, and mature stands, and amongst juvenile stands with distinctive structural differences [36]. Vegetation and topography impact on snow energy exchanges can thus result in de-synchronization of snowmelt in a basin [37,38] complicating the space-time patterns of snowmelt and therefore also on the separation of meltwater into surface runoff or infiltration. Further details regarding the complex interplay between physiographic variables sometimes exert contrasting influences on snowmelt runoff can be found in Smith et al. [35]. According to Smith et al. [35] most snowmelt studies exploring groundwater-related dynamics are made in small catchments (<50 ha) and have limited elevation ranges (<200 m) (such as [39]). The small scales have probably restricted the spatial variability in the timing and intensity of snowmelt inputs and associated impacts on groundwater and runoff response patterns. Few studies thus deal with groundwater response when meltwater from different part of a basin occur at different times (asynchronous) [35,40,41,42,43].

Laboratory and field studies on soil core scale consistently show decreased infiltration capacity with increasing ice content [44,45,46,47]. Some plot field studies also report reduced infiltration with increasing ice content [29,48], while others indicate no or minor impact [22,26,49]. Since seasonally frozen ground is mainly studied with focus on runoff processes, the effects on infiltration and recharge mostly have to be indirectly deduced. Catchment runoff studies often rely on numerical simulations and some studies suggested and successfully used a classification where the infiltration capacity of seasonally frozen ground was divided into different soil and frost types [50,51]. However a similar approach for a forested catchment found little improvement by including soil frost in the simulations [52]. In addition, no impact of frozen soil on runoff was found in a large catchment [23]. The importance of frost thus seems to decrease with increasing spatial scale and a number of possible reasons why this could be the case are suggested by Lundberg et al. [2]. Partitioning of meltwater during initiation of spring freshet into vertical versus lateral flux (due to intensity, timing and quantity) of melt seems to overwhelm the influence of topographic convergence on runoff source area dynamics in mountain areas [35]. The spatiotemporal distribution of snow and frost thus seems to lack influence on magnitude and timing of basin runoff but the limitations regarding the sizes and altitude ranges of presented studies mentioned above cannot be dismissed.

Additional knowledge regarding spatiotemporal variation in snow and frost processes is therefore needed in order to predict how global warming will influence these processes [2]. This in turn calls for appropriate knowledge on available measurement techniques to get information about relevant snow and soil frost parameters. The purpose with this paper is therefore to review different methods to measure properties of snow (depth, density, water equivalent, wetness, and layering) and soil frost (depth, water and ice content, permeability, and distance to groundwater) and to make recommendations for instrumentation of process studies aiming to improve knowledge regarding infiltration in seasonally frozen ground regions. Specific focus is on spatiotemporal distribution of snow and frost properties, which influence infiltration. Satellite based techniques can also be used to map snow and frost properties but they typically have low spatial and temporal resolution and are beyond the scope of this review. Snow and frost measurement techniques are evolving quickly so there is a risk for unavoidable omissions of some publications and this review is thus far from complete.

2. MEASUREMENT TECHNIQUES

There are existing reviews on ground based measurement techniques focusing on snow depth [53], SWE [54,55], snow density [56], forest snow [57,58,59], frost depth [60] and automated snow techniques [61]. However, no review focusing on the spatiotemporal variations in snow and frost properties seems to exist and the progress in measurement techniques is rapid. Current access to new cheap, wireless digital sensors allow continuous measurements of many meteorological and hydrological parameters [62,63].

First, systems to measure many relevant parameters are reviewed; electrical/dielectrical, tracer and temperature based techniques are, for example, deployed in both snow and frost studies. Then follow the review of techniques primarily used in snow studies and finally the review of methods usually used for soil frost recharge studies.

2.1. Snow Processes and Parameters and Associated Common Methodological Problems

Snow related processes that are essential in hydrological measurements include precipitation, snow depth and SWE. Besides these snow density, layering and albedo are important in some snow studies and applications. As forests effect frost and snow accumulation, vegetation cover needs special attention in snow measurements. In addition, snow packs are dynamic and particularly, the phase is important for snowmelt infiltration studies since rain on snow might refreeze at the base of the snowpack forming a more or less impermeable ice layer preventing infiltration [29], and there is an on-going trend to replace meteorological observers with automated systems, many of which lack precipitation discriminators [64].

All types of sensors located less than about 0.5 m under snow surface are likely to absorb more solar radiation than the snow itself, and this causes risk for local melt around the device and erroneous measurements [2,65]. However, thin, small and white installations cause small or negligible disturbances. If a part of the measuring device is buried in the snow and a part is extending above the snow surface there is also risk of air pockets forming around the device during high wind events by vibrations (particularly if the wind exposed area is large) causing similar problems.

2.2. Dielectrical and Electrical Techniques

The dielectrical techniques are based on variations in dielectric constant (permittivity) of the material, which differ markedly between liquid and frozen water. The permittivity (K) of wet and dry snow as well as of wet, dry and frozen soil can be calculated with a mixing model using the dielectric permittivity (ε) and the volume fractions (θ) of the different components [66]. For wet snow, the relation becomes

$$\varepsilon_{WS} = (\theta_W \varepsilon_W^{\alpha} + \theta_I \varepsilon_I^{\alpha} + \theta_A \varepsilon_A^{\alpha})^{\frac{1}{\alpha}}$$

where W, I, A denotes liquid water, ice and air, respectively. The contribution from air is small and is usually neglected. For frozen soil the volume fractions and the dielectric permittivity of the soil itself ($\theta_{Soil}\varepsilon^{\alpha}_{Soil}$) should be included [66]. A good fit over the whole range of liquid volume fraction ranging from 0 to 1 using the α-value 0.5 for soils was found [67], and this values has also been applied for snow [68]. For common frequencies the dielectric constant for liquid water is about 80, while values for ice and dry soil are about 3. Methods based on differences in dielectric properties between ice, water, soil and air have many different applications for studies of spatiotemporal variations in snow and ground frost. Radar systems, time domain reflectometry (*TDR*), 3-D tomography, LiDAR, and the "snow fork" are examples of such techniques.

2.2.1. LiDAR

Terrestrial laser scanning, also known as LiDAR or laser altimetry techniques, can give spatially detailed ground surface altitude information, and snow depth can be retrieved when the technique is applied on both bare and snow covered soil [69,70,71,72]. The development of LiDAR applications for snow depth mapping has exploded in recent years, and is now also used for investigating snow distribution below forest canopies [72]. The accuracy of the LiDAR technique is high but is hard to generalize since it depends on many factors; the measurement set-up (distance to the observed surface), the used method (beam divergence), the scanning speed, the quality check, etc. [71]. However, a skilled operator is required to perform data acquisition and post processing (to achieve trustworthy results). The purchase cost of the equipment is around 70,000–120,000 Euros [71].

Ground-based systems usually deliver millimeter-scale range accuracy and sub-meter point spacing over several kilometers for smooth and open terrain while the accuracy is lower in complex and forested terrain [53]. The technique has also been used to identify depressions exposed to concentrated infiltration in frozen soil [73].

2.2.2. Radar Systems

Ground penetrating radar, (GPR), also known as georadar, has been used for a variety of applications for snow soil and frost research, and exists in many different designs (systems with different antenna frequencies, with one or several frequencies, with one or several channels). The radar can be operated from vehicles, such as helicopters, airplanes or snow mobiles and can thus cover large lateral distances in short times, or it can be mounted stationary and monitor temporal changes [61]. *GPR*-techniques have been applied to determine the spatial distribution of snow depths [74,75,76,77,78], snow density [77,79], SWE [80], snow layering and stratigraphy [78,81,82], wet snow properties [83], soil frost depth [84,85,86,87], depths to thawing front and shallow freeze and thaw processes [85,86,88,89,90,91,92], soil water content [93,94] and depths to the water table [95,96].

Frequency-modulated continuous wave (FMWC) radar systems transmit a continuous signal where the frequency linearly increases over a time step and records the reflected waves, the minimum and maximum frequencies and the total sweep time. Usually, these systems generate frequencies of up to 40 GHz, resulting in a vertical resolution of 1–3 cm. These high radar-frequency systems are less expensive than impulse radars since the received signal only needs to be sampled with kilohertz frequency, i.e., several orders of magnitude lower than that of the transmitted signal. An overview of FMCW radar technology and its application for snow research can be found in Marshall et al. [97]. In these systems the received signal is mixed with a replica of the transmitted signal, (computer-generated or recorded by a radar antenna), and Fast Fourier Transform is used to determine the frequency difference between them. This difference along with the total sweep time and the frequency range (the difference between the maximum and minimum frequencies) are used to calculate the radar wave two-way travel time (TTWT) [98]. As for the impulse radar, snowpack depth is attained from the TTWT if the radar wave propagation velocity is known or estimated from independently measured snow density. This depth and the snow density are then finally used to calculate SWE. A limitation with this method is the requirement to perform measurements on dry snow or to determine and take into account the liquid water content (*LWC*). If the density varies substantially throughout the snowpack the accuracy of SWE estimates will be lower.

Snowpack stratigraphy can also be assessed by FMCW radar with high resolution [99]. When FMCW radar is combined with manual snow pit measurements, ice layers identified in the pit can be followed through the length of the line radar measurements. For dry snow covers with varying snow densities the propagation velocity can be calculated from the snow density for each layer and a good estimate of snow cover depth and snow density, and, therefore also SWE, can be acquired from the TTWT [100]. A discrete version of FMCW systems named step-frequency continuous wave (SFCW) radar systems, transmit a continuous signal which steps through a number of equidistant frequencies with constant step length, so the total sweep time is known [101]. SFCW radars have the same limitations as FMCW systems.

Field scale shallow soil moisture profiles using full-waveform inversion of GPR data were mapped [102], and Jadoon et al. [91] monitored temporal variations in freeze–thaw cycles on a snow covered agricultural field using a stationary mounted step-frequency radar and judged the method promising for real-time mapping of soil frost at the field scale. Steelman and Endres [85] used the development of high-frequency *GPR* direct ground wave propagation and could detect thinner frozen layers than when using the radar wave reflection from the frozen-unfrozen boarder. Steelman et al. [86] illustrated that this technique could be used to monitoring spatiotemporal dynamics of shallow freezing and thawing events.

Peak SWE is often the most important parameter for infiltration and recharge studies, and *GPR* measurement might then need to be performed on melting snowpacks. However, the traditional *GPR* technique is not well adjusted for measurements on melting snow, since SWE-values might be overestimated by about 20% for wet snowpacks [103]. Methods to improve the accuracy of SWE measurements using *GPR* on wet snow and to estimate snow wetness based on amplitude damping of the emitted pulse have been published [80,104]. However, the available methods and technologies for estimating snow properties such as density, wetness, and internal structure directly from GPR data, is still dependent on advanced methods of antenna configurations and analysis and have not yet been shown to be applicable for larger scale operational snow surveys. In the latter case, the application of GPR is probably still most efficiently made using manual snow depth and snow density observations as reference to calibrate the GPR data. The development and use of a stationary mounted upward looking GPR, placed at the ground at the beginning of the snow season to record temporal variations in snow-pack depth, layering and snow wetness are reported in literature [81,83,105,106]. An upward looking design as the one discussed above based on L-band FMCW radar for snow cover monitoring is presented by Okorn et al. [107].

2.2.3. TDR

In-situ time domain reflectometry (TDR) measurements are also based on differences in dielectrical constant between water, soil, air and liquid water. TDR has been used for registration of temporal changes in liquid water content (*LWC*) for frozen soils [46,67,108,109,110,111,112,113] and snow [46,114,115] as well as for density of dry snow [77]. The TDR technique determines dielectric permittivity (dielectric constant) of the material by measuring the delay in time between the emitted and the reflected electromagnetic wave using wave-guides (pair of stainless steel rods) installed in the material. The permittivity can then be converted to volumetric water content for soils and for snow to density (if the *LWC* is known) or to *LWC* (if the density is known or can be assumed to be constant) [114]. The main advantages of TDR techniques are high temporal resolution, rapidity acquisition (about 20 s) and the repeatability of the measurements [116]. A major disadvantage with the technique when applied to snow is the sensitivity to air gaps forming around the wave-guides. Wave-guides absorb more solar radiation than the surrounding snow and might melt the surrounding snow thus forming air gaps around the guides. TDR has been adapted to fit on mobile platforms, primarily for agricultural applications [116]. Recent TDR measurement of freezing and thawing curves on soils in western Canada showed significant hysteresis and mechanisms for this and possible influence on the interpretation of published datasets are discussed by He and Dyck [117]. The TDR technique is applied in the commercially available water content reflectometers (WCR) (Campbell Scientific Inc.) used by for example Iwata et al. [118]. A thermo-time domain reflectometry (T-TDR) probe monitors both soil temperature and *LWC* [119].

2.2.4. Additional Electrical Techniques

Impedance tomography has been applied to get a 3-D picture of the infiltration process [120,121] and electrical resistivity imaging (ERI) has been used to give spatial distribution data regarding infiltration [122,123,124]. An oscillating circuit and a sensing part embedded in the material are used by capacitance based techniques. The dielectric permittivity of the medium is determined by measuring the charge time of a capacitor, which uses the medium as a dielectric capacitor. Different dielectric operating frequency are applied to different media [116]. The nuclear magnetic resonance (NMR) method has been used to measure unfrozen water content in frozen soil [125]. The underlying principle is that hydrogen protons resonate when subjected to an oscillating magnetic field of the right frequency. Limitations with this technique are that it measures small sample sizes that it is difficult to sustain the desired temperature of the NMR device, and the device is expensive

[126]. Self-potential (streaming potential) theory has been applied for many hydrological applications for example for mapping of seepage in embankment dams [127,128,129]. A new theory and numerical model of electrical self-potential (SP) signals associated with unsaturated flow in melting snow have been developed and tested using laboratory column experiments [130]. The SP signals primarily depend on the temporal growth of snow porosity and meltwater flux, electrical conductivity and pH [130]. The authors conclude that non-intrusive SP measurements can serve as proxies for meltwater fluxes and that the acquisition methods could bridge the widely acknowledged gap in spatial scales between satellite and point measurements of snow properties. The technique has been applied to track meltwater fluxes in-situ snowpacks [131].

2.3. Temperature Based Techniques

Recent advances in fiber optics allow for rapid and accurate temperature measurements along lengthy fiber cables. High resolution temperature information (1 m, 10 s, and 0.1 °C) from fiber-optic distributed temperature sensing technology was for example employed by Tyler at el. [132] to map soil surface temperatures during melt out. Distributed temperature sensors (also called iButtons or Miniature Thermochron dataloggers) placed a few cm below or above the soil surface offer direct spatial information of soil temperatures and frozen ground and indirect information about presence or absence of snow, since diurnal temperature variations are much larger for bare than for snow covered ground [133,134]. iButtons were also used to get distributed data on the dates for snowpack basal ripening and for melt out [135] and to assess the likelihood of ice lenses formation in the snow pack [62]. Unfortunately these dataloggers are not waterproof, which can cause data loss, so one option is to seal them by coating them with plastic, and when comparing unmodified iButtons with iButtons modified in this way, Rznik and Alford [136] found small temperature differences (0–1.3 °C) and they conclude that coating iButtons with plastic is an reliable and affordable method to waterproof the data loggers. Thermistor strings are also used to maps snowpack temperature variations. Infrared thermal imager has been used for snow pit temperature measurements and to assess snow surface temperature [137]. However, it was pointed out [138] that during windy and cold conditions snow surfaces within a snow pit will quickly approach air temperatures thus limiting the use of infrared thermal imager for this type of application.

2.4. Tracer Techniques

Beside temperature, environmental tracers as well as artificial tracers are good tools to identify and quantify frost and snow processes and their impact on recharge rates and flow paths. A review of tracer studies aiming to increase the understanding of flow paths and residence times in northern catchments is presented by Tetzlaff et al. [139]. Artificial tracers can give information of a specific process at specific initial or boundary conditions of an experiment itself. Dye was used as a tracer to show that meltwater drip from trees seep more or less vertically through the snowpack to the soil but that water from different melt events follow different preferential flow channels [140]. Dyes were also used to visualize and quantify snowpack percolation [141]. Here, complex percolation patterns were observed by dye staining, highlighting the influence of ice layers and preferential flow paths in the snowpack. Similar findings were identified from the artificial application of rare earth elements to identify redistribution processes within the snowpack [142]. Ice layers within the snowpack acted as solute barriers and diurnal variability of melting processes were identified. Bromide in combination with electrical resistivity tomography has often been used to track infiltration pathways [20,122,124]. Salt tracers like bromide change the freezing point though and thus, their application as conservative tracers for snow and frost related processes are limited.

For more general applications, and on larger scales, environmental tracers have been used to estimate flow pathways in snowpacks, observe redistribution of water within the snowpack, mark the infiltration patters of melt water into soils, and quantify fractions of melt water in surface water or groundwater recharge [110,143,144,145,146,147,148]. Particularly, the stable environmental isotopes $\delta^{18}O$ and $\delta^{2}H$ are suited for providing information about meltwater infiltration in areas with frost and seasonal snow cover. Due to the temperature-isotope dependence, δ-values in snowfall and meltwater are lower compared to groundwater and annual mean of precipitation [149,150]. When snow reaches tree branches or the ground, the initial isotope signal can be modified by rain, condensation, sublimation, partial melting and percolation of meltwater. Meltwater isotopic signatures are for example influenced by the presence of forests, and Koeniger et al. [151] showed that heavy isotopes in the snowpack were progressively enriched as canopy stand density increased. During aging of the snow pack, isotope fractionation processes alter the isotope ratios of fresh snow [152]. For example, enrichment in $\delta^{18}O$ of meltwater can occur during percolation through the snow cover. Such enrichment can either be due to sublimation effects [150] or due to isotopic exchange between water and ice during percolation [153]. Both processes result in slopes of $\delta^{18}O/\delta^{2}H$ relationships that are significantly lower compared to the isotopic composition of precipitation. From the variability of the linear relationship between $\delta^{18}O$ and $\delta^{2}H$ due to water-ice interactions, Lee et al. [153] concluded on the melting history. To obtain snowmelt infiltration and recharge, melt water should therefore be collected and analyzed for water isotopic composition [149,154]. A solution is to use snow lysimeters like passive capillary samplers (PCS), earlier used to sample soil water. PCS were modified and placed at the snow soil interface to collect snowmelt water representative for the isotopic composition of the water infiltrating into the soil [155]. Penna et al. [156] state that PCS are low-cost, easy to install and can collect representative integrated snowmelt samples throughout the melt period or during single melt events. They are thus suitable for flow path determinations.

2.5. Techniques Focusing on Snow Parameters

2.5.1. Snow Precipitation

Most precipitation gauges collect less snowfall than the surrounding ground due to wind losses. Different types of wind shields and correction coefficients (functions of wind speed and air temperatures) have therefore been developed to compensate for this [157]. However, it is still a challenge to measure snow precipitation at wind exposed sites. Additional sources of errors are blockage of the gauge orifice by snow capping and accumulating on the side of the orifice walls. Heating the gauge eliminates capping and accumulation issues, however, heated gauges are prone to undercatch caused by updrafts over the gauge orifice and evaporation; particularly the latter should be avoided when using stable water isotopes as tracers. The survey by Nuti et al. [158] indicates that many automatic precipitation gauges varying in configuration, capacity, orifice area sensitivity and windshield are presently used for the measurement of solid precipitation. The use of a variety of gauge designs and approaches complicates intercomparison, and a harmonized measurement approach for snow research was called for [159]. Results from the WMO Solid Precipitation InterComparison (WMO-SPICE) to assess the quality of automated data [160] are now being published and readers are referred to other studies for details [156,158,161,162,163,164,165,166,167,168,169,170,171,172,173,174,175,176,177,178]. It is however worth mentioning that the two dimensional video disdrometers (2DVD) offer a novel technique which matches precipitation types based on images from two orthogonal cameras using supervised classification [179]. This allows for determination of both precipitation phase and precipitation density and thus of SWE for short time intervals. Examples of different precipitation types are shown in Figure 1.

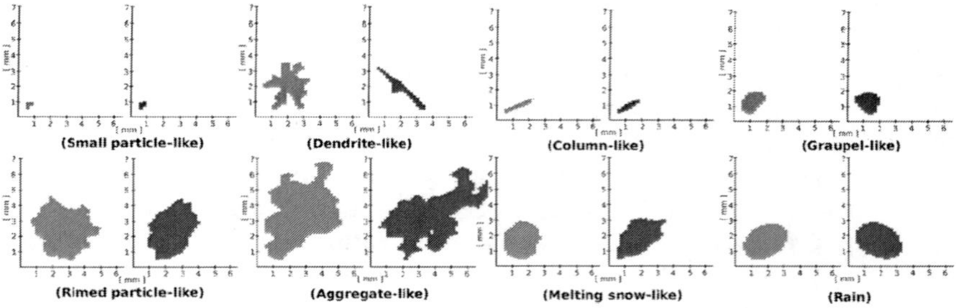

Figure 1. Examples of images from two orthogonal cameras showing the eight precipitation types identified by the 2DVD (with permission from [179]).

Most automated methods to discriminate between rain and snow use ((I) Doppler radar; (II) optical sensors; (III) hotplates; or (IV) two dimensional video disdrometers (2DVD)) have been introduced without an in-depth intercomparison, which raises the question of their quality, consistency, compatibility, and representativeness [158].

2.5.2. Snow Depth

The temporal and spatial evolution of snow depth in the early winter is especially important since shallow snow depths during the first cold spells often create thick frost. Ultrasonic sensors are extensively used to monitor temporal changes in snow depths for example at several SNOTEL sites [180], and they have an accuracy around ±1 cm [181]. SNOTEL is an automated system in Western US to measure snowpack and related meteorological parameters operated by the US Natural Resources Conservation Service (NRCS). Long-term trends (since 1928) of snow depths have been studied using snow stakes in the Austrian Alps [182]. Placed in view of time lapse cameras, they have been used to automatically record snow depth variation [183]. A snow depth probe combined with an inbuilt GPS giving positions with ±10 m (absolute value) and with ±2 m on repeated surveys can be ordered from the company SnowHydro [184]. By this technique several kilometers of observations with 2 m spacing can be obtained in just a few hours and downloaded to a computer at the end of the observations. However, information about how thick snow packs can be probed and how the device works with ice crust is lacking [184]. Detailed temperature profiles can be used to determine snow depth by analyzing the daily temperature amplitude (small variations within snow-pack, large above it). GPS receivers are today installed in numerous locations to provide precise positions for various scientific studies and this recent explosion of GPS networks means that daily data from thousands of GPS receivers are now available from public archives. Geodetic-quality GPS systems can be used to record average snow depths in an area of about 1000 m^2 around the GPS antenna in the depth range 0 to 150 cm [185,186,187,188].

2.5.3. Snow Water Equivalent (SWE)

SWE is the most essential parameter for assessing available water before spring melt which often is needed in hydrological forecasting related to hydropower, floods or recharge. SWE is the best indicator of the potential infiltration from snowmelt in spring, and the spatial variations in SWE give information about possible spatial variation in snowmelt volume. All methods to determine snow depths, mentioned above, can in combination with density be converted to SWE by multiplying with the density. The snow density can be measured (gravimetric snow sampling or retrieved from snow pits), estimated based on snow depth (maximum SWE in regions not exposed to midwinter melt [189]), on regional climate [190] or on season, snow depth, altitude, and location [191]. GPS depths were for example converted into SWE by SNOTEL site measured densities [188], however the spatial variation in snow density is large, see [192,193,194,195] for additional information regarding spatial and temporal variations in snow density. Temporal changes in SWE in one location can be recorded by snow pillows, weighing plates, attenuation of radioactivity or by attenuation of GPS signals or by repeated snow courses.

Snow courses: A snow course is a fixed transect in the terrain where snow depth and density are measured at fixed spatial intervals by snow tubes. Design criteria for them depend on study objectives and type of terrain [196]. Many evaluations of different snow tube designs are made and a recent comparison of two snow tube designs can be found in Dixon and Boon [197].

Snow pillows: A typical snow pillow is a big bladder filled with a non-toxic antifreeze liquid solution, installed under the snow, in level with the ground, with the weight of the snow (SWE) sensed by pressure transducers. They function well in fairly flat areas without freeze and thaw events. However, they are not suited for use in snowpacks with freeze thaw cycles forming ice sheets bridging the pillows, neither for dense forests nor steep terrain since it is difficult to dig foundations for them there. They are not suited for snowmelt infiltration studies since they block water transport into the soil. Furthermore, dark pillows should be avoided since they may absorb more solar radiation than the surrounding ground and therefore delay the onset of autumn snow accumulation [198,199]. There is also risk for leakage of antifreeze liquid. Recommendations for size, type of antifreeze liquid, etc. can be found in a Technical Report [200]. Impermeable weighing plates (SWE pressure sensors) suffer from similar weaknesses as snow pillows, but require less maintenance [201]. Tips on how to identify and correct SWE pressure sensor errors are presented in Johnston et al. [202]. Recent development and tests of different designs of permeable weighing plates better suited for infiltration studies can be found elsewhere [55,203].

Techniques based on radiation attenuation: That snowpack (more strictly hydrogen) attenuates radioactive radiation has long been used for SWE measurements [204] and at least three different applications have been used: (A) attenuation of radiation from an active radioactive source placed on the ground under the snowpack; (B) attenuation of the natural background radiation of the soil and rocks; and (C) attenuation of the natural cosmic radiation. Use of radioactive sources (active neutron sources) in a cased borehole, or access tube have been used previous to determine total water content in soils [205] but seems to have been more or less abandoned due to difficulties in gaining permission to handle, transport and leave radioactive sources in remote locations [206]. Soils and overburden emits natural gamma radiation and the strength depends on the type of soil and rock [207]. The attenuation of this radiation can after first measuring the background radiation without snow be related to SWE by a calibration function using the sensor CS 725 [208,209]. When the sensor is located about 3 m above the ground it measures an area of 50 to 100 m^2 (according to the manufacturer). This technique was tested and recommend [55,210] for non-alpine Norwegian regions (with melt and freeze cycles); however, the authors point out that it is not suited for SWE larger than 600 mm and that it should be connected to a power grid. A similar device run by solar power is presented by Martin et al. [209]. Cosmic gamma radiation also reaches the ground and incoming and attenuated cosmic radiation has been used to monitor SWE in Japan [211,212]. Between 1998 and 2004 about 40 Cosmic-ray snow sensors were installed in the mountainous regions of France giving satisfying SWE measurements [213]. However, local calibration seems to be needed and sometimes, the surrounding soil humidity has a strong influence on the SWE measurement and a specific sensor has been designed to solve this problem [213]. Today, a commercial product named SnowFox is accessible [214]. The cosmic radiation and the neutron probe techniques are similar so obstacles with identifying the measurement volume and with separating soil water and SWE could be expected. As mentioned earlier radar techniques can give line information about SWE.

2.5.4. SWE with Additional Snow Parameters

SAS2: Kinar and Pomeroy presented a non-destructive and time saving method based on acoustic reflectometry theory to determine SWE [215], and this method has later been upgraded to an electronic sensing system (SAS2) [216] to simultaneously measure LWC, snow density, and snow temperature and achieve images of the snowpack stratigraphy. Acoustic waves sent into and reflected from the snowpack by the SAS2 system and a sound propagation theory was used to acquire acoustic measurements of snow and images of the snowpack. Both stationary and portable versions of the SAS2 system were installed and tested and SWE could be determined quicker and more accurately than with a gravimetric snow tube [216].

SnowPack Analyzer/SnowPower: The SnowPack Analyzer system (also known as SnowPower) is based on low-frequency impedance measurements on a flat cable sensor for simultaneous measurement of snow density and liquid water content [217,218]. In combination with a snow depth sensor, it can also be used to measure SWE. The system measures the dielectric constant of snow at multiple frequencies in the kHz range using a newly developed low frequency impedance analyzer, and optionally also in the MHz range using an ordinary TDR analyzer. The dielectric constant of ice has a strong frequency variation in the kHz range, which enables determination of the volumetric content of both frozen and liquid water. The sensor consists of a 5 to 25 m long flat PVC-band cable, which can be mounted horizontally at some distance from the ground, to monitor conditions at a specific layer of the snow, or sloping from the ground to a position above the snow surface, to give integrated information over the entire snow profile. TDR measurements on the sensor cable may be further utilized to investigate the spatial variation of liquid water in the snowpack by inverse modeling of the Telegraph equation [219]. However, the low frequency measurements are enough for determination of the average density, SWE, and wetness; therefore the TDR measurements are not included in the later versions of the SnowPack Analyzer system. The sensor has been tested for instance in high-alpine Swiss test site [83,218], agricultural fields in Canada [217], mountain sites in Scandinavia, and forest sites in Germany and USA [220,221]. According to these tests the sensing system was quite robust with regard to the impedance analyzer and the horizontally mounted sensor cables. However, several tests indicated practical problems with the sloping installation such as wind-induced air pockets around the cable [218], running water along the sloping cable [83], and damages to the cable sensor [220]. The analysis if the sloping cable is also highly sensitive to the fraction of the cable covered by snow, which needs to be estimated for instance by an assumed geometry and snow depth observations. Due to these problems, the sloping cable installation is no longer recommended by the manufacturer (Wolfram Sommer, personal communication 2015). On the other hand, the field tests also showed that data from horizontal sensors mounted, either at the base of the snow pack, or at specific depths of the snowpack give reliable and useful information compared to manual snowpack measurements and numerical modeling of SWE, and LWC variation and snowmelt infiltration [83,218,220,221]. The Canadian study also accentuates the importance for the estimation accuracy on the choice of low frequency and the algorithm for the snow temperature extrapolation [217,218,221].

2.5.5. Snow Layering

Snow layering into different density layers has an effect on, for example, snow stability, which is important in avalanche research. Snow layering is important for spatial distribution of snowmelt and rain, which can be transported tens of meters along layers in the snow. Snow pit excavations are frequently used to determine vertical density variations and to identify snow layers and track water movement in snow [81,222]. How to dig and refill a snowpit without disturbing the snow is described in Williams et al. [223]. The penetrometer (SnowMicroPen) can be carried in a rucksack and can give extremely detailed information about snow structure profile (snow pack layering) [65,114,224]. A "snowguillotine" was used to produce high-resolution (1 cm^3), 3-D-information regarding meltwater flow through a snowpack by combining the measurements with image processing and geostatistical analysis [225]. The radar and acoustic techniques mentioned above can give line and area information about snow layers. Snow stratigraphy may also be observed using photographs of translucent snow profiles obtained by excavating a snow snowpit profile from two sides [226].

2.5.6. Liquid Water Content

The first snowmelt does not necessarily leave the snowpack but is held by capillary forces in the snow. Therefore, it is important to measure the temporal and spatial variations in liquid water content (LWC) to be able to know when and where snow is ripe (saturated LWC) and water is ready to leave the snowpack. Infrared radiation sensors can be used to measure snow surface temperature [227] and thus give information about if the snow is melting or not since melting snow has the constant temperature of 0 °C. The "snow fork" is a hand-held device which consists of a two-pronged wave guide that is inserted into the snow to determine its density (for dry snow) and wetness (if density is known) [228]. A similar design as the snow fork, said to have higher spatial resolution and accuracy, was presented by Kendra et al. [229]. A capacitive radio-frequency device was used to record snow density and LWC at several different levels [230]; a similar approach used a monopole antenna [231]. However, all these devices are not suited for stationary registration since they will absorb solar radiation creating air pockets around the probes. LWC measured by the snow fork was consistently 1.5 times greater than by the Denoth meter [232]. The spatial variations in LWC for snow and seasonally frozen ground are needed in order to assess if and how common preferential flow is. Radar, TDR, SnowPack Analyzer and electrical resistivity techniques mentioned before can also be used to monitor snowpack wetness.

2.5.7. Melt Water Collection and Separation

Many different types of lysimeters are used to collect meltwater [62,233,234], and the design is influenced by the purpose of the study. Design to measure snowmelt separation into surface, subsurface and deep percolation can be found in Bayard et al. [21]. Teflon-coated acid washed snowmelt lysimeters were used for measuring volumes, melt intensity and isotopic composition of the snowmelt [39]. Lysimeters in materials with low thermal conductivity and high albedo (plywood and white fiber glass) approximating the thermal characteristics of the snowpack are described by Winkler et al. [36]. Lysimeters are widely used in hydrograph separation studies either using volume weighted average value or sampling at different time steps throughout the melt event [235]. The earlier mentioned passive capillary samplers (PCS) can also be used.

2.5.8. Forest Snow Parameters

Forests reduce snowfall (evaporation and sublimation of canopy intercepted snow) and cause small-scale distribution of throughfall, thus causing uneven frost. Measurement techniques for canopy snow forests storage processes have been reviewed earlier [58,236]. Ideal canopy process studies should include continuous measurements of intercepted mass and throughfall with high time resolution and not disturb the meteorological conditions function and the method should work during periods of both melt and sublimation. However, all methods have individual drawbacks [57]. Winter season canopy forest sublimation has been determined as difference between peak snow accumulation in open and forested areas [237,238]. Sublimation/evaporation processes of canopy snow has been estimated by eddy covariance techniques [239,240] and by deuterium and oxygen 18 isotopes [151,241]. The eddy covariance technique has also been applied in studies of snow sublimation below forested canopies [242]. Canopy (tree crown) intercepted mass has been estimated using branch displacement by tensiometers [243], photos [240], digitized video images of branch deflection [244], tree weighing devices [58,236,237,245] and tree stem compression [246]. Throughfall has been measured using sealing of ground by tarpaulin [247], by weighing devices [58,248] and by non-weighing snow-melt lysimeters i.e., troughs equipped with tipping buckets [62]. The lysimeter troughs designed to record drip from tree crowns during rain on snow events seem well suited for the task for snowpacks <1 m. They were buried in the ground and stuffed with coniferous branches in order to prevent icing and blockages and the associated tipping bucket was enclosed and also buried in the ground to prevent freezing [62]. Frost measurements using radar techniques with a snow mobile in forests might be difficult due to poor accessibility. In special conditions, terrestrial laser scanning can also be used for detailed mapping of sub-canopy snow depth distribution [72].

2.6. Techniques Focusing on Soil and Groundwater Parameters

Several of the radar techniques are well suited to record spatial variations in frost depths, and depths to the thawing front, and a surface to surface application have also been used to map temporal variations in soil moisture [249]. The temperature techniques discussed above can also be used to assess frost depths [87]. Frost depth is important for the infiltration but even more crucial is the ice content of the frost (in reality it is the open pore space that matters). Most infiltration measurement methods deal with determining infiltration capacity (tension disks and pressure ring infiltrometers) while techniques to measure actual infiltration rates seem more rare. Infiltration rates may be estimated from measurements of increase in water content at various soil depths using combinations of dielectrical sensors. For example TDR-techniques, which can give unfrozen water content, and gamma or neutron scattering techniques, which gives total water content (liquid and frozen). Frost depth tubes filled with methylene blue solution [87], which changes color when frozen, are often used to monitor frost temporal development and their precision was, according to Iwata et al. [60] around ±3.5 cm, and thaw depths were overestimated by the tubes but could be calibrated using a linear equation.

As TDR are commonly used in hydrological studies to monitor soil moisture content, TDRs can provide additional information to many hydrological studies on soil freezing time or freezing front movement. Sun et al. [250] tested a frequency domain dielectrical sensor designed for use in access tubes to measure in-situ soil freeze–thaw cycle dynamics and concluded that the sensor was easier to operate than TDR-equipment, rapid, fairly accurate, non-destructive and more cost effective since a single sensor could be used at multiple sites. Yoshikawa and Overduin [251] who tested soil moisture probes based on electromagnetic properties in frozen soils showed that each sensor had a unique response to soil type and temperature changes, thus requiring unique calibrations. Wu et al. [252] presented a laboratory test using the complex resistivity method to monitor freeze–thaw transitions. They claim that their study provided a basis for exploring the potential of this method to monitoring spatiotemporal variations of freeze–thaw transitions over field-relevant scales. For neutron scattering [111,205] and gamma attenuation techniques a radioactive source is used (the gamma source is stronger) to measure the volumetric water content. The source is inserted into a vertical cased tube and winched up and down to quickly measure the moisture content at different depths. However, the use of radioactive sources requires training of staff, monitoring and regulation of shipping and handling [116] and has limited the use of the these methods. These techniques have been used in combination with TDR techniques where the latter technique is used to measure LWC and the total water content (ice and water) is measured by neutron [111,253,254] and gamma ray [126] scattering methods respectively. Errors for the gamma ray and TDR technique of 2% and 1% by volume (m^3/m^3) are reported [126].

Thermally insulated tensiometers designed for monitoring the potential of unfrozen soil below the frozen soil layer are described and used by several studies [14,29,118,255,256]. These tensiometers should be installed horizontally to minimize the soil disturbance. For recommendations on how to install and use piezometers and groundwater wells to monitor groundwater response to recharge see for example Rosenberry et al. [257]. For snow and frost applications, special attention must be paid to avoiding use of metal tubing, which might cause local enhanced frost around the observation wells. The soil freezing characteristic and soil moisture characteristic plays a key roles in soil frost modeling and a portable dielectric tube sensor along with a set of temperature sensors were successfully used for in situ determination of soil freezing characteristic and soil moisture characteristic [258].

3. SUMMARY AND FUTURE RECOMMENDATION ON MEASUREMENT TECHNIQUES

The spatial variability of both snow and soil frost seem to be the most distinguished requirement for enhancing our understanding of the snowmelt infiltration processes at the spatial scale important for runoff generation. Thus, measurement techniques that enable observations of the spatial distribution with high spatial variability are to be recommended foremost, in other words, space-continuous techniques such as ground penetrating radar have a high potential to provide new insights. On the other hand, the quantitative uncertainties are rather high in radar techniques, unless it is combined with other observation types to reduce these uncertainties. In the following, we summarize the pros and cons of the various measurement techniques included in this review:

Ultrasonic sensors and snow sticks (with or without time laps cameras) are inexpensive methods suited to monitor temporal changes in snow depths. GPS systems can be used to record average SWE over larger areas but with lower accuracy. There are many different methods used to map temporal variations in SWE (see Table 1) and they all have their individual advantages and disadvantages (see references for details) but methods that do not block the melt water entrance into the soil should be used for infiltration and recharge studies so snow pillows and impermeable electronic weighing devices should be avoided.

Table 1. Summary of methods to register temporal variations in snow water equivalent (SWE).

Method	Obs. Area *	Additional Snow Parameters	Refs
Snow pillows			
Bladders or Steel containers	≈3–10 m^2	-	[200,259]
Electronic weighing devices			
Impermeable Johnson	≈10 m^2	-	[198,260]
Permeable Johnson	≈1.5–9 m^2	-	[203]
Permeable Moen 2525	≈25 m^2	-	[55]
Gamma radiation attenuation techniques			
Artificial source	Point	-	[204]
Natural ground	≈75 m^2	-	[55,208,209]
Natural cosmic	≈75 m^2	-	[211–214] 2015, Hydroinnova
Stationary up-ward looking radar techniques			
Single frequency	Point	LWC, density, layers	[81,83,105,106]
Step frequency	Point	LWC, density, layers	[107]
Low Frequency Impedance Band (SnowPower)	≈5–25 m		[218,261]
SAS2 Electronic sensing system using acoustics	<45 m^3	LWC, density, temp, layers	[216]

* Observed area when applicable is given as observed distance or observed volume and for the Johnson et al. electronic devices is it the area of the weighing panels which are given, these panels are surrounded by eight equally sized panels.

Ground based radar comes in many different designs (systems with single or varying antenna frequencies, with one or several channels) and can be mounted stationary to record temporal changes or on a vehicle to assess spatial variations. Radars applying high frequencies can for example identify thin ice layers but have shallow penetration depths while radars with low frequencies can penetrate deeper but might fail to identify thin layers. Both the travel time for the radar pulse, and the attenuation of its amplitude could preferably be used to assess properties of the snow and the frozen soils. The recent development, of upward looking stationary mounted radars, has been used to successfully record temporal variations in snowpack properties (SWE, depth, LWC). Radars mounted on vehicles have fruitfully been used to map spatial distribution of snowpack properties (depth, SWE and ice layers) and frozen soil properties (depth of frozen soil, depth of thawing layer, soil LWC). The radar applications providing information of SWE, density and layers would have many applications. A combination of an upward looking stationary radar system, and another radar system for spatial surveys at (a) the time of maximum SWE; (b) towards the end of snowmelt; and (c) at the time of maximum thaw would give information about the temporal development and spatial distribution of snow, as well as spatial distribution of frost depth and thawing depth provided that soil conditions are favorable for radar measurements (preferably sandy soils, and not clay soils). As long as snow bulk properties are of interest, ground based radar systems could well be the optimal measurement techniques to collect snow information for studies on infiltration patterns in frozen soils. However, it must be noted that derivation of snow properties such as density and liquid water content from the radar signal without other reference data comes with large uncertainties, and it is recommended to always combine the radar data with for instance manual snow density and snow depth samples to reduce these uncertainties.

Time domain reflectometry (TDR) works well for determination of LWC in snow and soil, but large care has to be made to assure that airgaps are not formed around the probes. The thermo-time domain reflectometry (T-TDR) probe monitors both temperature and LWC. High resolution temperature information (1 m, 10 s, and 0.1°C) can be achieved from fiber-optic distributed temperature sensing technology and inexpensive coated iButtons and thermistor strings can also be useful to map soil and snow temperatures.

The low frequency impedance sensor system (SnowPack Analyzer/SnowPower) has the advantage over conventional TDR of determining both liquid and frozen water content, and the influencing area around the sensor is larger making it less sensitive to air gaps. On the other hand, the size of the sensor is both a problem and a potential strength of the system. The low frequency impedance measurements require the sensor cable to be at least about 5 m long. Low-frequency measurements will thus integrate over a larger volume, which might be preferable, but this length also causes problems with regard to the deployment in the snow, especially for sloping configurations. Thus far, the SnowPack Analyzer system seems to be most reliable and unique with regard to liquid water content measurements, and using the horizontal deployment of the sensor cable in the snow.

For flow path information some kind of tracer is recommended. Dye has been used to map meltwater paths along different flow paths in the snowpack and artificial application of rare earth elements were used to identify redistribution process within the snowpack. Electrical resistivity tomography in combination with bromide has often been used to track infiltration path ways. The stable environmental isotopes $\delta^{18}O$ and $\delta^{2}H$ are suited since the δ-values in snowfall and meltwater are lower compared to groundwater and annual mean of precipitation and thus giving information about snowmelt infiltration. However, meltwater isotopic signatures are influenced by phase change processes (freezing, melting and evaporation) and melt water leaving the snow pack should therefore be collected and analyzed for water isotopic composition by for example passive capillary samplers (PCS) or lysimeters. Design of lysimeters for collection and separation snowmelt into shallow flowing soil water and percolating water has to be adapted to the purpose and location of study. When water quality is of interest teflon-coated acid washed snowmelt lysimeters are recommended and material with thermal conductivity and albedo approximating those of the snowpack are recommended for use in the snowpack. The recent application of self-potential (SP) to map melt water flow paths seems to have large potential and so does the acoustic method, which also gives additional information.

Terrestial laser scanning technique (LiDAR) can preferably be used to identify possible locations for focused infiltration.

Recommended techniques for snowmelt infiltration studies:

- Electronic sensing system using acoustics, stationary upward mounted radars or low impedance band for temporal information about snow (LWC, SWE and layers).
- Mobile radars with different designs for spatial information about snow (LWC, SWE and layers) and soil (frost depth, water content, and thawing depth).
- Permeable weighing sensor for temporal variations in SWE.
- Natural or artificial tracers and Self-potential for flow paths in snow.
- TDR and Low-Frequency impedance sensor measurements for LWC in snow and soil.

- Neutron methods for total (liquid and frozen) water in soils.
- Fiber optics, thermistor strings or coated iButtons for temperature.
- Frost tubes for frost depth.
- Lysimeters designed for the study and SPC samplers to measure snowmelt.
- LiDAR to identify likely locations for focused infiltration.
- If precipitation gauges are used to measure snowfall, is it important to use a suitable design or to correct for wind losses.

4. RECOMMENDATION FOR FUTURE COLLABORATIVE RESEARCH

A collaborative research effort is recommended to obtain more information on snow and frost hydrology. Processes related to spatial and temporal variation of snow and soil frost and its interplay with recharge and runoff are crucial in hydrology, but relatively poorly understood and not well incorporated in hydrological models. With climate warming, changes in snow and frost occurrence will be a major change in hydrology of high latitudes and future climate change scenarios should urgently assess changes in hydrological process. The demand for such information is obviously needed for many sectors and communities (water supply, flood protection, agriculture, traffic, environmental protection, etc.) Information is needed on different scales and on different processes depending on application, e.g., small scale stability is crucial for avalanches, small scale accumulation of snow is needed for traffic and roof snow loads but large scale accumulation is more important for flood prediction and infiltration/overland flow partitioning (water quality), and recharge (groundwater renewal). Knowledge on spatial distributions of snow and frost will be especially urgent in regions that rely on snowmelt runoff and snow related groundwater renewal for water supply and irrigation. As soil frost and snowmelt have an effect on transport pathways, knowledge is also needed to assess climate change related changes in solutes transport such as DOC, nitrate, pesticide and phosphorous leaching. To cover the complexities linked to snow hydrology, experts with different specializations should work closely together to increase process understanding, share experiences in novel measurement techniques and integrate knowledge and raise awareness on the importance of snow hydrology.

We recommend a stronger future collaboration in hydrological research, data sharing and learning. Such joint research could focus on different type of catchments to cover the variability and gradients in climate, topography, geology and land cover. This should bring new insight into the importance of snow and seasonal soil frost processes on spatial and temporal distribution of snowmelt infiltration (and thus on the distribution of the nutrients and pollutants in the snowmelt) could be achieved with study design adjusted to the aim, the scale, and the means of the study using more than one technique to assess the spatiotemporal variations of the key parameters. Ground based radar, stationary mounted for temporal variations, and mounted on vehicles for spatial variations of snow and frost parameters is a useful tool in this context and the acoustics and the self-potential techniques, recently adjusted for snow measurements, seem very promising for this type of studies. Isotopic techniques and airborne methods linked to data assimilation could also be important. Collaboration across current experimental sites, establishing new sites and adding more detailed information on currently monitored sites is recommended. Some sites such as ICOS on green house gas fluxes currently monitor evapotranspiration and energy fluxes in seasonally snow covered regions (https://www.icos-ri.eu). For these sites hydrological studies could relatively simply be added to existing measurement set-ups to provide more detailed information on hydrological processes where all component to understand the hydrological dynamics and close the water balance. The ICOS network could be a model for future collaboration. As snow has wide implication on many sectors of society, experimental sites need to be established on a range of different land uses and soils.

Collaborative research and data sharing can have great potential to improve the knowledge in this field. To our knowledge, national monitoring networks on snow depth are under evaluation as new techniques have emerged. A common network on experts and experimental catchment could outline future recommendation regarding national monitoring efforts. The snow studies should be linked to catchments and groundwater hydrological studies with stronger links to experts on geophysics and airborne techniques to measure snow dynamics at appropriate spatial and temporal scales.

National monitoring set-ups could benefit from more detailed studies on snow where novel techniques that capture the complex temporal and spatial patterns of snow accumulation and melt as well as hydrological process of infiltration, overland flow, recharge and runoff generation are developed.

ACKNOWLEDGMENTS

Part of this work was made within the framework of the GENESIS project financed by the European Commission 7FP contract 226536 and by the Swedish Research Council Formas contract 242-2010-1187.

REFERENCES

1. Barnett, T.P.; Adam, C.J.; Lettenmaier, D.P. Potential impacts of a warming climate on water availability in snow-dominated regions. *Nature* **2005**, *438*, 303–309.
2. Lundberg, A.; Ala-Aho, P.; Eklo, O.; Klöve, B.; Kværner, J.; Stumpp, C. Snow and frost: Implications for spatiotemporal infiltration patterns—A review. *Hydrol. Process.* **2016**, *30*, 1230–1250.
3. Rasmus, S. *Snowpack Structure in Finland: Measurement and Modelling*; University of Helsinki: Helsinki, Finland, 2005.
4. Kreyling, J. Winter climate change: A critical factor for temperate vegetation performance. *Ecology* **2010**, *91*, 1939–1948.
5. Edwards, A.C.; Scalenghe, R.; Freppaz, M. Changes in the seasonal snow cover of alpine regions and its effect on soil processes: A review. *Quat. Int.* **2007**, *162*, 172–181.
6. EC. Directive 2000/60/ec of the european parliament and of the council establishing a framework for community action in the field of water policy. Available online: http://eur-lex.europa.eu/legal-content/EN/TXT/?uri=URISERV%3Al28002b (accessed on 20 Februay 2015).
7. EC. Directive 2007/60/ec on the assessment and management of flood risks. Available online: http://ec.europa.eu/environment/water/flood_risk/ (accessed on 20 Februay 2015).
8. EC. Directive 2006/118/ec of the european parliament and of the council on the protection of groundwater against pollution and deterioration. Available online: http://eur-lex.europa.eu/eli/dir/2006/118/oj (accessed on 20 Februay 2015).
9. Tague, C.; Grant, G.E. Groundwater dynamics mediate low-flow response to global warming in snow-dominated alpine regions. *Water Resour. Res.* **2009**, *45*, W07421.
10. Beniston, M.; Diaz, H.; Bradley, R. Climatic change at high elevation sites: An overview. *Clim. Chang.* **1997**, *36*, 233–251.
11. Pepin, N.; Losleben, M. Climate change in the colorado rocky mountains: Free air versus surface temperature trends. *Int. J. Climatol.* **2002**, *22*, 311–329.
12. Hardy, J.P.; Groffman, P.M.; Fitzhugh, R.D.; Henry, K.S.; Welman, A.T.; Demers, J.D.; Fahey, T.J.; Driscoll, C.T.; Tierney, G.L.; Nolan, S. Snow depth manipulation and its influence on soil frost and water dynamics in a northern hardwood forest. *Biogeochemistry* **2001**, *56*, 151–174.
13. Groffman, P.M.; Driscoll, C.T.; Fahey, T.J.; Hardy, J.P.; Fitzhugh, R.D.; Tierney, G.L. Colder soils in a warmer world: A snow manipulation study in a northern hardwood forest ecosystem. *Biogeochemistry* **2001**, *56*, 135–150.
14. Iwata, Y.; Hayashi, M.; Suzuki, S.; Hirota, T.; Hasegawa, S. Effects of snow cover on soil freezing, water movement, and snowmelt infiltration: A paired plot experiment. *Water Resour. Res.* **2010**, *46*, W09504.
15. Choi, G.; Robinson, D.A.; Kang, S. Changing northern hemisphere snow seasons. *J. Clim.* **2010**, *23*, 5305–5310.
16. Räisänen, J. Warmer climate: Less or more snow? *Clim. Dyn.* **2008**, *30*, 307–319.
17. Lundberg, A.; Beyerl, H. Ash on snow a tool—A tool to prevent flooding? *Nord. Hydrol.* **2001**, *32*, 195–214.

18. Popova, V. Winter snow depth variability over northern eurasia in relation to recent atmospheric circulation changes. *Int. J. Climatol.* **2007**, *27*, 1721–1733.
19. Brown, R.D.; Mote, P.W. The response of northern hemisphere snow cover to a changing climate. *J. Clim.* **2009**, *22*, 2124–2145.
20. Bayard, D.; Stähli, M. *Effects of Frozen Soil on the Groundwater Recharge in Alpine Areas*; Wiley: Chichester, UK, 2005; pp. 73–83.
21. Bayard, D.; Stähli, M.; Parriaux, A.; Flühler, H. The influence of seasonally frozen soil on the snowmelt runoff at two alpine sites in southern switzerland. *J. Hydrol.* **2005**, *309*, 66–84.
22. Nyberg, L.; Stähli, M.; Mellander, P.E.; Bishop, K.H. Soil frost effects on soil water and runoff dynamics along a boreal forest transect: 1. Field investigations. *Hydrol. Process.* **2001**, *15*, 909–926.
23. Shanley, J.B.; Chalmers, A. The effect of frozen soil on snowmelt runoff at sleepers river, vermont. *Hydrol. Process.* **1999**, *13*, 1843–1857.
24. Johnsson, H.; Lundin, L.-C. Surface runoff and soil water percolation as affected by snow and soil frost. *J. Hydrol.* **1991**, *122*, 141–159.
25. Stähli, M.; Jansson, P.E.; Lundin, L.C. Soil moisture redistribution and infiltration in frozen sandy soils. *Water Resour. Res.* **1999**, *35*, 95–103.
26. Sutinen, R.; Hänninen, P.; Venäläinen, A. Effect of mild winter events on soil water content beneath snowpack. *Cold Reg. Sci. Technol.* **2008**, *51*, 56–67.
27. Stein, J.; Proulx, S.; Lévesque, D. Forest floor frost dynamics during spring snowmelt in a boreal forested basin. *Water Resour. Res.* **1994**, *30*, 995–1007.
28. Jones, H.; Pomeroy, J. Early spring snowmelt in a small boreal forest watershed: Influence of concrete frost on the hydrology and chemical composition of streamwaters during rain-on-snow events. In Proceedings of the Eastern Snow Conference, Ottawa, ON, Canada, 17–19 May 2001; pp. 209–218.
29. Iwata, Y.; Nemoto, M.; Hasegawa, S.; Yanai, Y.; Kuwao, K.; Hirota, T. Influence of rain, air temperature, and snow cover on subsequent spring-snowmelt infiltration into thin frozen soil layer in northern Japan. *J. Hydrol.* **2011**, *401*, 165–176.
30. Espeby, B. Tracing the origin of natural-waters in a glacial till slope during snowmelt. *J. Hydrol.* **1990**, *118*, 107–127.
31. Flerchinger, G.; Lehrsch, G.; McCool, D. Freezing and thwing processes. In *Encyclopedia of Soils in the Environment*; Hillel, D., Ed.; Elsevier, Ltd.: Oxford, UK, 2005; pp. 104–110.
32. Ireson, A.M.; van der Kamp, G.; Ferguson, G.; Nachshon, U.; Wheater, H.S. Hydrogeological processes in seasonally frozen northern latitudes: Understanding, gaps and challenges. *Hydrogeol. J.* **2013**, *21*, 53–66.
33. Stadler, D.; Wunderli, H.; Auckenthaler, A.; Flühler, H.; Bründl, M. Measurement of frost-induced snowmelt runoff in a forest soil. *Hydrol. Process.* **1996**, *10*, 1293–1304.
34. Redding, T.; Devito, K. Aspect and soil textural controls on snowmelt runoff on forested boreal plain hillslopes. *Hydrol. Res.* **2011**, *42*, 250–267.
35. Smith, R.; Moore, R.; Weiler, M.; Jost, G. Spatial controls on groundwater response dynamics in a snowmelt-dominated montane catchment. *Hydrol. Earth Syst. Sci.* **2014**, *18*, 1835–1856.
36. Winkler, R.; Spittlehouse, D.; Golding, D. Measured differences in snow accumulation and melt among clearcut, juvenile, and mature forests in southern british columbia. *Hydrol. Process.* **2005**, *19*, 51–62.
37. Ellis, C.; Pomeroy, J.; Essery, R.; Link, T. Effects of needleleaf forest cover on radiation and snowmelt dynamics in the Canadian Rocky Mountains. *Can. J. For. Res.* **2011**, *41*, 608–620.
38. Jost, G.; Moore, R.D.; Smith, R.; Gluns, D.R. Distributed temperature-index snowmelt modelling for forested catchments. *J. Hydrol.* **2012**, *420*, 87–101.
39. Laudon, H.; Seibert, J.; Kohler, S.; Bishop, K. Hydrological flow paths during snowmelt: Congruence between hydrometric measurements and oxygen 18 in meltwater, soil water, and runoff. *Water Resour. Res.* **2004**, *40*, W03102.
40. Boyer, E.W.; Hornberger, G.M.; Bencala, K.E.; McKnight, D.M. Response characteristics of doc flushing in an alpine catchment. *Hydrol. Process.* **1997**, *11*, 1635–1647.

41. Deng, Y.; Flerchinger, G.; Cooley, K. Impacts of spatially and temporally varying snowmelt on subsurface flow in a mountainous watershed: 2. Subsurface processes. *Hydrol. Sci. J.* **1994**, *39*, 521–533.
42. Hinckley, E.L.S.; Barnes, R.T.; Anderson, S.P.; Williams, M.W.; Bernasconi, S.M. Nitrogen retention and transport differ by hillslope aspect at the rain-snow transition of the Colorado front range. *J. Geophys. Res. Biogeosci.* **2014**, *119*, 1281–1296.
43. Kuras, P.K.; Weiler, M.; Alila, Y. The spatiotemporal variability of runoff generation and groundwater dynamics in a snow-dominated catchment. *J. Hydrol.* **2008**, *352*, 50–66.
44. Burt, T.; Williams, P.J. Hydraulic conductivity in frozen soils. *Earth Surf. Process.* **1976**, *1*, 349–360.
45. Andersland, O.B.; Wiggert, D.C.; Davies, S.H. Hydraulic conductivity of frozen granular soils. *J. Environ. Eng.* **1996**, *122*, 212–216.
46. Kane, D.L.; Stein, J. Water movement into seasonally frozen soils. *Water Resour. Res.* **1983**, *19*, 1547–1557.
47. Van der Kamp, G.; Hayashi, M.; Gallen, D. Comparing the hydrology of grassed and cultivated catchments in the semi-arid Canadian prairies. *Hydrol. Process.* **2003**, *17*, 559–575.
48. Stadler, D.; Brundl, M.; Wunderli, H.; Auckenthaler, A.; Fluhler, H. Measurement of frost-induced snowmelt runoff in a forest soil [1996]. *Hydrol. Process.* **1997**, *11*, 427–427.
49. Zhao, L.; Gray, D.M.; Male, D.H. Numerical analysis of simultaneous heat and mass transfer during infiltration into frozen ground. *J. Hydrol.* **1997**, *200*, 345–363.
50. Gray, D.M.; Toth, B.; Zhao, L.T.; Pomeroy, J.W.; Granger, R.J. Estimating areal snowmelt infiltration into frozen soils. *Hydrol. Process.* **2001**, *15*, 3095–3111.
51. Gray, D.M.; Landine, P.G.; Granger, R.J. Simulating infiltration into frozen prairie soils in streamflow models. *Can. J. Earth Sci.* **1985**, *22*, 464–472.
52. Lindström, G.; Bishop, K.; Löfvenius, M.O. Soil frost and runoff at svartberget, northern Sweden—Measurements and model analysis. *Hydrol. Process.* **2002**, *16*, 3379–3392.
53. Deems, J.S.; Painter, T.H.; Finnegan, D.C. Lidar measurement of snow depth: A review. *J. Glaciol.* **2013**, *59*, 467–479.
54. Egli, L.; Jonas, T.; Meister, R. Comparison of different automatic methods for estimating snow water equivalent. *Cold Reg. Sci. Technol.* **2009**, *57*, 107–115.
55. Bache Stranden, H.; Lirhus Ree, B.; Moen, K. *Recommendations for Automatic Measurements of Snow Water Equivalent in Nve*; NVE Publisher: Oslo, Norway, 2015; p. 33.
56. Proksch, M.; Loewe, H.; Schneebeli, M. Density, specific surface area, and correlation length of snow measured by high-resolution penetrometry. *J. Geophys. Res. Earth Surf.* **2015**, *120*, 346–362.
57. Lundberg, A.; Halldin, S. Snow interception evaporation. Review of measurement techniques, processes, and models. *Theor. Appl. Climatol.* **2001**, *70*, 117–133.
58. Lundberg, A. Evaporation of intercepted snow—Review of existing and new measurement methods. *J. Hydrol.* **1993**, *151*, 267–290.
59. Lundberg, A.; Halldin, S. Snow measurement techniques for land-surface-atmosphere exchange studies in boreal landscapes. *Theor. Appl. Climatol.* **2001**, *70*, 215–230.
60. Iwata, Y.; Hirota, T.; Suzuki, T.; Kuwao, K. Comparison of soil frost and thaw depths measured using frost tubes and other methods. *Cold Reg. Sci. Technol.* **2012**, *71*, 111–117.
61. Lundberg, A.; Granlund, N.; Gustafsson, D. Towards automated 'ground truth' snow measurements—A review of operational and new measurement methods for Sweden, Norway, and Finland. *Hydrol. Process.* **2010**, *24*, 1955–1970.
62. Floyd, W.; Weiler, M. Measuring snow accumulation and ablation dynamics during rain-on-snow events: Innovative measurement techniques. *Hydrol. Process.* **2008**, *22*, 4805–4812.
63. Pohl, S.; Garvelmann, J.; Wawerla, J.; Weiler, M. Potential of a low-cost sensor network to understand the spatial and temporal dynamics of a mountain snow cover. *Water Resour. Res.* **2014**, *50*, 2533–2550.
64. Feiccabrino, J.; Gustafsson, D.; Lundberg, A. Surface-based precipitation phase determination methods in hydrological models. *Hydrol. Research* **2013**, *44*, 44–57.

65. Schneebeli, M.; Coleou, C.; Touvier, F.; Lesaffre, B. Measurement of density and wetness in snow using time-domain reflectometry. *Ann. Glaciol.* **1998**, *26*, 69–72.
66. Roth, K.; Schulin, R.; Fluhler, H.; Attinger, W. Calibration of time domain reflectometry for water-content measurement using a composite dielectric approach. *Water Resour. Res.* **1990**, *26*, 2267–2273.
67. Watanabe, K.; Wake, T. Measurement of unfrozen water content and relative permittivity of frozen unsaturated soil using NMR and TDR. *Cold Reg. Sci. Technol.* **2009**, *59*, 34–41.
68. Lundberg, A. Laboratory calibration of TDR-probes for snow wetness measurements. *Cold Reg. Sci. Technol.* **1997**, *25*, 197–205.
69. Egli, L.; Jonas, T.; Grunewald, T.; Schirmer, M.; Burlando, P. Dynamics of snow ablation in a small alpine catchment observed by repeated terrestrial laser scans. *Hydrol. Process.* **2012**, *26*, 1574–1585.
70. Grunewald, T.; Schirmer, M.; Mott, R.; Lehning, M. Spatial and temporal variability of snow depth and ablation rates in a small mountain catchment. *Cryosphere* **2010**, *4*, 215–225.
71. Prokop, A. Assessing the applicability of terrestrial laser scanning for spatial snow depth measurements. *Cold Reg. Sci. Technol.* **2008**, *54*, 155–163.
72. Revuelto, J.; López-Moreno, J.I.; Azorin-Molina, C.; Vicente-Serrano, S.M. Canopy influence on snow depth distribution in a pine stand determined from terrestrial laser data. *Water Resour. Res.* **2015**, *51*, 3476–3489.
73. Kværner, J.; Eklo, O.M.; Solbakken, E.; Solberg, I.; Sorknes, S. An integrated approach for assessing influence of agricultural activities on pesticides in a shallow aquifer in south-eastern Norway. *Sci. Total Environ.* **2014**, *499*, 520–532.
74. Marchand, W.D.; Bruland, O.; Killingtveit, A. Improved measurements and analysis of spatial snow cover by combining a ground based radar system with a differential global positioning system receiver. *Nord. Hydrol.* **2001**, *32*, 181–194.
75. Richardson, C.; Holmlund, P. Spatial variability at shallow snow-layer depths in central Dronning Maud Land, East Antarctica. *Ann. Glaciol.* **1999**, *29*, 10–16.
76. McGrath, D.; Sass, L.; O'Neel, S.; Arendt, A.; Wolken, G.; Gusmeroli, A.; Kienholz, C.; McNeil, C. End-of-winter snow depth variability on glaciers in Alaska. *J. Geophys. Res. Earth Surf.* **2015**, *120*, 1530–1550.
77. Previati, M.; Godio, A.; Ferraris, S. Validation of spatial variability of snowpack thickness and density obtained with GPR and TDR methods. *J. Appl. Geophys.* **2011**, *75*, 284–293.
78. Singh, K.K.; Datt, P.; Sharma, V.; Ganju, A.; Mishra, V.D.; Parashar, A.; Chauhan, R. Snow depth and snow layer interface estimation using ground penetrating radar. *Curr. Sci.* **2011**, *100*, 1532–1539.
79. Forte, E.; Dossi, M.; Colucci, R.R.; Pipan, M. A new fast methodology to estimate the density of frozen materials by means of common offset GPR data. *J. Appl. Geophys.* **2013**, *99*, 135–145.
80. Sundström, N.; Gustafsson, D.; Kruglyak, A.; Lundberg, A. Field evaluation of a new method for estimation of liquid water content and snow water equivalent of wet snowpacks with GPR. *Hydrol. Res.* **2013**, *44*, 600–613.
81. Heilig, A.; Schneebeli, M.; Eisen, O. Upward-looking ground-penetrating radar for monitoring snowpack stratigraphy. *Cold Reg. Sci. Technol.* **2009**, *59*, 152–162.
82. Harper, J.T.; Bradford, J.H. Snow stratigraphy over a uniform depositional surface: Spatial variability and measurement tools. *Cold Reg. Sci. Technol.* **2003**, *37*, 289–298.
83. Mitterer, C.; Heilig, A.; Schweizer, J.; Eisen, O. Upward-looking ground-penetrating radar for measuring wet-snow properties. *Cold Reg. Sci. Technol.* **2011**, *69*, 129–138.
84. Ma, Y.; Zhang, Y.; Zubrzycki, S.; Guo, Y.; Farhan, S.B. Hillslope-scale variability in seasonal frost depth and soil water content investigated by GPR on the southern margin of the sporadic permafrost zone on the Tibetan plateau. *Permafr. Periglac. Process.* **2015**, *26*, 321–334.
85. Steelman, C.M.; Endres, A.L. Evolution of high-frequency ground-penetrating radar direct ground wave propagation during thin frozen soil layer development. *Cold Reg. Sci. Technol.* **2009**, *57*, 116–122.

86. Steelman, C.M.; Endres, A.L.; van der Kruk, J. Field observations of shallow freeze and thaw processes using high-frequency ground-penetrating radar. *Hydrol. Process.* **2010**, *24*, 2022–2033.
87. Butnor, J.R.; Campbell, J.L.; Shanley, J.B.; Zarnoch, S.J. Measuring soil frost depth in forest ecosystems with ground penetrating radar. *Agric. For. Meteorol.* **2014**, *192*, 121–131.
88. Westermann, S.; Wollschlager, U.; Boike, J. Monitoring of active layer dynamics at a permafrost site on svalbard using multi-channel ground-penetrating radar. *Cryosphere* **2010**, *4*, 475–487.
89. Wollschlager, U.; Gerhards, H.; Yu, Q.; Roth, K. Multi-channel ground-penetrating radar to explore spatial variations in thaw depth and moisture content in the active layer of a permafrost site. *Cryosphere* **2010**, *4*, 269–283.
90. Pan, X.; You, Y.; Roth, K.; Guo, L.; Wang, X.; Yu, Q. Mapping permafrost features that influence the hydrological processes of a thermokarst lake on the Qinghai-Tibet plateau, China. *Permafr. Periglac. Process.* **2014**, *25*, 60–68.
91. Jadoon, K.Z.; Weihermller, L.; McCabe, M.F.; Moghadas, D.; Vereecken, H.; Lambot, S. Temporal monitoring of the soil freeze-thaw cycles over a snow-covered surface by using air-launched ground-penetrating radar. *Remote Sens.***2015**, *7*, 12041–12056.
92. Brosten, T.R.; Bradford, J.H.; McNamara, J.P.; Gooseff, M.N.; Zarnetske, J.P.; Bowden, W.B.; Johnston, M.E. Estimating 3D variation in active-layer thickness beneath arctic streams using ground-penetrating radar. *J. Hydrol.***2009**, *373*, 479–486.
93. Huisman, J.; Hubbard, S.; Redman, J.; Annan, A. Measuring soil water content with ground penetrating radar. *Vadose Zone J.* **2003**, *2*, 476–491.
94. Ardekani, M.R.M. Off- and on-ground GPR techniques for field-scale soil moisture mapping. *Geoderma* **2013**, *200*, 55–66.
95. Doolittle, J.A.; Jenkinson, B.; Hopkins, D.; Ulmer, M.; Tuttle, W. Hydropedological investigations with ground-penetrating radar (GPR): Estimating water-table depths and local ground-water flow pattern in areas of coarse-textured soils. *Geoderma* **2006**, *131*, 317–329.
96. Hengari, G.M.; Hall, C.R.; Kozusko, T.J.; Bostater, C.R. Use of ground penetrating radar for determination of water table depth and subsurface soil characteristics at kennedy space center. In Proceedings of the Earth resources and Environmental Remote Sensing/GIS Applications IV, Dresden, Germany, 23 September 2013; Volume 8893.
97. Marshall, H.-P.; Schneebeli, M.; Koh, G. Snow stratigraphy measurements with high-frequency FMCW radar: Comparison with snow micro-penetrometer. *Cold Reg. Sci. Technol.* **2007**, *47*, 108–117.
98. Yankielun, N.; Rosenthal, W.; Davis, R.E. Alpine snow depth measurements from aerial FMCW radar. *Cold Reg. Sci. Technol.* **2004**, *40*, 123–134.
99. Koh, G.; Yankielun, N.E.; Baptista, A.I. Snow cover characterization using multiband fmcw radars. *Hydrol. Process.***1996**, *10*, 1609–1617.
100. Richardson, C. *Spatial Distribution of Snow in Antarctica and Other Glaciers Studies Using Ground-Penetrating-Radar*; Stockholm University: Stockholm, Sweden, 2001.
101. Iizuka, K.; Freundorfer, A.P.; Wu, K.H.; Mori, H.; Ogura, H.; Nguyen, V.K. Step-frequency radar. *J. Appl. Phys.* **1984**, *56*, 2572–2583.
102. Minet, J.; Wahyudi, A.; Bogaert, P.; Vanclooster, M.; Lambot, S. Mapping shallow soil moisture profiles at the field scale using full-waveform inversion of ground penetrating radar data. *Geoderma* **2011**, *161*, 225–237.
103. Lundberg, A.; Thunehed, H. Snow wetness influence on impulse radar snow surveys theoretical and laboratory study. *Nord. Hydrol.* **2000**, *31*, 89–106.
104. Sundström, N.; Kruglyak, A.; Friborg, J. Modeling and simulation of GPR wave propagation through wet snowpacks: Testing the sensitivity of a method for snow water equivalent estimation. *Cold Reg. Sci. Technol.* **2012**, *74–75*, 11–20.
105. Heilig, A.; Eisen, O.; Schneebeli, M. Temporal observations of a seasonal snowpack using upward-looking GPR. *Hydrol. Process.* **2010**, *24*, 3133–3145.
106. Schmid, L.; Heilig, A.; Mitterer, C.; Schweizer, J.; Maurer, H.; Okorn, R.; Eisen, O. Continuous snowpack monitoring using upward-looking ground-penetrating radar technology. *J. Glaciol.* **2014**, *60*, 509–525.

107. Okorn, R.; Brunnhofer, G.; Platzer, T.; Heilig, A.; Schmid, L.; Mitterer, C.; Schweizer, J.; Eisen, O. Upward-looking l-band FMCW radar for snow cover monitoring. *Cold Reg. Sci. Technol.* **2014**, *103*, 31–40.
108. He, H.; Dyck, M.F.; Si, B.C.; Zhang, T.; Lv, J.; Wang, J. Soil freezing-thawing characteristics and snowmelt infiltration in cryalfs of Alberta, Canada. *Geoderma Reg.* **2015**, *5*, 198–208.
109. Oberdoerster, C.; Vanderborght, J.; Kemna, A.; Vereecken, H. Investigating preferential flow processes in a forest soil using time domain reflectometry and electrical resistivity tomography. *Vadose Zone J.* **2010**, *9*, 350–361.
110. Murray, C.D.; Buttle, J.M. Infiltration and soil water mixing on forested and harvested slopes during spring snowmelt, turkey lakes watershed, central Ontario. *J. Hydrol.* **2005**, *306*, 1–20.
111. Kahimba, F.; Ranjan, R.S. Soil temperature correction of field TDR readings obtained under near freezing conditions. *Can. Biosyst. Eng.* **2007**, *49*, 19–25.
112. Stähli, M.; Stadler, D. Measurement of water and solute dynamics in freezing soil columns with time domain reflectometry. *J. Hydrol.* **1997**, *195*, 352–369.
113. Christensen, A.F.; He, H.L.; Dyck, M.F.; Turner, E.L.; Chanasyk, D.S.; Naeth, M.A.; Nichol, C. In situ measurement of snowmelt infiltration under various topsoil cap thicknesses on a reclaimed site. *Can. J. Soil Sci.* **2013**, *93*, 497–510.
114. Schneebeli, M.; Johnson, J.B. A constant-speed penetrometer for high-resolution snow stratigraphy. *Ann. Glaciol.* **1998**, *26*, 107–111.
115. Waldner, P.A.; Schneebeli, M.; Schultze-Zimmerman, U.; Fluhler, H. Effect of snow structure on water flow and solute transport. *Hydrol. Process.* **2004**, *18*, 1271–1290.
116. Lekshmi, S.S.U.; Singh, D.N.; Baghini, M.S. A critical review of soil moisture measurement. *Measurement* **2014**, *54*, 92–105.
117. He, H.; Dyck, M. Application of multiphase dielectric mixing models for understanding the effective dielectric permittivity of frozen soils. *Vadose Zone J.* **2013**, *12*.
118. Iwata, Y.; Hayashi, M.; Hirota, T. Comparison of snowmelt infiltration under different soil-freezing conditions influenced by snow cover. *Vadose Zone J.* **2008**, *7*, 79–86.
119. Tian, Z.; Heitman, J.; Horton, R.; Ren, T. Determining soil ice contents during freezing and thawing with thermo-time domain reflectometry. *Vadose Zone J.* **2015**, *14*.
120. Gnecchi, J.A.G.; Patino, A.M.; Paramo, F.L.; Anguiano, A.D.T.; Espinoza, D.L. Investigation of wetting front propagation dynamics using soil impedance measurements: Implications for modelling and irrigation scheduling. *Water Resour. Manag.* **2015**, *29*, 197–210.
121. Gnecchi, J.G.; Chávez, A.G.-T.; Campos, G.C.; Peregrino, V.O.; Pineda, E.M. Soil water infiltration measurements using electrical impedance tomography. *Chem. Eng. J.* **2012**, *191*, 13–21.
122. French, H.; Binley, A. Snowmelt infiltration: Monitoring temporal and spatial variability using time-lapse electrical resistivity. *J. Hydrol.* **2004**, *297*, 174–186.
123. Berthold, S.; Bentley, L.R.; Hayashi, M. Integrated hydrogeological and geophysical study of depression-focused groundwater recharge in the Canadian prairies. *Water Resour. Res.* **2004**, *40*, W06505.
124. French, H.K.; Hardbattle, C.; Binley, A.; Winship, P.; Jakobsen, L. Monitoring snowmelt induced unsaturated flow and transport using electrical resistivity tomography. *J. Hydrol.* **2002**, *267*, 273–284.
125. Watanabe, K.; Mizoguchi, M. Amount of unfrozen water in frozen porous media saturated with solution. *Cold Reg. Sci. Technol.* **2002**, *34*, 103–110.
126. Zhou, X.; Zhou, J.; Kinzelbach, W.; Stauffer, F. Simultaneous measurement of unfrozen water content and ice content in frozen soil using gamma ray attenuation and TDR. *Water Resour. Res.* **2014**, *50*, 9630–9655.
127. Boleve, A.; Revil, A.; Janod, F.; Mattiuzzo, J.L.; Fry, J.J. Preferential fluid flow pathways in embankment dams imaged by self-potential tomography. *Near Surf. Geophys.* **2009**, *7*, 447–462.
128. Ikard, S.J.; Revil, A.; Jardani, A.; Woodruff, W.F.; Parekh, M.; Mooney, M. Saline pulse test monitoring with the self-potential method to nonintrusively determine the velocity of the pore water in leaking areas of earth dams and embankments. *Water Resour. Res.* **2012**, *48*, W04201.

129. Panthulu, T.V.; Krishnaiah, C.; Shirke, J.M. Detection of seepage paths in earth dams using self-potential and electrical resistivity methods. *Eng. Geol.* **2001**, *59*, 281–295.
130. Kulessa, B.; Chandler, D.; Revil, A.; Essery, R. Theory and numerical modeling of electrical self-potential signatures of unsaturated flow in melting snow. *Water Resour. Res.* **2012**, *48*, W09511.
131. Thompson, S.; Kulessa, B.; Essery, R.; Lüthi, M. The electrical self-potential method is a non-intrusive snow-hydrological sensor. *Cryosphere Discuss.* **2015**, *9*, 4437–4457.
132. Tyler, S.W.; Burak, S.A.; McNamara, J.P.; Lamontagne, A.; Selker, J.S.; Dozier, J. Spatially distributed temperatures at the base of two mountain snowpacks measured with fiber-optic sensors. *J. Glaciol.* **2008**, *54*, 673–679.
133. Reusser, D.E.; Zehe, E. Low-cost monitoring of snow height and thermal properties with inexpensive temperature sensors. *Hydrol. Process.* **2011**, *25*, 1841–1852.
134. Lundquist, J.D.; Lott, F. Using inexpensive temperature sensors to monitor the duration and heterogeneity of snow-covered areas. *Water Resour. Res.* **2008**, *44*, W00D16.
135. Schmid, M.O.; Gubler, S.; Fiddes, J.; Gruber, S. Inferring snowpack ripening and melt-out from distributed measurements of near-surface ground temperatures. *Cryosphere* **2012**, *6*, 1127–1139.
136. Roznik, E.A.; Alford, R.A. Does waterproofing thermochron ibutton dataloggers influence temperature readings? *J. Therm. Biol.* **2012**, *37*, 260–264.
137. Shea, C.; Jamieson, B.; Birkeland, K.W. Use of a thermal imager for snow pit temperatures. *Cryosphere* **2012**, *6*, 287–299.
138. Schirmer, M.; Jamieson, B. Limitations of using a thermal imager for snow pit temperatures. *Cryosphere* **2014**, *8*, 387–394.
139. Tetzlaff, D.; Buttle, J.; Carey, S.K.; McGuire, K.; Laudon, H.; Soulsby, C. Tracer-based assessment of flow paths, storage and runoff generation in northern catchments: A review. *Hydrol. Process.* **2015**, *29*, 3475–3490.
140. Bründl, M.; Schneebeli, M.; Flühler, H. Routing of canopy drip in the snowpack below a spruce crown. *Hydrol. Process.* **1999**, *13*, 49–58.
141. Campbell, F.M.A.; Nienow, P.W.; Purves, R.S. Role of the supraglacial snowpack in mediating meltwater delivery to the glacier system as inferred from dye tracer investigations. *Hydrol. Process.* **2006**, *20*, 969–985.
142. Lee, J.; Nez, V.E.; Feng, X.; Kirchner, J.W.; Osterhuber, R.; Renshaw, C.E. A study of solute redistribution and transport in seasonal snowpack using natural and artificial tracers. *J. Hydrol.* **2008**, *357*, 243–254.
143. Buttle, J.M.; Sami, K. Recharge processes during snowmelt—An isotopic and hydrometric investigation. *Hydrol. Process.* **1990**, *4*, 343–360.
144. Stähli, M.; Bayard, D.; Wydler, H.; Flühler, H. Snowmelt infiltration into alpine soils visualized by dye tracer technique. *Arct. Antarct. Alp. Res.* **2004**, *36*, 128–135.
145. Stumpp, C.; Hendry, M.J. Spatial and temporal dynamics of water flow and solute transport in a heterogeneous glacial till: The application of high-resolution profiles of $\delta^{18}O$ and $\delta^{2}H$ in pore waters. *J. Hydrol.* **2012**, *438*, 203–214.
146. Dietermann, N.; Weiler, M. Spatial distribution of stable water isotopes in alpine snow cover. *Hydrol. Earth Syst. Sci.***2013**, *17*, 2657–2668.
147. Bengtsson, L.; Lepistö, A.; Saxena, R.K.; Seuna, P. Mixing of meltwater and groundwater in a forested basin. *Aqua Fenn.* **1991**, *21*, 3–12.
148. Mueller, M.H.; Alaoui, A.; Külls, C.; Leistert, H.; Meusburger, K.; Stumpp, C.; Weiler, M.; Alewell, C. Tracking water pathways in steep hillslopes by $\delta^{18}O$ soil depth profiles of soil water. *J. Hydrol.* **2014**, *519*, 340–352.
149. Rodhe, A. Snowmelt dominated systems. *Isot. Tracers Catchment Hydrol.* **1998**, *21*, 391–433.
150. Clark, I.; Fritz, P. *Environmental Isotopes in Hydrogeology*; Lewis Publishers: Boca Raton, NY, USA, 1997.
151. Koeniger, P.; Hubbart, J.A.; Link, T.; Marshall, J.D. Isotopic variation of snow cover and streamflow in response to changes in canopy structure in a snow-dominated mountain catchment. *Hydrol. Process.* **2008**, *22*, 557–566.

152. Herrmann, A.; Lehrer, M.; Stichler, W. Isotope input into runoff systems from melting snow covers. *Nord. Hydrol.* **1981**, *12*, 309–318.
153. Lee, J.; Feng, X.; Faiia, A.M.; Posmentier, E.S.; Kirchner, J.W.; Osterhuber, R.; Taylor, S. Isotopic evolution of a seasonal snowcover and its melt by isotopic exchange between liquid water and ice. *Chem. Geol.* **2010**, *270*, 126–134.
154. Cooper, L. Isotopic Fractionation in Snow Cover. In *Isotope Tracers in Catchment Hydrology*; Elsevier Science B.V.: Amsterdam, The Netherlands, 1998.
155. Penna, D.; Ahmad, M.; Birks, S.J.; Bouchaou, L.; Brencic, M.; Butt, S.; Holko, L.; Jeelani, G.; Martinez, D.E.; Melikadze, G.; et al. A new method of snowmelt sampling for water stable isotopes. *Hydrol. Process.* **2014**, *28*, 5637–5644.
156. Garrett, T.J.; Yuter, S. Observed influence of riming, temperature, and turbulence on the fallspeed of solid precipitation. *Geophys. Res. Lett.* **2014**, *41*, 6515–6522.
157. Ma, Y.; Zhang, Y.; Yang, D.; Farhan, S.B. Precipitation bias variability *versus* various gauges under different climatic conditions over the Third Pole Environment (TPE) region. *Int. J. Climatol.* **2014**, *26*, 321–334.
158. Nuti, R.; Wong, K. *Instruments and Observing Methods, Cimo Survey on National Summaries of Methods and Instruments for Solid Precipitation Measurement at Aws*; World Meteorological Organization: Geneva, Switzerland, 2010; p. 57.
159. Young, K.; Bolton, R.; Killingtveit, A.; Yang, D. Assessment of precipitation and snowcover in northern research basins. *Nord. Hydrol.* **2006**, *37*, 377–391.
160. WMO. Spice: Wmo Solid Precipitation Intercomparison Experiment (2012–2015). Available online: http://www.wmo.int/pages/prog/www/IMOP/intercomparisons/SPICE/SPICE.html (accessed on 20 Februay 2015).
161. Boudala, F.S.; Isaac, G.A.; Rasmussen, R.; Cober, S.G.; Scott, B. Comparisons of snowfall measurements in complex terrain made during the 2010 winter Olympics in Vancouver. *Pure Appl. Geophys.* **2014**, *171*, 113–127.
162. Boudala, F.S.; Rasmussen, R.; Isaac, G.A.; Scott, B. Performance of hot plate for measuring solid precipitation in complex terrain during the 2010 Vancouver winter Olympics. *J. Atmos. Ocean. Technol.* **2014**, *31*, 437–446.
163. Lee, J.-E.; Jung, S.-H.; Park, H.-M.; Kwon, S.; Lin, P.-L.; Lee, G. Classification of precipitation types using fall velocity-diameter relationships from 2D-video distrometer measurements. *Adv. Atmos. Sci.* **2015**, *32*, 1277–1290.
164. Battaglia, A.; Rustemeier, E.; Tokay, A.; Blahak, U.; Simmer, C. Parsivel snow observations: A critical assessment. *J. Atmos. Ocean. Tech.* **2010**, *27*, 333–344.
165. Yang, D. Double fence intercomparison reference (DFIR) vs. Bush gauge for "true" snowfall measurement. *J. Hydrol.* **2014**, *509*, 94–100.
166. Yang, D.; Simonenko, A. Comparison of winter precipitation measurements by six tretyakov gauges at the valdai experimental site. *Atmos. Ocean* **2014**, *52*, 39–53.
167. Grazioli, J.; Tuia, D.; Monhart, S.; Schneebeli, M.; Raupach, T.; Berne, A. Hydrometeor classification from two-dimensional video disdrometer data. *Atmos. Meas. Tech.* **2014**, *7*, 2869–2882.
168. Zhang, L.; Zhao, L.; Xie, C.; Liu, G.; Gao, L.; Xiao, Y.; Shi, J.; Qiao, Y. Intercomparison of solid precipitation derived from the weighting rain gauge and optical instruments in the interior Qinghai-Tibetan Plateau. *Adv. Meteorol.* **2015**.
169. Rasmussen, R.M.; Hallett, J.; Purcell, R.; Landolt, S.D.; Cole, J. The hotplate precipitation gauge. *J. Atmos. Ocean. Technol.* **2011**, *28*, 148–164.
170. Theriault, J.M.; Rasmussen, R.; Ikeda, K.; Landolt, S. Dependence of snow gauge collection efficiency on snowflake characteristics. *J. Appl. Meteorol. Climatol.* **2012**, *51*, 745–762.
171. Theriault, J.M.; Rasmussen, R.; Petro, E.; Trepanier, J.-Y.; Colli, M.; Lanza, L.G. Impact of wind direction, wind speed, and particle characteristics on the collection efficiency of the double fence intercomparison reference. *J. Appl. Meteorol. Climatol.* **2015**, *54*, 1918–1930.
172. Rasmussen, R.; Baker, B.; Kochendorfer, J.; Meyers, T.; Landolt, S.; Fischer, A.; Black, J.; Thériault, J.; Kucera, P.; Gochis, D.; et al. How well are we measuring snow; the NOAA/FAA/NCAR winter precipitation test bed. *Bull. Am. Meteorol. Soc.* **2010**, *936*, 811–829.

173. Wolff, M.; Isaksen, K.; Braekkan, R.; Alfnes, E.; Petersen-Overleir, A.; Ruud, E. Measurements of wind-induced loss of solid precipitation: Description of a norwegian field study. *Hydrol. Res.* **2013**, *44*, 35–43.
174. Wolff, M.A.; Isaksen, K.; Petersen-Overleir, A.; Odemark, K.; Reitan, T.; Braekkan, R. Derivation of a new continuous adjustment function for correcting wind-induced loss of solid precipitation: Results of a norwegian field study. *Hydrol. Earth Syst. Sci.* **2015**, *19*, 951–967.
175. Yuter, S.E.; Kingsmill, D.E.; Nance, L.B.; Loeffler-Mang, M. Observations of precipitation size and fall speed characteristics within coexisting rain and wet snow. *J. Appl. Meteorol. Climatol.* **2006**, *45*, 1450–1464.
176. Naaim-Bouvet, F.; Bellot, H.; Nishimura, K.; Genthon, C.; Palerme, C.; Guyomarc'h, G.; Vionnet, V. Detection of snowfall occurrence during blowing snow events using photoelectric sensors. *Cold Reg. Sci. Technol.* **2014**, *106*, 11–21.
177. Colli, M.; Lanza, L.G.; Rasmussen, R.; Thériault, J.M. Tthe collection efficiency of shielded and unshielded precipitation gauges- part II: Modeling particle trajectories. *J. Hydrometeorol.* **2015**, 231–243.
178. Kane, D.L.; Stuefer, S.L. Reflecting on the status of precipitation data collection in Alaska: A case study. *Hydrol. Res.***2015**, *46*, 478–493.
179. Grazioli, J.; Tuia, D.; Monhart, S.; Schneebeli, M.; Raupach, T.; Berne, A. Hydrometeor classification from two-dimensional video disdrometer data. *Atmos. Meas. Tech.* **2014**, *7*, 2869–2882.
180. Serreze, M.C.; Clark, M.P.; Armstrong, R.L.; McGinnis, D.A.; Pulwarty, R.S. Characteristics of the western United States snowpack from snowpack telemetry (snotel) data. *Water Resour. Res.* **1999**, *35*, 2145–2160.
181. Brazenec, W.A. *Evaluation of Ultrasonic Snow Depth Sensors for Automated Surface Observing Systems (Asos)*; Colorado State University: Fort Collins, CO, USA, 2005.
182. Schöner, W.; Auer, I.; Böhm, R. Long term trend of snow depth at sonnblick (austrian alps) and its relation to climate change. *Hydrol. Process.* **2009**, *23*, 1052–1063.
183. Garvelmann, J.; Pohl, S.; Weiler, M. From observation to the quantification of snow processes with a time-lapse camera network. *Hydrol. Earth Syst. Sci.* **2013**, *17*, 1415–1429.
184. SnowHydro. Gps Depth Probe Snowhydro. Available online: http://www.snowhydro.com/products/column2.html (accessed on 20 Februay 2015).
185. Boniface, K.; Braun, J.J.; McCreight, J.L.; Nievinski, F.G. Comparison of snow data assimilation system with gps reflectometry snow depth in the western United States. *Hydrol. Process.* **2015**, *29*, 2425–2437.
186. Gutmann, E.D.; Larson, K.M.; Williams, M.W.; Nievinski, F.G.; Zavorotny, V. Snow measurement by GPS interferometric reflectometry: An evaluation at niwot ridge, colorado. *Hydrol. Process.* **2012**, *26*, 2951–2961.
187. Larson, K.M.; Gutmann, E.D.; Zavorotny, V.U.; Braun, J.J.; Williams, M.W.; Nievinski, F.G. Can we measure snow depth with GPS receivers? *Geophys. Res. Lett.* **2009**, *36*, L17502.
188. McCreight, J.L.; Small, E.E.; Larson, K.M. Snow depth, density, and swe estimates derived from GPS reflection data: Validation in the western U.S. *Water Resour. Res.* **2014**, *50*, 6892–6909.
189. Lundberg, A.; Richardson-Näslund, C.; Andersson, C. Snow density variations: Consequences for ground-penetrating radar. *Hydrol. Process.* **2006**, *20*, 1483–1495.
190. Sturm, M.; Taras, B.; Liston, G.E.; Derksen, C.; Jonas, T.; Lea, J. Estimating snow water equivalent using snow depth data and climate classes. *J. Hydrometeorol.* **2010**, *11*, 1380–1394.
191. Jonas, T.; Marty, C.; Magnusson, J. Estimating the snow water equivalent from snow depth measurements in the Swiss Alps. *J. Hydrol.* **2009**, *378*, 161–167.
192. Bormann, K.J.; Westra, S.; Evans, J.P.; McCabe, M.F. Spatial and temporal variability in seasonal snow density. *J. Hydrol.* **2013**, *484*, 63–73.
193. Zhong, X.; Zhang, T.; Wang, K. Snow density climatology across the former USSR. *Cryosphere* **2014**, *8*, 785–799.

194. Rees, A.; English, M.; Derksen, C.; Toose, P.; Silis, A. Observations of late winter canadian tundra snow cover properties. *Hydrol. Process.* **2014**, *28*, 3962–3977.
195. Proksch, M.; Rutter, N.; Fierz, C.; Schneebeli, M. Intercomparison of snow density measurements: Bias, precision and spatial resolution. *Cryosphere Discuss.* **2015**, *9*, 3581–3616.
196. Hydrometry. *Manual Measurement of Snow Water Equivalent*; European Committee for Standardization: Brussels, Belgium, 2014; SIS/TK 432.
197. Dixon, D.; Boon, S. Comparison of the snowhydro snow sampler with existing snow tube designs. *Hydrol. Process.***2012**, *26*, 2555–2562.
198. Johnson, J.B.; Schaefer, G.L. The influence of thermal, hydrologic, and snow deformation mechanisms on snow water equivalent pressure sensor accuracy. *Hydrol. Process.* **2002**, *16*, 3529–3542.
199. Trabant, D.; Clagett, G. Measurement and evaluation of snowpacks. *Cold Reg. Hydrol. Hydraul.* **1990**, 39–93.
200. CEN/TR; 15996. Hydrometry. In *Measurement of Snow Water Equivalent Using Snow Mass Registration Devices*; European Committee for Standardization: Brussels, Belgium, 2010.
201. Johnson, J.B.; Gelvin, A.; Schaefer, G. An engineering design study of electronic snow water equivalent sensor performance. In Proceedings of the 75th Annual Western Snow Conference, Kailua-Kona, HI, USA, 16–19 April 2007; pp. 23–30.
202. Johnson, J.B.; Marks, D. The detection and correction of snow water equivalent pressure sensor errors. *Hydrol. Process.* **2004**, *18*, 3513–3525.
203. Johnson, J.B.; Gelvin, A.B.; Duvoy, P.; Schaefer, G.L.; Poole, G.; Horton, G.D. Performance characteristics of a new electronic snow water equivalent sensor in different climates. *Hydrol. Process.* **2015**, *29*, 1418–1433.
204. Pomeroy, J.; Gray, D. *Snow Cover—Accumulation, Relocation and Management*; National Hydrology Institute NHRI: Saskatoon, Canada, 1995; p. 134.
205. Daniel, J.A.; Staricka, J.A. Frozen soil impact on ground water-surface water interaction. *J. Am. Water Resour. Assoc.***2000**, *36*, 151–160.
206. Dobriyal, P.; Qureshi, A.; Badola, R.; Hussain, S.A. A review of the methods available for estimating soil moisture and its implications for water resource management. *J. Hydrol.* **2012**, *458*, 110–117.
207. Desilets, D.; Zreda, M.; Ferré, T. Nature's neutron probe: Land surface hydrology at an elusive scale with cosmic rays. *Water Resour. Res.* **2010**, *46*, W11505.
208. Choquette, Y.; Ducharme, P.; Rogoza, J. Cs725, an accurate sensor for the snow water equivalent and soil moisture measurements. In Proceedings of the International Snow Science Workshop, Grenoble, France, 7–11 October 2013.
209. Martin, J.P.; Houdayer, A.; Lebel, C.; Choquette, Y.; Lavigne, P.; Ducharme, P. An unattended gamma monitor for the determination of Snow Water Equivalent (SWE) using the natural ground gamma radiation. In Proceedings of the Nuclear Science Symposium Conference, Dresden, Germany, 19–25 October 2008; pp. 983–988.
210. Bache Stranden, H.; Lirhus Ree, B.; Møen, K.M. *Måling av snø ved Hjelp Avgammasensor*; Norges Vassdrags-og Energidirektorat: Oslo, Norway, 2014; p. 46.
211. Kodama, M. Continuous monitoring of snow water equivalent using cosmic-ray neutrons. *Cold Reg. Sci. Technol.* **1980**, *3*, 295–303.
212. Kodama, M.; Nakai, K.; Kawasaki, S.; Wada, M. Application of cosmic-ray neutron measurements to the determination of the snow-water equivalent. *J. Hydrol.* **1979**, *41*, 85–92.
213. Paquet, E.; Laval, M.T. Operation feedback and prospects of edf cosmic-ray snow sensors. *Houille Blanche* **2006**, *2*, 113–119.
214. SnowFox. Swe Probe. Facta Sheet. Available online: http://hydroinnova.com/reso_product.html (accessed on 20 Februay 2015).
215. Kinar, N.; Pomeroy, J. Determining snow water equivalent by acoustic sounding. *Hydrol. Process.* **2007**, *21*, 2623–2640.
216. Kinar, N.J.; Pomeroy, J.W. Measurement of the physical properties of the snowpack. *Rev. Geophys.* **2015**, *53*, 481–544.

217. Niang, M.; Bernier, M.; Stacheder, M.; Brandelik, A.; Van Bochove, E. Influence of snow temperature interpolation algorithm and dielectric mixing-model coefficient on density and liquid water content determination in a cold seasonal snow pack. *Subsurf. Sens. Technol. Appl.* **2006**, *7*, 1–22.
218. Stähli, M.; Stacheder, M.; Gustafsson, D.; Schlaeger, S.; Schneebeli, M.; Brandelik, A. A new *in situ* sensor for large-scale snow-cover monitoring. *Ann. Glaciol.* **2004**, *38*, 273–278.
219. Schlaeger, S. A fast tdr-inversion technique for the reconstruction of spatial soil moisture content. *Hydrol. Earth Syst. Sci. Discuss.* **2005**, *9*, 481–492.
220. Heggli, A. Data analysis of the snow pack analyzing system tested by UCAR. In Proceedings of the 81st Annual Western Snow Conference, Jackson Hole, WY, USA, 15–18 April 2013.
221. Gustafsson, D.; Sommer, W.; Ahlberg, J. Measurement of liquid water content in snow and its application in snow hydrological modelling. In Proceedings of the XXVI Nordic Hydrological Conference: Hydrology: From Research to Water Management, Riga, Latvia, 9–11 August 2010; p. 34.
222. Eiriksson, D.; Whitson, M.; Luce, C.H.; Marshall, H.P.; Bradford, J.; Benner, S.G.; Black, T.; Hetrick, H.; McNamara, J.P. An evaluation of the hydrologic relevance of lateral flow in snow at hillslope and catchment scales. *Hydrol. Process.* **2013**, *27*, 640–654.
223. Williams, M.W.; Seibold, C.; Chowanski, K. Storage and release of solutes from a subalpine seasonal snowpack: Soil and stream water response, niwot ridge, Colorado. *Biogeochemistry* **2009**, *95*, 77–94.
224. Schneebeli, M.; Pielmeier, C.; Johnson, J.B. Measuring snow microstructure and hardness using a high resolution penetrometer. *Cold Reg. Sci. Technol.* **1999**, *30*, 101–114.
225. Williams, M.W.; Erickson, T.A.; Petrzelka, J.L. Visualizing meltwater flow through snow at the centimetre-to-metre scale using a snow guillotine. *Hydrol. Process.* **2010**, *24*, 2098–2110.
226. Gustafsson, D.; Waldner, P.A.; Stähli, M. Factors governing the formation and persistence of layers in a subalpine snowpack. *Hydrol. Process.* **2004**, *18*, 1165–1183.
227. Semadeni-Davies, A.; Lundberg, A.; Bengtsson, L. Radiation balance of urban snow: A water management perspective. *Cold Reg. Sci. Technol.* **2001**, *33*, 59–76.
228. Sihvola, A.; Tiuri, M. Snow fork for field determination of the density and wetness profiles of a snowpack. *IEEE Trans. Geosci. Remote Rens.* **1986**, *Ge-24*, 717–721.
229. Kendra, J.R.; Ulaby, F.T.; Sarabandi, K. Snow probe for in situ determination of wetness and density. *IEEE Trans. Geosci. Remote Rens.* **1994**, *32*, 1152–1159.
230. Denoth, A. An electronic device for long-term snow wetness recording. *Annales of Glaciology* **1994**, *19*, 104–106.
231. Denoth, A. Monopole-antenna: A practical snow and soil wetness sensor. *IEEE Trans. Geosci. Remote Rens.* **1997**, *35*, 1371–1375.
232. Williams, M.W.; Rikkers, M.F.; Pfeffer, T.; Sommerfeld, R. Comparison of snow liquid water measurements with a denoth meter and finnish snow-fork. In Proceedings of the AGU Fall Meeting, Niwot Ridge, CO, USA, 14–16 October 1996.
233. Whitaker, A.C.; Sugiyama, H. Seasonal snowpack dynamics and runoff in a cool temperate forest: Lysimeter experiment in Niigata, Japan. *Hydrol. Process.* **2005**, *19*, 4179–4200.
234. French, H.K.; van der Zee, S. Field observations of small scale spatial variability of snowmelt drainage and infiltration. *Nord. Hydrol.* **1999**, *30*, 161–176. [Google Scholar]
235. Laudon, H.; Hemond, H.F.; Krouse, R.; Bishop, K.H. Oxygen 18 fractionation during snowmelt: Implications for spring flood hydrograph separation. *Water Resour. Res.* **2002**, *38*, 1258.
236. Hedstrom, N.R.; Pomeroy, J.W. Measurements and modelling of snow interception in the boreal forest. *Hydrol. Process.* **1998**, *12*, 1611–1625.
237. Nakai, Y. *An Observational Study on Evaporation from Intercepted Snow on Forest Canopies*; Kyoto University: Kyoto, Japan, 1996; (In Japanese, tables and figures in English).
238. Lundberg, A.; Koivusalo, H. Estimating winter evaporation in boreal forests with operational snow course data. *Hydrol. Process.* **2003**, *17*, 1479–1493.

239. Harding, R.J.; Pomeroy, J.W. The energy balance of a winter boreal landscape. *J. Clim.* **1996**, *9*, 2778–2787.
240. Nakai, Y.; Sakamoto, T.; Terajima, T.; Kitamura, K.; Shirai, T. The effect of canopy-snow on the energy balance above a coniferous forest. *Hydrol. Process.* **1999**, *13*, 2371–2382.
241. Claassen, H.C.; Downey, J.S. A model for deuterium and oxygen-18 isotope changes during evergreen interception of snowfall. *Water Resour. Res.* **1995**, *31*, 601–618.
242. Reba, M.L.; Pomeroy, J.; Marks, D.; Link, T.E. Estimating surface sublimation losses from snowpacks in a mountain catchment using eddy covariance and turbulent transfer calculations. *Hydrol. Process.* **2012**, *26*, 3699–3711.
243. Hancock, N.H.; Crowther, J.M. A technique for the direct measurement of water storage on a forest canopy. *J. Hydrol.***1979**, *41*, 105–122.
244. Bründl, M.; Bartelt, P.; Schneebeli, M.; Flühler, H. Measuring branch defection of spruce branches caused by intercepted snow load. *Hydrol. Process.* **1999**, *13*, 2357–2369.
245. Montesi, J.; Elder, K.; Schmidt, R.A.; Davis, R.E. Sublimation of intercepted snow within a subalpine forest canopy at two elevations. *J. Hydrometeorol.* **2004**, *5*, 763–773.
246. Martin, K.A.; Van Stan, J.T.; Dickerson-Lange, S.E.; Lutz, J.A.; Berman, J.W.; Gersonde, R.; Lundquist, J.D. Development and testing of a snow interceptometer to quantify canopy water storage and interception processes in the rain/snow transition zone of the north cascades, Washington, USA. *Water Resour. Res.* **2013**, *49*, 3243–3256.
247. Lundberg, A.; Calder, I.; Harding, R. Evaporation of intercepted snow: Measurement and modelling. *J. Hydrol.* **1998**, *206*, 151–163.
248. Storck, P.; Lettenmaier, D.P.; Bolton, S.M. Measurement of snow interception and canopy effects on snow accumulation and melt in a mountainous maritime climate, Oregon, United States. *Water Resour. Res.* **2002**, *38*, 1223.
249. Cassiani, G.; Giustiniani, M.; Ferraris, S.; Deiana, R.; Strobbia, C. Time-lapse surface-to-surface GPR measurements to monitor a controlled infiltration experiment. *Boll. Geofis. Teor. Appl.* **2009**, *50*, 209–226.
250. Sun, Y.; Cheng, Q.; Xue, X.; Fu, L.; Chai, J.; Meng, F.; Lammers, P.S.; Jones, S.B. Determining *in-situ* soil freeze-thaw cycle dynamics using an access tube-based dielectric sensor. *Geoderma* **2012**, *189*, 321–327.
251. Yoshikawa, K.; Overduin, P.P. Comparing unfrozen water content measurements of frozen soil using recently developed commercial sensors. *Cold Reg. Sci. Technol.* **2005**, *42*, 250–256.
252. Wu, Y.; Hubbard, S.S.; Ulrich, C.; Wullschleger, S.D. Remote monitoring of freeze-thaw transitions in arctic soils using the complex resistivity method. *Vadose Zone J.* **2013**, *12*.
253. Yi, J.; Zhao, Y.; Shao, M.a.; Zhang, J.; Cui, L.; Si, B. Soil freezing and thawing processes affected by the different landscapes in the middle reaches of Heihe river basin, Gansu, China. *J. Hydrol.* **2014**, *519*, 1328–1338.
254. Hayhoe, H.N.; Bailey, W.G. Monitoring changes in total and unfrozen water-content in seasonally frozen soil using time domain reflectometry and neutron moderation techniques. *Water Resour. Res.* **1985**, *21*, 1077–1084.
255. Hirota, T.; Iwata, Y.; Hayashi, M.; Suzuki, S.; Hamasaki, T.; Sameshima, R.; Takayabu, I. Decreasing soil-frost depth and its relation to climate change in Tokachi, Hokkaido, Japan. *J. Meteorol. Soc. Jpn.* **2006**, *84*, 821–833.
256. Iwata, Y.; Hirota, T. Monitoring over-winter soil water dynamics in a freezing and snow-covered environment using a thermally insulated tensiometer. *Hydrol. Process.* **2005**, *19*, 3013–3019.
257. Rosenberry, D.O.; LaBaugh, J.W.; Hunt, R.J. Use of monitoring wells, portable piezometers, and seepage meters to quantify flow between surface water and ground water. *Field Techniques for Estimating Water Fluxes between Surface Water and Ground Water*; U.S. Geological Survey: Reston, VA, USA, 2008. 4-D2.
258. Cheng, Q.; Sun, Y.; Xue, X.; Guo, J. *In situ* determination of soil freezing characteristics for estimation of soil moisture characteristics using a dielectric tube sensor. *Soil Sci. Soc. Am. J.* **2014**, *78*, 133–138.

259. Sorteberg, H.; Engeset, R.; Udnaes, H. A national network for snow monitoring in Norway: Snow pillow verification using observations and models. *Phys. Chem. Earth* **2001**, *26*, 723–729.
260. Johnson, J.B. A theory of pressure sensor performance in snow. *Hydrol. Process.* **2004**, *18*, 53–64.
261. Niang, M.; Bernier, M.; Van Bochove, E.; Durand, Y. Precision of estimated snow water equivalent (swe) derived from the new sensor snowpower in Quebec (Canada). *Houille Blanche* **2006**, 128–133.

CHAPTER 14

A Monte Carlo-Based Approach for Groundwater Chemistry Inverse Modeling

Walt W. McNab Jr.

Kennedy/Jenks Consultants, San Francisco, USA

ABSTRACT

Inverse geochemical modeling of groundwater entails identifying a set of geochemical reactions which can explain observed changes in water chemistry between two samples that are spatially related in some sense, such as two points along a flow pathway. A common inversion approach is to solve a set of simultaneous mass and electron balance equations involving water-rock and oxidation-reduction reactions that are consistent with the changes in concentrations of various aqueous components. However, this mass-balance approach does not test the thermodynamic favorability of the resulting model and provides limited insight into the model uncertainties. In this context, a Monte Carlo-based forward-inverse modeling method is proposed that generates probability distributions for model parameters which best match the observed data using the Metropolis-Hastings search strategy. The forward model is based on the well-vetted PHREEQC geochemical model. The proposed modeling approach is applied to two test applications, one involving an inverse modeling example supplied with the PHREEQC code that entails groundwater interactions with a granitic rock mineral assemblage, and the other concerning the impact of fuel hydrocarbon bioattenuation on groundwater chemistry. In both examples, the forward-inverse approach is able to approximately reproduce observed water quality changes invoking mass transfer reactions that are all thermodynamically favorable.

Keywords: Inverse Model, Geochemical Model, Monte Carlo Simulation

1. INTRODUCTION

Inverse process models attempt to estimate model parameter values based on changes in observed data between two or more data sets (in contrast to forward process models, which predict the values of variables based on assumed model parameters). In this context, the compositional differences between two groundwater samples—an initial water and a final water, should contain the information necessary for an inverse model to unravel candidate causative reaction and mixing histories. The results of such an inverse model may not necessarily be a uni- que problem exacerbated by uncertainties associated with the groundwater chemistry data itself (e.g., analytical uncertainty, representativeness of samples, incomplete analyses suites).

Different strategies can be employed to address the inverse problem and associated uncertainties. Simple mass and charge balance constraints can be used to determine which reactions are mathematically consistent with the data [1], although these do not directly test whether or not any particular reaction is thermodynamically favorable. This approach is used in codes such as NETPATH [2] [3] and PHREEQC [4]. PHREEQC's inverse modeling capability is designed to work with uncertain input in the form of analytical uncertainties associated with each component. Consequently, PHREEQC is able to generate families of inverse models, with associated maxima and minima that can feasibly be contributed by each reaction. Prior studies employing the mass and charge balance inverse modeling approach include [5] , who used PHREEQC's inverse modeling capability to study the mobilization of arsenic under reducing conditions in groundwater in a shallow aquifer in Arkansas, USA, as well as [6] , who developed a geographical information system (GIS)—an assisted approach coupling a groundwater flow model, and an inverse geochemical model to quantify phase mass transfers involving calcite, dolomite, gypsum, CO_2 and fluorite between two points along a groundwater flow path in an Indian aquifer.

An alternative strategy is to utilize a forward-inverse approach that entails running many forward models to identify those reactions that best match the observed data. Least-squares error minimization or similar approaches can be invoked for parameter estimation, a strategy employed by generic inverse models such as CXTFIT [7] [8] and PEST [9]. In the specific context of inverse geochemical modeling, [10] employed a parameter fitting strategy as part of a reactive transport modeling study entailing iron and manganese reduction in coastal aquifers, for example. Nonetheless, while optimization methods can be used to develop a best-fit specific inverse model set from forward models—one that is inherently thermodynamically favored—the uniqueness of the result is still not assured. The optimization routine employed may simply identify a local minimum of whatever error function is being minimized and thus may fail to return better estimates of a global minimum, or, alternatively, multiple sets of objective function minima encompassing a broad sample of the parameter space. Additional information can be gleaned from the data by conducting Monte Carlo simulations to systematically sample the parameter space. However, such a brute-force Monte Carlo approach can be inefficient in identifying an optimal parameter set which reproduces the measured data with sufficient fidelity.

To address these issues, a modified Monte-Carlo-based geochemical modeling strategy is proposed to constrain the parameter space search. The strategy is based on the Metropolis-Hastings algorithm [11] , a Markov-Chain-based technique for approximating probability distributions of parameter values that minimize or maximize a multi-parameter objective function. The Metropolis-Hastings Monte Carlo (MHMC) approach is particularly useful for understanding inverse model parameter probability distributions in cases where those distributions are poorly understood a priori. This is often an issue for groundwater geochemical models, where various aqueous components and phases can interact with one another through a large number of aqueous complexation reactions, electron transfer reactions, surface reactions, and other mechanisms, sometimes obscuring simple relationships between model input and model output. Reference [12] employed a MHMC strategy to quantify water-rock reactions along a flow path in a carbonate system with simple mineralogy. In this current study, a more general geochemical modeling approach is considered, with example applications that include inverse modeling of groundwater interactions with a granitic rock mineral assemblage and an assessment of the impact of fuel hydrocarbon bioattenuation on groundwater chemistry.

2. METHODS

The MHMC strategy involves the sequential generation of sets of proposed parameter values, submission of those values as input into a forward model, and quantification of the resulting goodness-of-fit of the model output to the data through a scoring scheme. The approach is iterative in nature in that the current score for a given iteration is compared with the score from the prior run; if the goodness of fit using the current parameter set represents an improvement, the proposal is automatically accepted. Proposals that score poorly in comparison are not necessarily rejected, however, provided that the proposal scores above some threshold fraction (typically chosen at random) of the prior score. This strategy prevents the algorithm from prematurely converging on local minima or maxima, allowing for more complete exploration of the parameter space but with a bias towards more promising parameter combinations. The parameter set values themselves are selected as part of a Markov Chain sequence; each new proposal represents only relatively minor changes in postulated values in comparison to the prior set. Rejected proposals result in a subsequent new set of posited parameter values using the previous "seed" values, whereas the parameter values associated with accepted proposals are assigned as the new seed values for the next iteration.

The overall algorithm for the application of the MHMC strategy to the generation of an inverse geochemical model is summarized in Figure 1. A given inverse problem is first defined by an initial, or ambient water composition, and a final composition. One or more aqueous components must be present in both solutions to permit differences to be quantified. Reactions that could explain these differences can include: 1) mass transfer between the solution and mineral or gas phases; 2) electron transfer (i.e., oxidation-reduction); 3) ion exchange; 4) surface complexation; 5) evaporation. In addition, mixing of a third water composition—added to the ambient water composition and thus contributing to the final composition can also be considered. Specific parameters that define the nature of these reactions but may represent unknown quantities include masses of mineral phases or gases initially present in the model and their dissolution equilibrium constants (if subject to some uncertainty), the quantity of an ion exchanger present (the equivalent of cation exchange capacity), the amount of water evaporated from the ambient solution composition, and the quantity of any third water composition mixed with the ambient water. Once constraints on each of the parameter ranges are prescribed, a generator algorithm can define the input for a forward geochemical simulation as a trial for a given proposal.

After a proposal is run as a forward geochemical model, the resulting concentration predictions must be compared to the corresponding measured concentrations in the final water composition for each of the selected metrics. A useful definition of a proposal's score is a lumped sum of the squares of the errors for each metric component, defined logarithmically:

$$s = \sum_{i=1}^{N} w_i \ln\left(\frac{c_{i,sim}}{c_{i,obs}}\right)^2 \tag{1}$$

where N is the number of water quality metrics, c_{sim} and c_{obs} the simulated and observed concentrations, respectively, and w a weighting factor introduced to provide some flexibility in emphasizing the importance of matching certain metrics with respect to others (any such intentional bias will be application-specific).

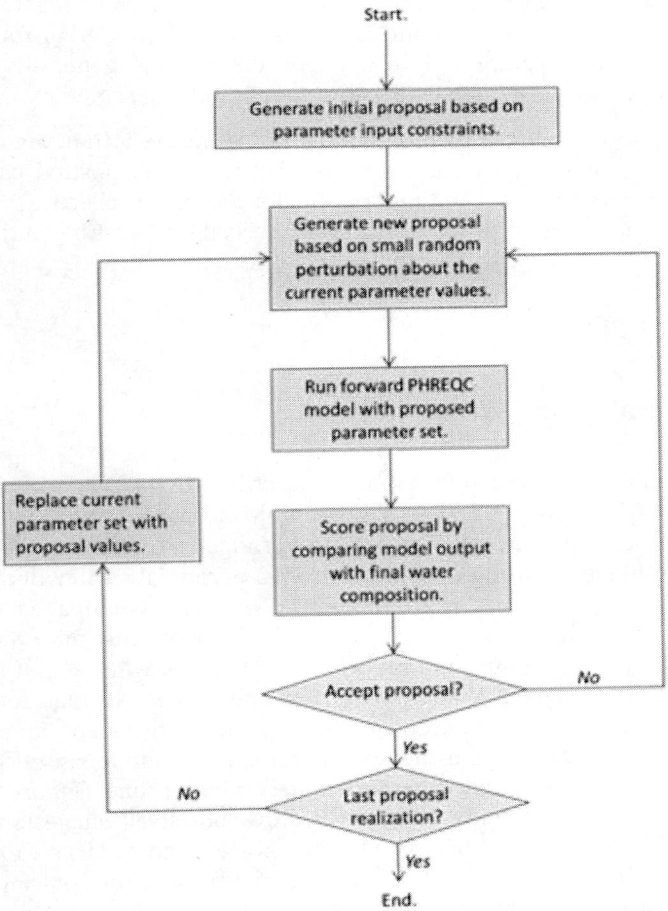

Figure 1. The Metropolis-Hasting Monte Carlo (MHMC) algorithm, as applied to an inverse groundwater geochemical problem.

The MHMC algorithm accepts any proposal yielding a score that is less than that of the prior proposal, with scores that do not meet this requirement accepted only conditionally. Specifically, proposals with higher scores than the prior trial are accepted when the inequality,

$$r \leq \left(\frac{s_{j-1}}{s_j} \right)^{\alpha} \quad (2)$$

is satisfied, where r is a random number selected between 0 and 1, S_{j-1} and s_j the prior and current scores. The exponent α serves as a means to throttle the proposal acceptance rate and is selected by trail-and-error for specific applications; an acceptance rate between 0.3 and 0.7 generally leads to quicker convergence to viable distributions of model parameter values.

PHREEQC was employed as the forward model in the MHMC algorithm. A python script was used to automate the generation of posited parameter sets for each iteration, write the input file for PHREEQC, execute PHREEQC, and subsequently read, process, score, and record the resulting output.

3. RESULTS

3.1. Sierra Spring Water

A geochemical inverse model example described by [13] is provided with the PHREEQC software package. The example consists of two sets of spring water data collected in the Sierra Nevada range in California, USA: water from an ephemeral springs and water from a perennial spring. Both waters are assumed to have been in contact with a mineral assemblage consisting of silica, plagioclase (as $Na_{0.62}Ca_{0.38}Al_{1.38}Si_{2.62}O_8$), biotite mica (as $KMg_3AlSi_3O_{10}(OH)_2$), kaolinite, montmorillonite (as $Ca_{0.17}Al_{2.33}Si_{3.67}O_{10}(OH)_2$), calcite, halite, gypsum, and CO_2. The perennial spring sample is characterized by higher concentrations of major cations and anions (e.g., Na, Ca, Mg, HCO_3, SO_4) than the ephemeral spring sample, an observation consistent with presumed differences in residence time (Figure 2). The inverse modeling capability of the PHREEQC code itself suggests that the compositional differences between the two waters can be largely explained by additional dissolution of CO_2, calcite, in the perennial spring water, coupled with the precipitation of kaolinite and silica. Because PHREEQC's inverse model considers compositional uncertainty in the measured data, two different possible inverse models are indicated (Figure 3).

The MHMC inverse model for the Sierra Spring water problem assumes a set of phases included in PHREEQC's thermodynamic database (phreeqc.dat) that approximately correspond to the mineralogy specified in the example problem. This assemblage includes calcium montmorillonite (characterized by the same stoichiometry as the PHREEQC inverse modeling example), kaolinite, and chalcedony as the silica phase. A K- mica phase, $KAl_3Si_3O_{10}(OH)_2$, was used as a surrogate for the biotite mica assumed by the PHREEQC inverse model example (for which an equilibrium constant was not specified). Unlike the example mica, this K-mica phase does not contain any magnesium, so magnesium was not included as a metric for scoring the proposal. Plagioclase was modeled as an ideal solid solution consisting of albite, $NaAlSi_3O_8$, and anorthite, $CaAl_2Si_2O_8$, end-members, with an initial Na:Ca ratio of approximately 2:1. In addition to this Na:Ca ratio, other uncertain model parameters included initial masses of kaolinite and K-mica as well as equilibrium constants for calcium montmorillonite and K-mica. Search ranges for the initial mineral masses were lognormally distributed between 10^{-6} and 10^{-3} moles/kgw, while uncertainties in the equilibrium constants were addressed by specifying a log saturation index constraint for mineral precipitation/dissolution in the PHREEQC input files between −0.25 and +0.25.

A total of 5000 trial proposals were evaluated, requiring several minutes of run time on a Windows-based personal computer, with the score contribution weighting factors for all components set equal to 1.0 except for Na and pH, for which weighting factors were set to 5.0 and 10.0 to improve the rate of convergence. Approximately 37 percent of the proposals were accepted. Matches of the MHMC inverse algorithm results to the water quality data are shown in Figure 2. The mineralogical changes associated with the best-case proposals (e.g., those with the lowest composite scores) for both scenarios are compared to the PHREEQC inverse model in Figure 3. The model matches the concentration data well except for the concentration of potassium, which remains unchanged because the K-mica phase is not favored to dissolve, a reflection of its particular stoichiometry and dissolution equilibrium constant, and magnesium, which is excluded. Mass transfer reactions are qualitatively consistent with the results of PHREEQC's inverse model in a broad sense in that halite, gypsum, calcite, gas-phase CO_2, and plagioclase all dissolve into solution by approximately corresponding amounts. However, the remaining silicate phases exhibit somewhat different behavior in that Ca-montomorillonite is favored to dissolve. This dissolution reaction contributes to the precipitation of both kaolinite and chalcedony in quantities that exceed those yielded by the PHREEQC inverse model results.

While the results of the only the best-fit proposal are shown in Figure 1 and Figure 2, the MHMC approach is designed to furnish approximations to the probability distributions for individual constituents about the optimal solution. The distribution of initial calcite mass among the best-scoring 100 proposals is shown in Figure 4 (all of the initial calcite mass in all of these proposals dissolves completely in the respective forward models). The probability distribution used by the proposal generator assumed, purely as a trial basis, that the initial calcite mass ranges between 10^{-6} and 10^{-3} moles/kilogram of water, characterized by a lognormal distribution (log mean = 3.2×10^{-5} mol/kgw). The MHMC-derived distribution shown in Figure 4 differs appreciably from this prior distribution and thus represents a refined estimate over the postulated initial range of reacted calcite mass.

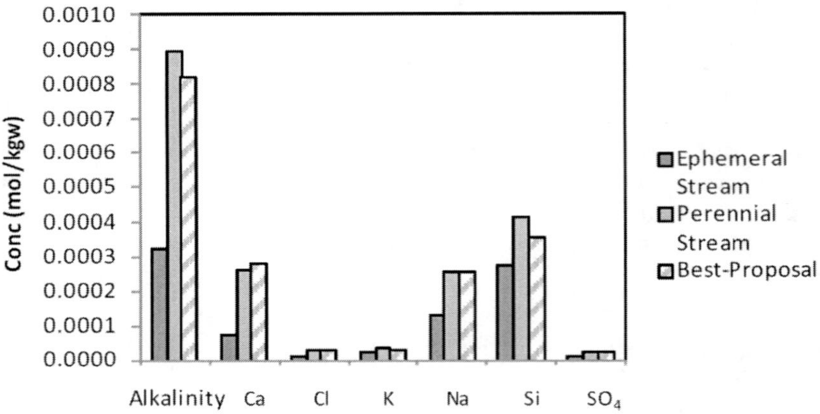

Figure 2. Measured concentrations of groundwater solution constituents compared with forward-model best-proposal results (i.e., minimal weighted error scenario) of the MHMC algorithm.

3.2. Fuel Hydrocarbon Bioattenuation and Trace Element Mobilization

A second demonstration application concerns the impact of fuel hydrocarbon bioattenuation on groundwater chemistry at an anonymous site in Montana, USA, where impacts to groundwater by diesel from a leaking above-ground tank are known to have occurred. Inverse modeling entailed using a limited site quality data set to assess consistency of the data with bioattenuation and to quantify hydrocarbon mass reduction to support remedial planning. Groundwater quality data measured in site groundwater wells includes manganese, ferrous iron, sulfate, sulfide, methane, bicarbonate alkalinity, chloride, and trace elements of potential environmental concern, in addition to dissolved-phase diesel fuel, reported as total petroleum hydrocarbons (TPH). In contrast, general indicators of groundwater quality, such as major cations and pH, have not been subject to monitoring.

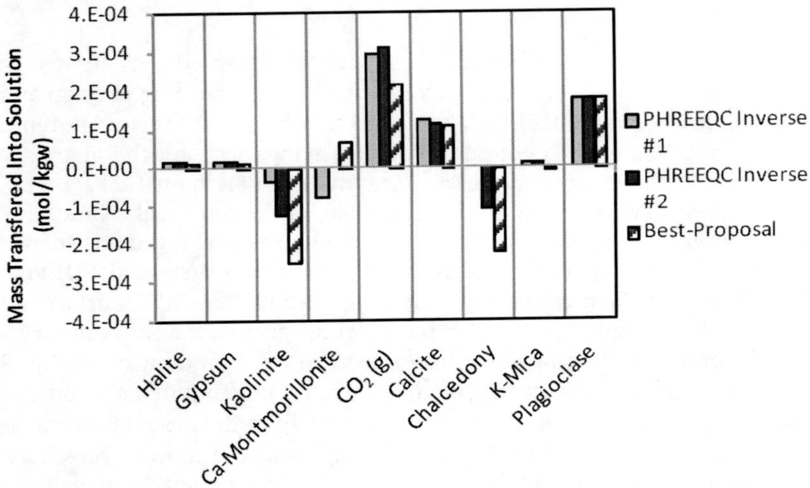

Figure 3. Mass transfer reactions consistent with observed differences between the Ephemeral and Perennial spring water compositions. The two PHREEQC inverse model results reflect ideal mass balances, within prescribed analytical uncertainties, but are not explicitly constrained by thermodynamic considerations. The best-proposal MHMC result—the minimum-error forward model realization, is thermodynamically valid but does not guarantee mass balance with respect to the data.

Processes that are commonly observed to occur in association with bioattenuating organic matter or fuel hydrocarbons, such as reduction of ferric iron and sulfate as well as methanogenesis [14] [15] , are suggested by spatial variability in historic site groundwater data. In addition, arsenic and barium concentrations are also locally elevated in site groundwater in apparent association with dissolved fuel hydrocarbons. The MHMC algorithm was used to identify a set of mass and electron transfer reactions that could consistently explain the observed differences in groundwater chemistry between hydrocarbon-impacted and non-impacted (i.e., "background") groundwater from a recent (2013) groundwater sampling event. The background groundwater composition was defined by the median concentrations of the constituents of interest (bicarbonate alkalinity, arsenic, barium, iron, methane, sulfate, and sulfide) among those groundwater samples characterized by the absence of TPH below the detection limit of 10 μg/L, representing a total of twelve wells. The impacted groundwater sample composition was defined by the median concentrations of the same constituents in six wells characterized by TPH concentrations in the uppermost quartile of wells with positive TPH detections, equivalent to a threshold concentration of approximately 300 μg/L.

Both dissolved iron and manganese were detected in nearly all groundwater samples collected from the site, including background samples. Consequently, PHREEQC was used to set the initial redox potential to reflect equilibrium with goethite, FeOOH (manganese geochemistry was not included in the inverse modeling assessment). Ferrihydrite, $Fe(OH)_3$, is an alternative ferric oxide mineral choice, although trial simulations indicated it was much less effective in matching the observed data. Goethite was also designated as a reactive hydrous ferric oxide surface (HFO) vis-à-vis PHREEQC's surface chemistry modeling feature. The HFO surface was assumed to be in equilibrium with background groundwater chemistry as an initial condition. This specification fixed the total amount of arsenic in the system, with the majority of the arsenic mass existing in a sorbed state on the HFO (arsenic was modeled as existing only in the arsenate form, i.e., the +V redox state). Additional phases included in the model were barite ($BaSO_4$), calcite ($CaCO_3$), and mackinawite (FeS). Diesel was represented in the model by a mean stoichiometry of $C_{12}H_{23}$.

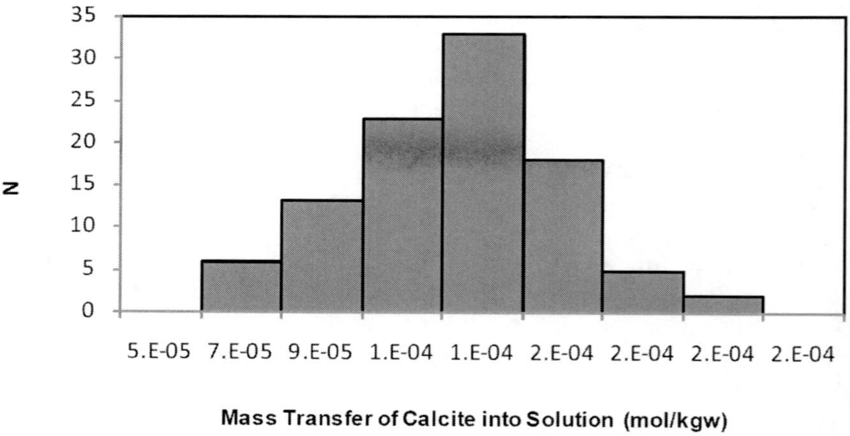

Figure 4. Representation of the probability distribution of calcite mass transferred into solution for the Sierra spring water example, implied by the best 100 scoring proposals among those accepted from the 5000-trial MHMC simulation.

A total of 5000 proposals were generated and assessed using the MHMC algorithm, of which 1816 proposals, or approximately 36 percent, were accepted. A comparison of median measured background and impacted biodegradation constituents and the forward-model output generated by the best-scoring proposal is shown in Figure 5. The best-scoring model results match the differences in the observed data well in a semi-quantitative sense; mismatches are likely attributable to spatial variability in both background groundwater composition and the extent to which various bioattenuation mechanisms vary locally. Nonetheless, despite the lack of a comprehensive suite of analytes fully characterizing the groundwater chemistry at the site, the MHMC results indicate that the available bioattenuation indicator data alone is consistent with the transformation of a particular quantity of hydrocarbon material, assuming equilibrium constant values that are consistent with the thermodynamic data.

The sensitivity of the accepted proposal scores to two key parameters—the initial mass of diesel and the initial mass of goethite are shown in Figure 6 and Figure 7, respectively. The initial mass of diesel added is well-constrained among the best-scoring proposals, indicating the threshold mass required to bring about removal of the majority, but not the entirety, of sulfate via reduction. In contrast, the initial mass of goethite, while clearly exerting an effect on the proposal score through impacts on dissolved iron, sulfide (via FeS precipitation), and arsenic concentrations, is characterized by a wider distribution of values for best-scoring proposals.

In addition to the posited initial masses, equilibrium dissolution constants (represented in proxy by fixing the logarithm of the saturation index for the mineral phase) for mineral phases with presumably uncertain composition—hydrous ferric oxide and iron sulfide were also treated as unknowns and thus allowed to vary within certain ranges in the proposal parameter sets. Among the top ten scoring proposals, log saturation indices for the iron oxyhydroxide phase, goethite), ranged between +1.2 and +1.8; that of the best proposal was approximately +1.5. In effect, this represents an intermediate solubility between that of goethite and the more soluble phase $Fe(OH)_3$, as represented in PHREEQC's database but not considered in the inversion. Saturation indices among the top ten proposals for FeS phase (mackinawite) ranged between +1.0 and +1.4, with that of the best proposal at approximately +1.3. Analogous to the iron oxyhydroxide phase, this value corresponds to an intermediate solubility between mackinawite and the more soluble amorphous FeS precipitate phase represented in the phreeqc.dat database of the PHREEQC code but not included in the model.

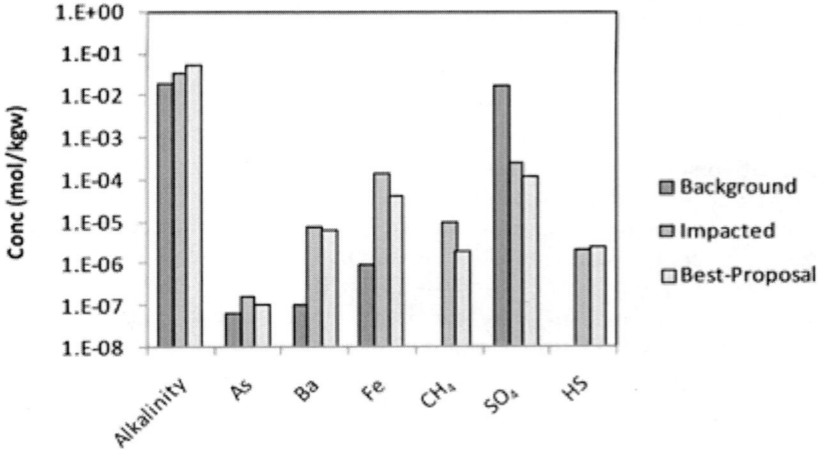

Figure 5. Comparison of median background, impacted and modeled concentrations (best-proposal score MHMC result) for fuel hydrocarbon bioattenuation indicators in groundwater.

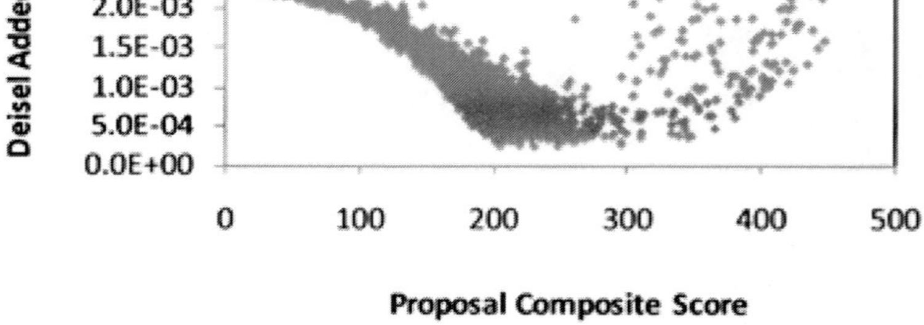

Figure 6. Association of proposal composite score with posited initial diesel (idealized as $C_{12}H_{23}$) mass loading to the modeled solution, per accepted MHMC proposal.

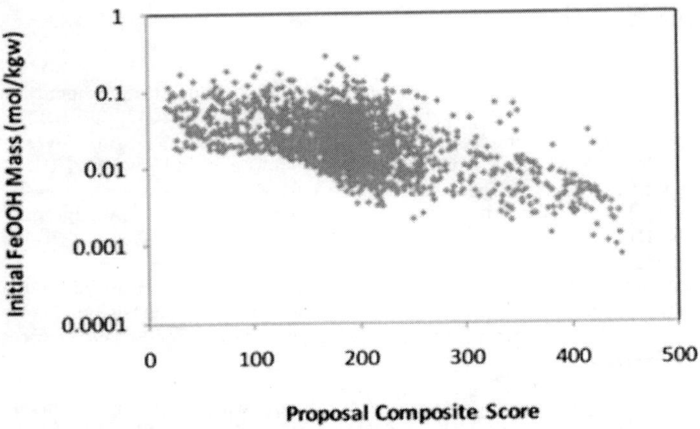

Figure 7. Association of proposal composite score with posited initial FeOOH mass in contact with solution, per accepted MHMC proposal.

4. CONCLUSION

The goal of the MHMC strategy is to identify approximate probability distributions of model parameters that best reproduce observed data. In principle, employing ever larger numbers of iterations should yield increasingly accurate approximations, assuming that the underlying forward model is itself applicable. For the geochemical inverse modeling approach proposed in this study, the forward model PHREEQC has already been well vetted elsewhere in the literature. Therefore, the best-scoring proposals generated by the MHMC algorithm, employing a large number of iterations, can be regarded as providing insights that are consistent with respect to mass balance, charge balance, electron transfer, and thermodynamic considerations. In this context, the MHMC approach serves to safeguard against inverse models that contain spurious mass transfer reaction sets that should not otherwise occur, a threat that exists with approaches such as PHREEQC's built-in inverse modeling tool. However, some limitations of the MHMC approach, as proposed, should be considered. Clearly, the validity of its application to any specific problem should always be assessed on a case-by-case basis, requiring consideration of mineral phase selection and the reliability of thermodynamic data. Second, the geochemical inverse modeling does not take into account the reaction path followed by the system as one water composition is transformed into another; only the initial and final solution states are addressed (*i.e.*, reaction kinetics are not addressed).

REFERENCES

1. Zhu, C. and Anderson, G. (2002) Environmental Applications of Geochemical Modeling. Cambridge University Press, Cambridge. http://dx.doi.org/10.1017/CBO9780511606274
2. Plummer, L.N., Prestemon, E.C. and Parkhurst, D.L. (1991) An Interactive Code (NETPATH) for Modeling NET Geochemical Reactions along a Flow PATH. USA Geological Survey Water-Resources Investigations Report, 91- 4078.
3. Plummer, L.N., Prestemon, E.C. and Parkhurst, D.L. (1994) An Interactive Code (NETPATH) for Modeling NET Geochemical Reactions along a Flow PATH, Version 2.0. USA Geological Survey Water-Resources Investigations Report, 94-4169.
4. Parkhurst, D.L. and Appelo, C.A.J. (1999) User's Guide to PHREEQC (Version 2): A Computer Program for Speciation, Batch-Reaction, One-Dimensional Transport, and Inverse Geochemical Calculations. USA Geological Survey Water-Resources Investigations Report, 99-4259.
5. Sharif, M.U., Davis, R.K., Steele, K.F., Kim, B., Kresse, T.M. and Fazio, J.A. (2008) Inverse Geochemical Modeling of Groundwater Evolution with Emphasis on Arsenic in the Mississippi River Valley Alluvial Aquifer, Arkansas (USA). Journal of Hydrology, 350, 41-55. http://dx.doi.org/10.1016/j.jhydrol.2007.11.027
6. Dhiman, S.D. and Keshari, A.K. (2006) GIS Assisted Inverse Geochemical Modeling for Plausible Phase Transfers in Aquifers. Environmental Geology, 50, 1211-1219. http://dx.doi.org/10.1007/s00254-006-0293-2
7. Toride, N., Leij, F.J. and van Genuchten, M.Th. (1995) The CXTFIT Code for Estimating Transport Parameters from Laboratory or Field Tracer Experiments, Version 2.0. USA Salinity Laboratory, Riverside.
8. Tang, G., Mayes, M.A., Parker, J.C. and Jardine, P.M. (2010) CXTFIT/Excel—A Modular Adaptable Code for Parameter Estimation, Sensitivity Analysis and Uncertainty Analysis for Laboratory or Field Tracer Experiments. Computers & Geosciences, 36, 1200-1209. http://dx.doi.org/10.1016/j.cageo.2010.01.013
9. Tonkin, M. and Doherty, J. (2008) Calibration-Constrained Monte Carlo Analysis of Highly-Parameterized Models Using Subspace Techniques. Water Resources Research, 45, 1-17.
10. Wang, Y. and Van Cappellen, P. (1996) A Multicomponent Reactive Transport Model of Early Diagenesis: Application to Redox Cycling in Coastal Marine Sediments. Geochimica et Cosmochimica Acta, 60, 2993-3014. http://dx.doi.org/10.1016/0016-7037(96)00140-8
11. Metropolis, N., Rosenbluth, A.W., Rosenbluth, M.N., Teller, A.H. and Teller, E. (1953) Equations of State Calculations by Fast Computing Machines. Journal of Chemical Physics, 21, 1087-1092. http://dx.doi.org/10.1063/1.1699114
12. McNab, W.W., Ramirez, A.L. and Johnson, J.W. (2013) Quantifying Reactive Chemistry along an Injected CO_2 Flow Path at the Field Scale Using a Monte Carlo Simulation Approach. International Journal of Greenhouse Gas Control, 16, S194-S202. http://dx.doi.org/10.1016/j.ijggc.2013.01.017
13. Garrels, R.M. and Mackenzie, F.T. (1967) Origin of the Chemical Composition of Some Springs and Lakes. In: Stumm, W., Ed., Equilibrium Concepts in Natural Water Systems, Chap. 10, American Chemical Society, Washington DC, 222-242. http://dx.doi.org/10.1021/ba-1967-0067.ch010
14. Chapelle, F.H. and Lovley, D.R. (1990) Rates of Bacterial Metabolism in Deep Coastal-Plain Aquifers. Applied and Environmental Microbiology, 56, 1865-1874.
15. Murphy, E.M. and Schramke, J.A. (1998) Estimate of Microbial Rates in Groundwater by Geochemical Modeling Constrained with Stable Isotopes. Geochimica et Cosmochimica

CHAPTER 15

Equivalent Porous Media (EPM) Simulation of Groundwater Hydraulics and Contaminant Transport in Karst Aquifers

Reza Ghasemizadeh[1], Xue Yu[1], Christoph Butscher[2], Ferdi Hellweger[1], Ingrid Padilla[3], Akram Alshawabkeh[1*]

[1]*Department of Civil and Environmental Engineering, Northeastern University, Boston, Massachusetts 02115, United States of America,*

[2]*Department of Engineering Geology, Institute of Applied Geosciences, Karlsruhe Institute of Technology, 76131, Karlsruhe, Germany,*

[3]*Department of Civil Engineering and Surveying, University of Puerto Rico, Mayaguez, Puerto Rico 00682, United States of America*

AUTHOR CONTRIBUTIONS

Conceived and designed the experiments: RG XY CB FLH IYP ANA. Performed the experiments: RG. Analyzed the data: RG XY CB FLH. Contributed reagents/materials/analysis tools: RG XY CB FLH IYP ANA. Wrote the paper: RG XY CB FLH. Initiated this work: IYP ANA.

ABSTRACT

Karst aquifers have a high degree of heterogeneity and anisotropy in their geologic and hydrogeologic properties which makes predicting their behavior difficult. This paper evaluates the application of the Equivalent Porous Media (EPM) approach to simulate groundwater hydraulics and contaminant transport in karst aquifers using an example from the North Coast limestone aquifer system in Puerto Rico. The goal is to evaluate if the EPM approach, which approximates the karst features with a conceptualized, equivalent continuous medium, is feasible for an actual project, based on available data and the study scale and purpose. Existing National Oceanic Atmospheric Administration (NOAA) data and previous hydrogeological U. S. Geological Survey (USGS) studies were used to define the model input parameters. Hydraulic conductivity and specific yield were estimated using measured groundwater heads over the study area and further calibrated against continuous water level data of three USGS observation wells. The water-table fluctuation results indicate that the model can practically reflect the steady-state groundwater hydraulics (normalized RMSE of 12.4%) and long-term variability (normalized RMSE of 3.0%) at regional and intermediate scales and can be applied to predict future water table behavior under different hydrogeological conditions. The application of the EPM approach to simulate transport is limited because it does not directly consider possible irregular conduit flow pathways. However, the results from the present study suggest that the EPM approach is capable to reproduce the spreading of a TCE plume at intermediate scales with sufficient accuracy (normalized RMSE of 8.45%) for groundwater resources management and the planning of contamination mitigation strategies.

1. INTRODUCTION

Karst aquifers account for 25% of groundwater resources in the world and 40% in the US [1]. They are formed when the dissolution process in primarily soluble carbonate rocks creates complex networks of preferential flow pathways, such as solutionary fractures and conduits, within the rock matrix (karstification). Conduits are crucial for groundwater flow and contaminant transport in karst aquifers [2], but their distribution is often unknown, thus limiting the applicability and validity of the numerical models that require detailed data on conduits [3]. Subsurface flow within the aquifer ranges from laminar to turbulent, with laminar flow in the rock matrix and predominantly turbulent flow in conduits, depending on flow velocities [4]. Karst areas include swallets, sinkholes, infiltrating streams, and other highly porous surface features that limit the availability of surface water, making groundwater the primary water resource for domestic, agricultural, and industrial utilization.

Traditional simulation of groundwater hydrodynamics with numerical models based on Darcy's law may not be directly applicable for modeling flow in karst [5, 6]. Such models are typically used for laminar groundwater flow regime and slow groundwater velocity conditions and their application in karst aquifers require extra attention. In an Equivalent Porous Media (EPM) approach for karst groundwater systems, the default assumption is that carbonate aquifers' behavior is equivalent to porous media for both flow and transport. Also known as single continuum porous equivalent approach (SCPE), heterogeneous continuum approach, smeared conduit approach, or single continuum approach [7, 8, 9, 10], it is the simplest distributed modeling approach for karst aquifers. For the reasons outlined above, its ability to simulate groundwater flow in karst, however, is limited.

The EPM approach assumes that the rock matrix including fractures and conduit networks can be represented by an equivalent porous medium with equivalent hydraulic conductivity in a certain area [11, 12]. In highly karstified aquifers, however, the contaminant transport may depend primarily on the karst conduit network rather than matrix hydraulic conductivity. EPM models often do not distinctly account for preferential flow; instead, they approximate the overall local conductivity of the matrix as well as possible fractures and conduit networks with an enhanced equivalent conductivity [5, 13]. Despite this limitation, it can lead to representative results depending on the degree of aquifer karstification and the scale of the modeling effort. Generally, the EPM approach is more suitable for regional scales rather than local and intermediate scales [5, 14].

Scanlon et al. [5] evaluated the accuracy of two different EPM approaches, lumped parameter and distributed parameter, for simulating regional scale (330 km^2) groundwater flow in the highly karstified Barton Springs Edwards aquifer in Austin, Texas. Both models simulated the temporal variations in spring discharge fairly accurate, but the latter approach reproduced the effect of pumping on groundwater levels, evaluated the potentiometric surface at different times, and spring discharge more accurately. According to their study, it is practical to simulate karstic aquifers as equivalent porous media for groundwater flow but not for contaminant transport (especially at local scales) or to delineate protection zones for wells or springs because transport velocities may be substantially underestimated.

Using both porous medium and conduit network simulations, Worthington [6] investigated high-permeability interconnected conduits in the limestone aquifer at Mammoth Cave (Kentucky, USA) at a regional scale (area of 258 km^2) both experimentally and numerically to explore if the aquifer behaves more similar to a porous medium or to a karst aquifer with flow both in the rock matrix and karst conduits. While the underground drainage pattern was comparatively well-known from previous tracer studies, that study identified a significant difference in the simulated aquifer behavior with and without considering conduits; especially at local scales, the EPM approach was inadequate to simulate solute transport, represent spring hydrographs, and to interpret tracer test results.

Using MODFLOW 2000, Putnam and Long [15] developed a regional EPM flow model for the Madison and Minnelusa aquifers (area of 2590 km^2) to synthesize available hydrogeological data and quantify the regional water budget. Saller et al. [16] further developed the above model by including high conductivity zones, representing karst conduits in the EPM model, which resulted in a better match with the observed transient well data; however, the simulated spring discharge was not improved and was similar to the original EPM.

If sufficient geologic data are available so that the location and characteristics of conduit networks within karst aquifers are fairly well known, some investigators have used MODFLOW to simulate those networks (on a scale order of less than 100 m to several kilometers) with various techniques, including lines of drain cells and internal sinks [17, 18] and lines of high hydraulic conductivity cells [19]. Although not directly comparable, results from above studies may be applicable to other karst aquifers on a case-by-case basis. Yet there remain questions as to how reliable and to what temporal resolution the EPM approach can replicate groundwater level fluctuation in karst or whether EPMs can simulate the long-term fate and transport of point source contamination.

The present study evaluates the utility of the Equivalent Porous Media (EPM) approach to simulate dynamic groundwater hydraulics and contaminant transport in highly permeable karst unconfined aquifers using an example from the North Coast limestone aquifer system in Puerto Rico at an intermediate scale (area of 132.5 km^2 and scale order of 10 kilometers). The three dimensional finite difference model MODFLOW 2005 [20] was used for simulating flow, and the transient transport model MT3DMS [21] was used to simulate the spreading of a point source TCE contamination at a Superfund site. The main goal of this study is to evaluate if the EPM approach is suitable and practical for the assessment of major contaminated sites in karst areas, based on available data and the study's scale and purpose.

2. MATERIALS AND METHODS

2.1 Ethics Statement

No specific permits were required for the described field studies, nor for the locations and activities. The studied location is not privately-owned or protected in any way, and the field studies did not involve endangered or protected species.

2.2 Study Site

The study was carried out at a test site within the north coast region of Puerto Rico. Karst limestone features cover 27.5% of the land surface of Puerto Rico (Fig 1). The Northern Karst Belt covers an area of approximately 2000 km^2 (16% of the island) and elevates at heights ranging from about 300 m in the mountainous areas to sea level at the coast (Fig 1) [22]. This aquifer system is the largest aquifer of the island and contains the island's most extensive freshwater aquifer, accounting for 80% of northern Puerto Rico's groundwater resources [23].

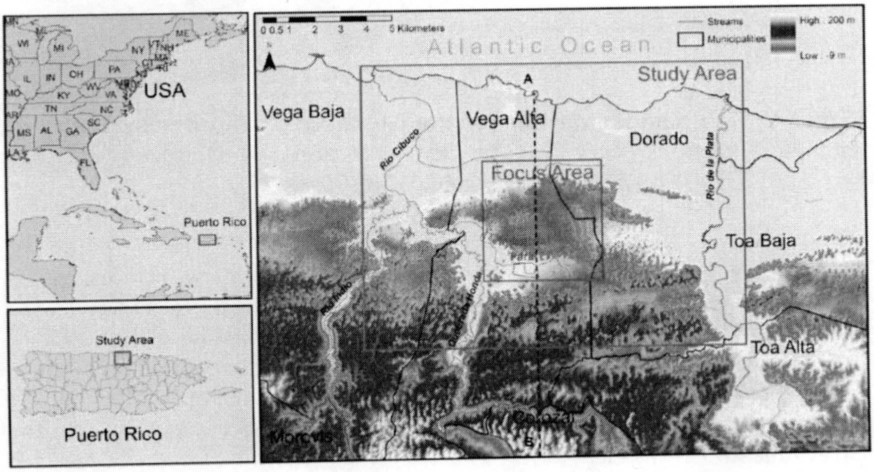

Figure 1. Location of study site in northern Puerto Rico. The map includes study area boundary, municipalities, major streams, focus area, Industrial Park, land surface elevation, and position of cross-section A-B. https://doi.org/10.1371/journal.pone.0138954.g001

2.3 Geology and hydrology.

The North Coast limestone aquifer system of Puerto Rico is sub-divided into four generalized hydrogeological units: the upper water-table aquifer, the middle confining unit, the lower confined aquifer, and the basal confining unit [24]. The thickness of the karst aquifer system increases from the south (Southern Karst Uplands) to the central part of the Northern Karst Belt (Karst Upland Plateau), and subsequently decreases to the coastal plane in the north (c.f. Fig 2).

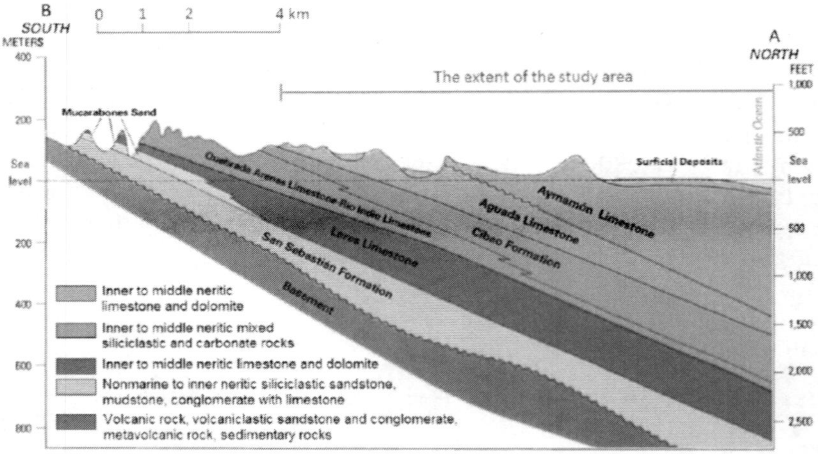

Figure 2. Geologic formations of the North Coast Aquifer system. The geologic sections are along cross-section A-B (see Fig 1). https://doi.org/10.1371/journal.pone.0138954.g002

The aquifer system consists of a series of Tertiary limestone formations containing eogenetic karst [25, 26, 27]. The term eogenetic refers to the early stage of karst porosity evolution at low burial, involving surface processes and meteoric diagenesis [28, 29]. The limestone formations include, from north to south and stratigraphically from top to bottom, the Aymamon limestone, the Aguada limestone, the Cibao Formation, and the Lares limestone (Fig 2).

The Tertiary limestone units are locally covered with Quaternary soils such as clay, blanket sand deposits, marine terrain deposits and stream valley alluvium [30, 31]. In the area of Vega Alta, where the present study was carried out, the aquifer system does not contain the Lares limestone formation. Due to the high annual rainfall and the warm climate of the area (average annual temperature about 25°C), the limestone units and soils are highly prone to erosional processes, and climatic conditions are favorable for karst development. The karst conduit systems typically occupy only a small portion of the total aquifer porosity [9]. Nevertheless, the karst conduit system may have a major impact on the hydraulic behavior of the total karst system. The frequency of sinkholes is < 1% in most locations and is much less in Vega Alta than in other municipalities of the north coast. As a result, point recharge into the aquifer is minor, leading to relatively low groundwater level fluctuations [32].

The limestone formations in northern Puerto Rico are slightly or moderately karstified [33]. Generally, the hydraulic conductivities of these units decrease exponentially from a maximum of about 2,040 m/day in the upper Aymamon limestone to a minimum of about 0.04 m/day in the basal Lares limestone [32]. In Vega Alta area, a first set of horizontal hydraulic conductivities (K values) and their vertical distribution over the model layers were defined by Sepúlveda [34] based on slug tests data conducted in 23 multiport wells with different monitoring ports at various depths. The slug tests indicated that K values ranged from 0.06 to 305 m/d, with generally higher values near the water table. The karst upland plateau in Vega Alta encompasses numerous closely spaced deep sinkholes and few gentle depressions in Aguada limestone and Aymamon limestone, respectively. Hydrological investigations [32, 35] demonstrate highly variable porosity and high permeability contrasts in short distances reflecting the variable distribution of conduit porosity in the north coast and suggest that water-bearing conduits are present between Rio Camuy and the Aguadilla area.

The presence of caverns in test wells drilled in Hatillo and Isabela in the north and northwest of Puerto Rico has been verified by Rodriguez and Hartley [36]. Using aerial photographs and geophysical methods, Rodriguez and Richards [37] investigated a few sites in northwestern Puerto Rico to detect the presence of conduits and probability of conduit-controlled groundwater flow based on presence of natural potential anomalies. The exact locations of highly transmissive features such as cavities, conduits and fractures in central northern aquifers, however, are uncertain; and no dye tracer or cave diving evidence exists for locations and properties of subsurface conduits discharging groundwater into the northern coastal plain or the submarine outcrop of the aquifer [38]. Rodriguez [39] classified the springs in the north coast limestone based on their hydrodynamics and suggested that some springs are interconnected by an integrated conduit network. Mackovic spring (with limited discharge data) in Vega Alta exhibits little short-term response to rainfall, while Maguayo spring (with flow range of 147 to 2,789 m3/d) in Dorado exhibits a strong short-term response to rainfall and therefore they are known as diffuse-flow and conduit-type minor springs, respectively (see Fig 3).

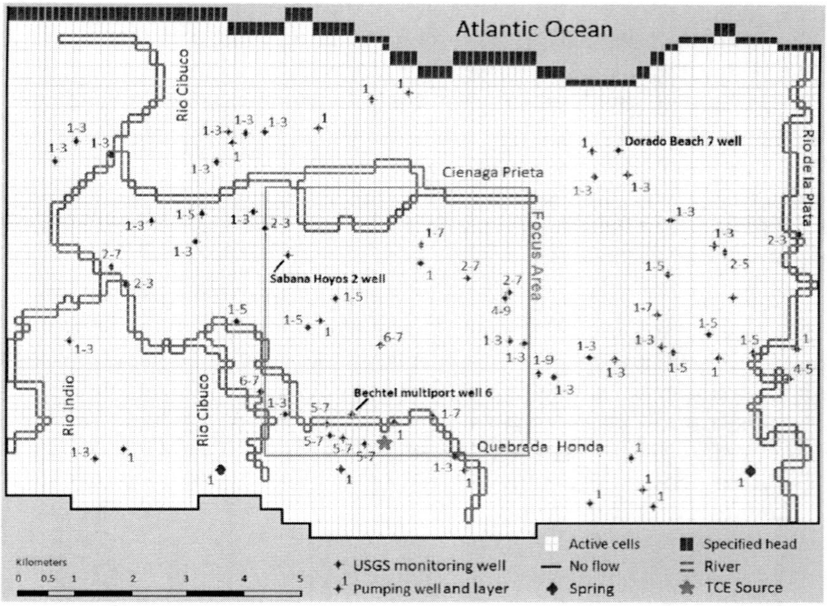

Figure 3. The model domain and defined boundaries. The domain extends beyond the two rivers and consists of focus area, location of TCE source, springs, USGS monitoring wells, and groundwater extraction wells. The well labels indicate the range of layers from which the groundwater was extracted. https://doi.org/10.1371/journal.pone.0138954.g003

The rainfall in Puerto Rico varies geographically and seasonally, with February being the driest month and September and October the wettest months [40]. There is a dry period starting in December and usually ending in March or April, followed with a spring rainfall period in April and May, an erratic, semidry period in June and July, and a wet season from August through November [33]. The average annual rainfall in the area of the North Coast limestone aquifer system is about 180 cm (NOAA). With a tropical climate and a high average level of solar radiation, the area experiences high evapotranspiration and receives relatively low rainfall intensity and duration, especially in winters. The average annual evapotranspiration is about 110 cm, or 63% of the average annual rainfall [33]. In Vega Alta area, assuming a minimal runoff, a preliminary estimate of annual aquifer recharge rates for watersheds in Puerto Rico was obtained by Giusti [33] from a correlation of evapotranspiration vs. rainfall (up to 84 cm recharge for a mean annual precipitation of 200 cm). Gómez and Torres-Sierra [41] estimated the spatial distribution of recharge for Vega Alta aquifer and developed a 2-dimensional groundwater model to assess the impact of additional withdrawals on groundwater balance.

2.4 Contamination concerns.

The focus area of the present study was the unconfined Vega Alta limestone aquifer, which is a part of regional North Coast limestone aquifer. The aquifer consists of Aymamon and Aguada limestone (the two top layers in Fig 2) and is underlain by Cibao Formation in the south and saltwater-freshwater interface in the center and to the north. It is located in the Vega Alta and Dorado municipalities in the central part of the north coast of Puerto Rico between the municipalities Vega Baja and Toa Baja. The study area is bounded by the USGS Vega Alta quadrangle and the low permeability Cibao Formation to the south, the Atlantic Ocean coast to the north, the Rio de la Plata River to the west, and the Rio Indio River to the east (Fig 1).

Because of the high aquifer productivity, among other reasons, many pharmaceutical, chemical, and manufacturing industries have settled in the North Coast of Puerto Rico, with subsequent growth in population. Many of these industries rely on the use of hazardous materials, which can enter the karst groundwater from accidental spills and deliberate disposals. Many sinkhole depressions have also been used as clandestine waste disposal sites. Water quality surveys in Puerto Rico have shown a vast contamination of the North Coast limestone aquifer [42]. Contamination with chlorinated chemicals, which have been measured in a large percentage of sample wells, is of great concern. Extensive contamination has resulted in the closure of 41% of drinking water supply wells by 1987 [43]. Since then, there have been more closures, and recently, groundwater has been avoided as a primary source of public water supply in northern Puerto Rico [23].

Up to 49 thousand cubic meters per day of groundwater was extracted in 1991 from 20 public supply wells in Vega Alta and Dorado for public water supply and private usage (industrial, commercial, and agricultural) [40]. Being among the most productive aquifers of the island, this aquifer is at the same time one of the most contaminated. Volatile Organic Compounds (VOCs) used in the metal, electronic, and dry cleaning industries have been leaching from beneath an Industrial Park (see Fig 1) for decades. It was first considered contaminated in 1983 when water samples from 17 of 90 wells collected at the Vega Alta Public Supply Wells site (Latitude: 18.41806, Longitude: -66.33028) were identified to contain high concentrations of methylene chloride, extractable organic compounds, trichloroethylene (TCE) and other VOCs [44].

The US Environmental Protection Agency (EPA) was concerned about health effects of this contamination, and the contaminated wells were immediately shut down by the Puerto Rico Aqueduct and Sewer Authority (PRASA) being responsible for the operation and maintenance of the public water supply system in Vega Alta. In 1984, the site was included in the Superfund National Priorities List (NPL) as "Vega Alta Public Supply Wells" Superfund site. The Vega Alta landfill and the Industrial Park have been recognized as possible sources for TCE and other VOCs. The Vega Alta limestone aquifer was estimated to contain 5900 kg (in 1990) and 5800 kg (in 1992) of TCE, and solute TCE influx into the aquifer was estimated at 9.98 kg/yr under long term average net recharge rates [22]. The drinking water supply of about 40,000 people in Vega Alta municipality depends upon the health of the highly vulnerable karst aquifer. Establishing a thorough understanding of the hydrogeology of this major karst aquifer system is important for water resources and groundwater management and also provides the foundation for predicting the contamination migration.

2.5 Model Setup and Calibration

To illustrate the application, performance, and limitations of the EPM modeling approach in evaluating the release and transport of contamination in karst aquifers, a MODFLOW model was developed and paired with MT3DMS; both were implemented in Visual MODFLOW Standard 2011.1 [45]. The variably-spaced grid is composed of 76 rows and 113 columns, encompassing a total area of 132.5 km^2 (Fig 3). The grid is oriented south-north in order to align the model columns with the dominant coastward direction of groundwater flow. The rectangular cells have a maximum size ($\Delta X \times \Delta Y$) of 236×248 m^2 and a minimum size of 118×124 m^2. The final grid consists of 8588 cells in each layer, with the finest refinement around the contamination source (the eastern side of the Industrial Park) and water extraction wells. The unconfined aquifer was subdivided into 12 layers with basis elevations of -3.05, -15.24, -22.86, -30.48, -38.1, -45.72, -53.34, -60.96, -68.58, -76.2, -86.87, and -97.54 m a.s.l, whereas the last elevation corresponds to the base of the karst aquifer (lower model boundary). The subdivision is based on similar hydraulic conductivities and initial TCE concentrations in these layers [46]. The saturation thickness of the top layer varies as a function of the water table.

Transient simulations are needed to analyze time dependent variables such as the response of water table in aquifer systems to transient signals like seasonal precipitation pulses. The transient flow simulations are run using monthly rainfall between February 1983 and December 2011 (29 years), and also using a refined daily rainfall between June 2004 and December 2005 (19 months) under the same conditions to evaluate the performance of the flow model in a greater temporal resolution.

The Atlantic Ocean coastline in the north is defined as a specified (constant) head boundary of the model with zero head at the sea-level. Predominantly south to north groundwater flow, suggested by historical water level data, implies no-flow boundary conditions along the western and eastern boundaries (c.f., Fig 3). These assumptions are consistent with previous modeling efforts [34, 41, 46] and are constant throughout the simulations. The unconfined aquifer bottom is underlain by the Cibao formation toward the south and by a static saltwater interface (groundwater with estimated dissolved solids exceeding 35,000 mg/L) toward the northern part of the area. In the Vega Alta karst system, the southern Cibao formation is almost impermeable with very low effective porosity due to the existence of clay layers and the absence of karst features. Therefore, a no-flow boundary for each layer is used as the southern and bottom boundary of the model, as well as the northern model boundary (Atlantic Ocean) in all layers except for the topmost two layers in the north.

Recharge is a key component of the hydrological balance in groundwater systems, and it is often difficult to quantify this parameter since it depends on variety of complex factors such as runoff and evapotranspiration. Due to the relatively low relief (maximum altitude of around 150 m a.s.l.) in the Vega Alta area, the average rainfall depends only little on the altitude (slightly greater values at higher elevations), which is confirmed by NOAA data. Therefore the model assumes that the rainfall is independent of the elevation, and rainfall data are only taken from Dorado rain gauge in the study area for model setup. Diffuse infiltration of rainfall through permeable outcrops is the major source of recharge to the aquifer. The net transient recharge is spatially distributed in six zones over the aquifer with rates calculated by multiplying the monthly rainfall to the calibrated discrete aerially-weighted conversion factors of 3%, 6%, 9%, 16%, 25%, and 31% of the net rainfall for the different geologic formations. The highest recharge rates correspond to the limestone outcrops and medium and lower recharge rates for areas with thick blanket deposits and coastal clay deposits, respectively. Aerial recharge was considered as specified flux (Neumann) boundary condition.

The major streams flowing through the study area are Rio de la Plata (the longest in the island) and Rio Cibuco with drainage areas of 227 and 519 square kilometers, respectively (c.f., Fig 1). At base flow conditions, river stages measured at the intersections with highway 2 correspond to 1.52 and 0.91 m a.s.l respectively. Both water level stages are assumed constant. The rivers were implemented as a transfer (Cauchy) boundary condition in the topmost layer of the model, with riverbed conductance values used for river cells ranging from 4.645 to 371.61 m^2/day, according to the studies of Monroe [30] and Gómez and Torres-Sierra [41].

There were 72 private and public pumping wells in the Vega Alta study area. Historically, well closure was the major reason for the decline in groundwater withdrawals, and the lost water public supply was compensated from surface-water sources [47]. Withdrawal rates and filter depth data were taken from field measurement by Sepúlveda [34] and PRASA before 1992, and were assumed constant between 1992 and 2011. The transient model requires initial conditions for the hydraulic head and contaminant distribution. A map of water levels measured in February 1983 [46] is used to define initial conditions of the transient flow model. The effects of uncertainty in initial conditions on simulation results decrease as the simulation progresses. Therefore, simulations were given sufficient "spin-up" time before the period being modeled to limit errors associated with possibly erroneous initial conditions.

Measured water table heads of February 1983 [48] in 70 wells were used for steady-state calibration to estimate the initial conditions for subsequent transient simulation. Parameters used for steady-state flow calibration are net recharge and hydraulic conductivity. These values are slightly adjusted and refined during the manual steady-state calibration processes to match modeled and spatially observed hydraulic heads (Fig 4).

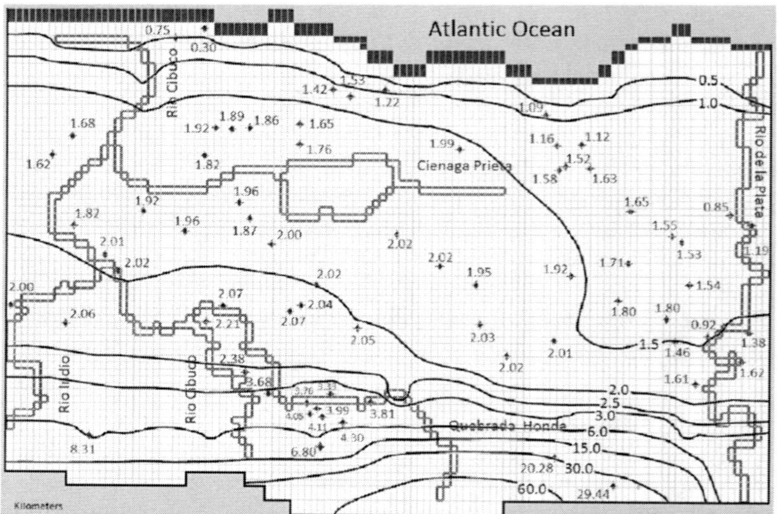

Figure 4. Calibrated steady-state water table contours for February 1983. The simulated groundwater levels were used as the initial conditions for the transient flow model. The monitoring well labels represent the measured groundwater heads (m above sea level). https://doi.org/10.1371/journal.pone.0138954.g004

Transmissivity within the upper limestone aquifer is largely regulated by the thickness of the freshwater lens that thins landward and coastward [24]. The distribution of hydraulic conductivities first estimated by Sepúlveda [34] were further adjusted for each layer in this model (Fig 5A), with lower values in the Cibao formation and higher values in the Aguada and Aymamon formations. The highest K values in the middle of model domain correspond to the upland plateau characterized by outcrops and closely spaced sinkholes which are surrounded by low rolling hills.

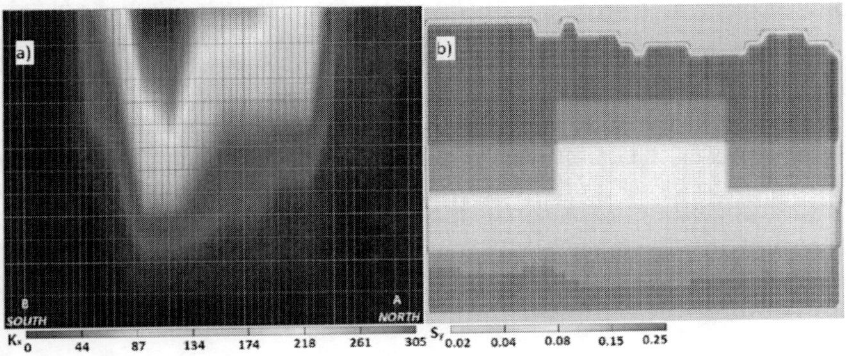

Figure 5. Variability of aquifer hydraulic parameters. (a) Vertical distribution of calibrated hydraulic conductivity (m/day) along cross section A-B (the plot is 50 times exaggerated in the vertical axis); and (b) specific yield of all model layers. https://doi.org/10.1371/journal.pone.0138954.g005

The same model parameters (hydraulic conductivity and recharge) used for steady-state calibration were used for the transient calibration to adjust the storage parameters. In the north of the karst upland plateau, with alluvial aquifer and beach deposits, transient calibration resulted in a specific yield (S_y) of 0.25. The limestone aquifers of the karst upland plateau were calibrated with a S_y of 0.08; to the north, west and east of the plateau, where it transitions to the alluvial aquifer, all limestone units were calibrated with S_y of 0.15. The south of the karst upland plateau with clay lenses was calibrated with S_y of 0.02 after a transition area with S_y of 0.04 (c.f., Fig 5B).

A relatively sharp interface between fresh and saline water was estimated for the freshwater lens with a depth to the interface about 40 times of height of freshwater and the saltwater zone can extend up to 7 km inland of the Atlantic Ocean coastline [24]. The thickest part of the freshwater lens (exceeding 100 m) generally lies along a line where the interface with an approximate slope of 2% intersects the base of the aquifer (i.e. southern Cibao Formation).

Commonly, the frequency distribution of the hydraulic conductivity is closer to log-normal than normal [49] in karst aquifers, but could be positively or negatively skewed [29]. The distribution of fracture aperture widths is generally assumed to follow a log-normal [50] or power-law distribution. Conduits spacing should therefore also follow a similar distribution, however, if present, a single conduit can dominate an aquifer [51]. The frequency distribution of the hydraulic conductivity of the present study is presented in Fig 6. Hydrogeological data sets often display a positive skewed frequency distribution, as hydraulic conductivity did in the present study (Fig 6A). The EPM favors a right-skewed distribution toward higher equivalent conductivity values, trying to approximate the individual highly-permeable pathways with an equivalent higher conductivity value in the model cells, which can be better seen in the logarithmic frequency distribution log(K) (Fig 6B). However, one should be aware that these plots do not represent actual field conditions with a possible conduit network, which can impose higher log(K) and potentially result in a log-normal distribution.

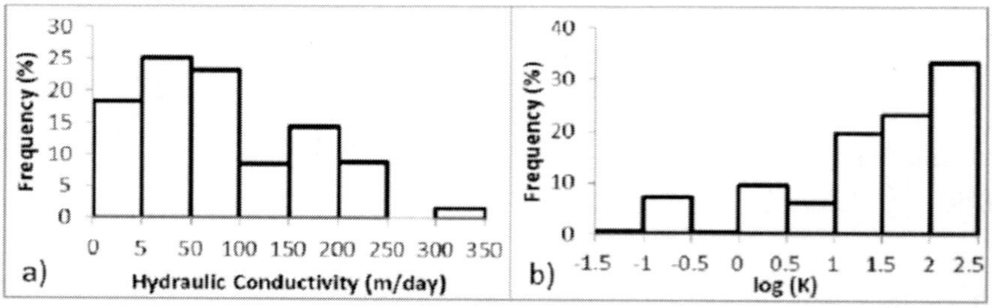

Figure 6. Frequency distribution of calibrated hydraulic conductivity. (a) Calibrated hydraulic conductivity is positively skewed; and (b) Logarithmic hydraulic conductivity log(K) is negatively skewed. https://doi.org/10.1371/journal.pone.0138954.g006

The transient transport model was run using monthly rainfall between January 1990 and January 2011. For the initial conditions, the transport model uses the head distribution in January 1990 computed by the flow model, and the TCE distribution in January 1990, the latter generated by linear interpolation of TCE concentration data measured in wells of the contaminated area [46]. Except for the TCE source cell near the eastern part of the Industrial Park (c.f., Fig 3) in the uppermost model layer, the TCE recharge concentration (C_{RCH}) in the study area was set to zero. The TCE mass present as free phase in the source zone or adsorbed onto the media provides ongoing sources of dissolved contamination in the aqueous phase. The mixing of the dissolved TCE in the source zone is a function of the local groundwater flow field within the TCE entrapment zone. Although the influx concentration may vary depending on hydrological conditions, it is considered constant in the transport simulations. The subsequent development of the dissolved TCE plume is controlled by the regional groundwater flow field and various processes such as advection, dispersion, diffusion, adsorption, and a minimal decay.

Since chlorinated organic compounds tend to decay very slowly in groundwater, TCE was assumed conservative with a minimal first-order decay constant of 1.95×10^{-6} d^{-1} (a half-life of 975 years). A linear distribution coefficient (k_d) of 0.2 l/kg (corresponding to a retardation factor of 3.95) was used in this study based on theoretical methods and field measurements on distribution coefficient for sorption of TCE in soil water systems with different organic carbon content [52].

Manual calibration of the transport model was performed based on comparing simulated and measured TCE concentrations between January 1990 and January 2008. Groundwater samples were collected quarterly from monitoring wells in the focus area by USGS and EPA. The environmental and hydrogeologic data were collected from recent studies and US Geological Survey [34, 46]. All data were compiled into a sustainable database in the Puerto Rico Testsite for Exploring Contamination Threats (PROTECT) Center in Northeastern University for current and future study.

2.6 Model Evaluation

To compare the closeness of modeled values to measured values, the model's estimation errors were calculated using two measures of root mean square error (RMSE) and mean absolute error (MAE). Also linear regression analysis was used to evaluate the dependency of water level fluctuations to precipitation events.

3. RESULTS AND DISCUSSION

One of the main objectives of the study is to evaluate the performance and applicability of the EPM approach to simulate flow and transport for a highly permeable karst aquifer. For this reason, efforts have been undertaken to challenge model results and thoroughly evaluate them against given data from the study site.

3.1 Flow Model

The steady-state calibration of the groundwater flow model was accomplished based on minimizing the MAE and RMSE of the simulated potentiometric surface map (Fig 4). Using average conditions in February 1983, RMSE of the model is 1.0889 m and MAE is 0.5230 m for 70 steady-state calibration wells (Fig 7A), representing about 1.7% and 3.6% of the measured water level variations across this unconfined aquifer. The number of available measured heads for transient calibration were much fewer than steady-state data, and the only three water level stations used for transient model calibration (operated by USGS Caribbean Water Science Center) are Dorado Beach #7, and Sabana Hoyos #2 and Bechtel #6 multiport wells (see Fig 3). The long term transient calibration results are presented in Fig 7B as a scatter plot diagram. Based on a statistical analysis of transient model results, the calculated low errors of 0.2049 m (MAE) and 0.2673 m (RMSE) indicate satisfactory simulation of water table variations in those stations. The linear trend line of data (R^2 = 0.6867) is slightly misaligned from the expected 45° line. A Pearson correlation coefficient (R) of 0.8186 between simulated and observed hydraulic heads indicates generally good temporal correlation between observed and simulated water tables. Taking into consideration the used equivalent porous media approach, these values suggest satisfying results for the model calibration. Higher values of hydraulic head are slightly underestimated and lower heads are slightly overestimated by the model. This suggests that the recharge in dry seasons is overestimated, while it is underestimated in wet seasons.

The simulated steady-state water balance in the aquifer demonstrates that the total aquifer recharge mostly originates from precipitation (98.69%) and river infiltration (1.31%). Groundwater flow in the study area discharges mostly into the Atlantic Ocean (19.62%) in the uppermost two layers, and extraction wells (74.41%). Except the two fourth order springs (together discharging 1.05%), no major springs have been identified in the Vega Alta area (Fig 3), and surface rivers drain only 4.92% of the total groundwater discharge. As net recharge rates increased (decreased) or pumping rate decreased (increased) during the transient simulations, discharge rates to the rivers and to the Atlantic Ocean increased (decreased). Due to absence of natural drainage features (such as wetlands or lagoons); in Vega Alta area more groundwater is available for withdrawals than in adjacent aquifers [53]. As a result, the impact of excessive pumping from this productive unconfined aquifer on water table elevations and flow patterns are significant. The head contours are locally deviated around highly productive wells and the steady-state flow is diverged eastward close to Rio de la Plata River (see Fig 4). During extreme rainfall events, the surface rivers discharge to aquifer because the hydraulic gradient towards the aquifer increases.

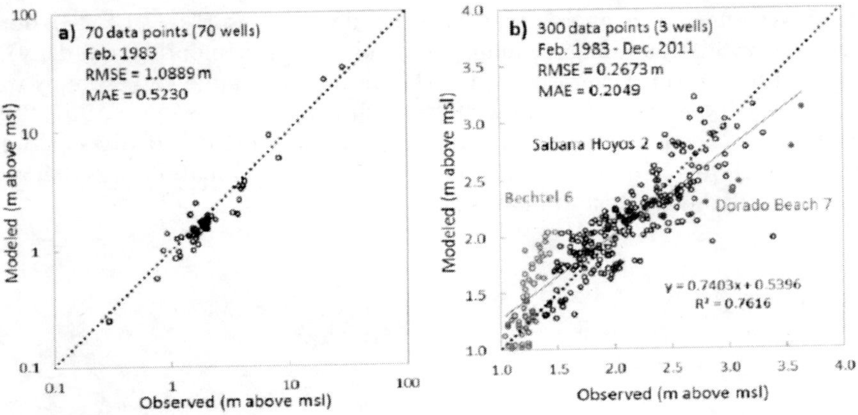

Figure 7. Scatter calibration diagrams of simulated versus observed hydraulic heads. (a) Steady-state calibration of the flow model (70 monitoring wells); and (b) Transient calibration of the flow model over three decades (3 monitoring wells). https://doi.org/10.1371/journal.pone.0138954.g007

As the distance from the coast increases (>6700 m), the Aguada formation at the bottom of the aquifer becomes shallower and reduced in thickness. As a result, the depth-averaged storativity and conductivity values decrease and the average groundwater level rises faster relative to sea level (Fig 8). The magnitude of seasonal fluctuations depends upon the distance from the constant head boundary condition in the north. The low hydraulic gradient in the central part of the model reflects high hydraulic conductivities (c.f., Fig 5B), while slightly steeper hydraulic gradients reflect the relatively lower hydraulic conductivity in the northern model part. The steepest hydraulic gradient in south reflects the lowest values of hydraulic conductivity in the Vega Alta aquifer, which is a typical relationship between different zones in karst aquifers [5]. The water-table gradient is almost flat (0.57 to 0.76 meters per kilometer) in the Aymamon limestone. It increases to the south in the Aguada limestone (2.85 to 3.79 meters per kilometer) toward the outcrop areas of the Cibao formation (model boundary).

Fig 9 shows the long term variation of the water table at three USGS wells and the water table rise of about 0.76 m at Sabana Hoyos #2 well over the last three decades. The rise is due to greater mean annual rainfall and the shutdown of some of the extraction. The simulated transient heads agree well with observed values, validating the assumption of constant withdrawal rates. This also suggests the reliability of EPM approach for replicating annual groundwater level fluctuations in northern Puerto Rico karst system, which agrees well with similar studies [5, 7, 12, 13] but provides validation in higher temporal resolutions (i.e. monthly and daily). Intermediate-term (several years) variation in water table corresponds to wet years, dry years and groundwater withdrawal activities. After 1967, the next severe dry period of the twentieth century in Puerto Rico was from 1993 to 1995 [54], when the annual rainfall accumulation was below normal with the greatest rainfall deficit.

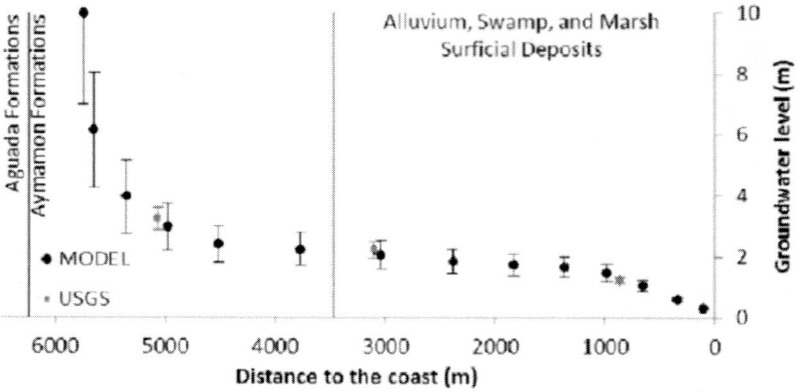

Figure 8. Simulated and measured groundwater levels vs. variable distance to coast. The mean and standard deviation for water level fluctuations in 3 USGS monitoring wells (red) and simulated points along the cross section A-B (see Fig 1). Distance dividers indicate the limit of surficial deposits and the Aguada-Aymamon intersection. https://doi.org/10.1371/journal.pone.0138954.g008

Precipitation is often higher in winter than in summer. Therefore, the karst storage may not be fully recharged in summer and can demonstrate seasonal water table variations. The study results reveal a strong relationship between rainfall and water table fluctuations, demonstrating that the groundwater table was mostly recharged by the rainfall. In addition to climatic conditions, water table fluctuation also responds to the closing of contaminated public water supply wells and the operation of new ones between February 1983 and April 1992, after which there have not been more wells reported as closed. The results illustrate that the water table level achieved its overall minimum in 1995 and maximum in 2011.

Monthly rainfall intensities varied between 0 and 42.67 cm/month, with an average intensity of about 14.73 cm/month. The water table responds soon after the occurrence of intense rainfalls which suggests rapid infiltration and limited delayed recharge through vadose zone. The match between the observed and simulated response of the water table to rainfall events can be well reproduced by the model (Fig 9). For example, short-term seasonal groundwater levels in 2005 varied up to a maximum of 0.6 m according to both measured and simulated data (Fig 9). Analyses of different recharge events suggest that aquifer replenishment in the model responds greatly to rainfalls of any magnitude. It can be observed in the model results that the increasing response to short-term precipitation events, such as storms, is faster than the groundwater decline following the storm.

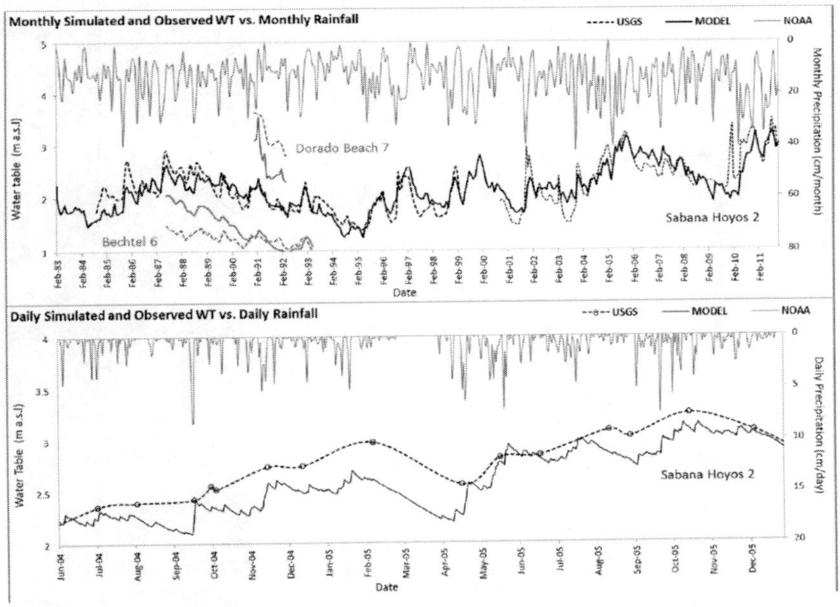

Figure 9. Monthly and daily averaged hydrogeologic conditions in Vega Ata aquifer. Simulated (solid line) and observed (dotted line) monthly water level hydrographs of three USGS wells vs. precipitation between 1983–2011 (upper); simulated and observed daily water levels in one well between June 2004 –December 2005 (lower). https://doi.org/10.1371/journal.pone.0138954.g009

Water table recovery, which is important in aquifer protection and sustainable water resource management, is affected by factors including precipitation rate, recharge and discharge rates, porosity, topography, and geologic structure. The water table recovery after each precipitation event was faster in the wet season and is observed several hours or several days after rainfall events in the study area. The linear correlation of monthly rainfall intensities with the water table recovery observed by Chuanmao [55] is acknowledged by the model results of the present study in the Vega Alta aquifer. It is thus possible to estimate the water table raise in each month based on monthly rainfall intensity. For example, the linear regression equation for water table recovery at Sabana Hoyos #2 well (R^2 = 0.9132) is shown on Fig 10, where ΔWT is monthly recovery of water table (m), and P is the monthly precipitation (cm/ month).

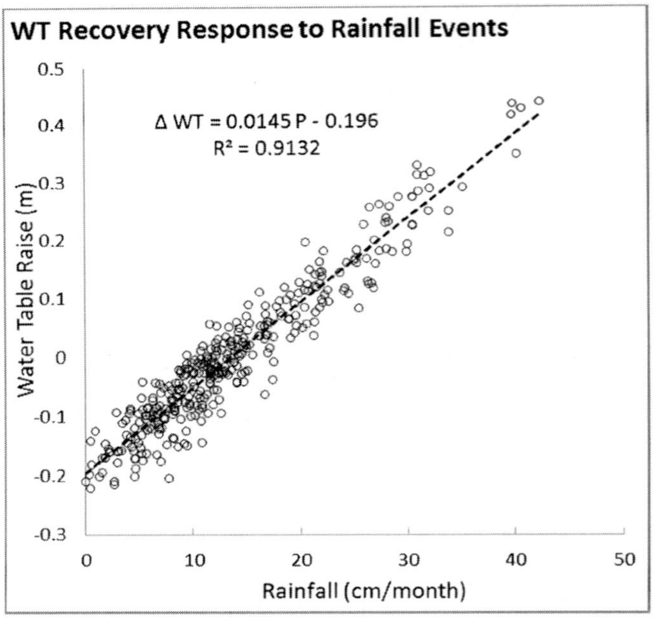

Figure 10. Scatter diagram of simulated water level recovery as a linear function of rainfall. On average, a monthly rainfall less (more) than 13.5 cm would lower (increase) the water table in USGS Sabana Hoyos #2 well. https://doi.org/10.1371/journal.pone.0138954.g010

The hydrogeologic effects of subsurface conduit networks cannot be reproduced using an EPM modeling approach, nor can be well detected using limited number of head monitoring wells in this study. The developed flow model for Vega Alta karst aquifer adequately simulates the spatial distribution and high-resolution temporal variation of groundwater levels over 21 years. It also uses monthly-refined rainfall, and captures the water table fluctuations in a larger temporal resolution for a shorter time period using daily-refined rainfall. Other flow modeling studies in northern Puerto Rico (e.g. Sepúlveda, 1996; Cherry, 2001) only demonstrated EPM's ability to regenerate steady-state regional hydrogeology, or transient simulations using very coarse rainfall resolution (e.g. annual).

3.2 Contaminant Transport Model

The calibration was accomplished by adjusting C_{RCH} value to 1,700 μg/l in the TCE source cell. Based on this value, the average TCE influx from the source area can be calculated by multiplying C_{RCH} with the recharge rate and the TCE source cell size, resulting in 7.035 kg/yr TCE influx into the unconfined aquifer. To control the mixing of the TCE in the water, uniform scale-dependent longitudinal, transverse, and vertical dispersivity values of 3, 0.3, and 0.03 m, respectively, were resulted during the calibration for all model layers, with assuming a typical molecular diffusion coefficient k_m of 10^{-5} cm^2/s. The calibrated model agrees well with observations in 10 monitoring wells in the area, and the correlation coefficient between simulated and measured TCE concentrations is 0.857 (Fig 11). Besides the limited field data used, the model simulation of concentrations produced an overall RMSE of 59.10 μg/l (normalized RMSE = 8.45%). This correlation suggests that the use of an EPM approach to simulate solute transport in an intermediate scale for this eogenetic karst limestone aquifer is justified. As noted by other studies [5], EPMs are not suitable for transport modeling in a local scale, where in-depth knowledge of distribution patterns of fracture and conduit networks is required to describe the complex transport processes.

TCE concentrations were measured at several observation wells in discrete sampling zones at different depths with the deepest at -82.6 m a.s.l. Percentile curve for the simulated TCE concentrations show good agreement with that for the TCE measurements distributed over the focus area (Fig 12). High degrees of spatial and vertical variation in TCE concentrations indicate the necessity in three dimensional simulating of the complex karst environment. The zone of high TCE concentrations extends to the depth of the deepest wells in the area (-61 m a.s.l).

In northern Puerto Rico, 56.6% of samples had TCE concentrations above detection limit (DL) and 11.5% of those samples had TCE concentrations above maximum contamination level (MCL) of 5 μg/l established by EPA. The mean, median, standard deviation, and coefficients of variation (CV %) for all measured TCE concentrations were 83.7 μg/l, 46.3 μg/l, 112.6 μg/l, and 41.1%, respectively, indicating that concentrations were substantially skewed (γ: 2.78) toward lower values. The sampling frequency and distribution have decreased significantly over recent years causing spatial uncertainties [56, 57]. Historically no significant seasonal fluctuation of TCE concentrations was observed in the taken samples (Yu et al., 2015); therefore in 2007, EPA proposed to reduce the sampling frequency from quarterly to semi-annually or annually. No TCE concentration was reported for Mackovic Spring in Vega Alta or Maguayo Spring in Dorado, both located outside of the focus area. Variability of spatially detected concentrations is commonly associated with complex subsurface hydrodynamics that affect storage and mobility of contaminants.

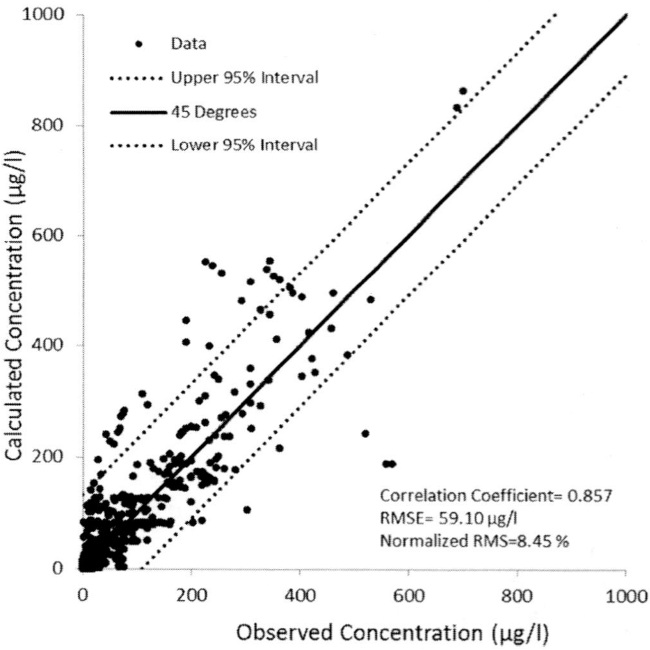

Figure 11. Transient calibration diagram for the transport model. Scatter diagram of simulated vs. observed TCE concentrations at various depths in 10 monitoring wells between January 1990 and January 2008. https://doi.org/10.1371/journal.pone.0138954.g011

The initial distribution of TCE concentration and the simulated spreading of TCE plume in the focus area at a depth of -30.48 m a.s.l between 1993 and 2008 are presented in Fig 13. The presented observation data are interpolated values collected from monitoring zones between depths of -25.91 to -36.58 m a.s.l. The spreading of the plume is simply observed by the outermost concentration contour which corresponds to the MCL (5 µg/l). The dominant flow path in the focus area is slightly northeast and downward towards the deeper aquifer, and concentrations directly south of the Industrial Park are negligible.

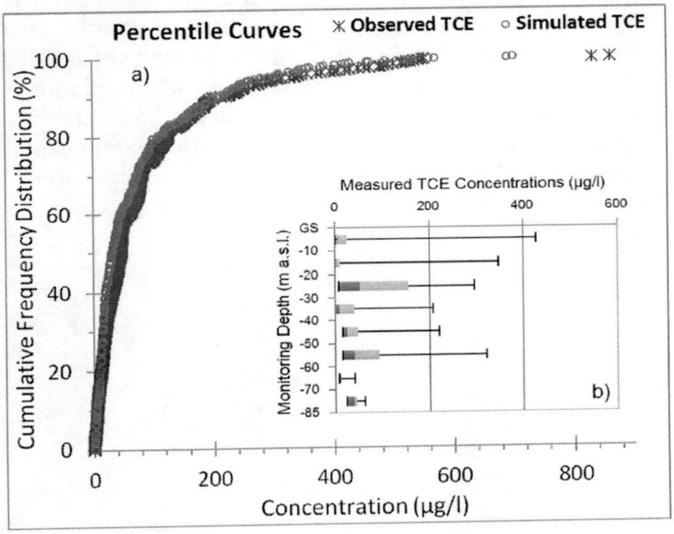

Figure 12. Cumulative frequency and vertical distribution of detected TCEs in groundwater. (a) Cumulative quantile plots of observed and simulated TCE concentrations were obtained from historical data and model simulations, respectively. (b) Box plots for the depth distribution of observed TCE concentrations indicate 5%, 25%, 50%, 75%, and 95% quantiles (GS: ground surface).

Due to the very low transport velocities, the TCE plume is very unlikely to reach the Atlantic Ocean. Sensitivity analysis suggests that the plume would spread further downward and to the north if no sorption of TCE was accounted for. The vertical hydraulic gradient becomes smaller toward deeper parts of the aquifer and toward the Atlantic Ocean. Accordingly, the simulation indicates a slightly downward and eastward transport in addition to the general northward movement of the TCE plume especially in the shallower parts of the aquifer.

The closure of some public water supply wells on the right side of focus area until 1992 has pushed the slight northeast groundwater direction further to the east. The unknown preferential flow paths and complex network of fractures, conduits and channels in this karst aquifer causes irregular spreading patterns and affects the contaminant dispersal and storage mechanisms. Measured TCE concentrations in different wells within the focus area had different temporal patterns reflecting high subsurface anisotropy and heterogeneity and different attenuation rates (also see Yu et al., 2015). In some wells, high variations of TCE concentrations were observed between consecutive monitoring events that are not captured by the model, e.g. after the plume had progressed enough (2005), the TCE concentrations in one site north of the plume started to increase gradually over time which was not seen by the model. The highest TCE concentration occurs directly downstream of the source cell and as the distance to the source increases, generally the TCE concentration decreases.

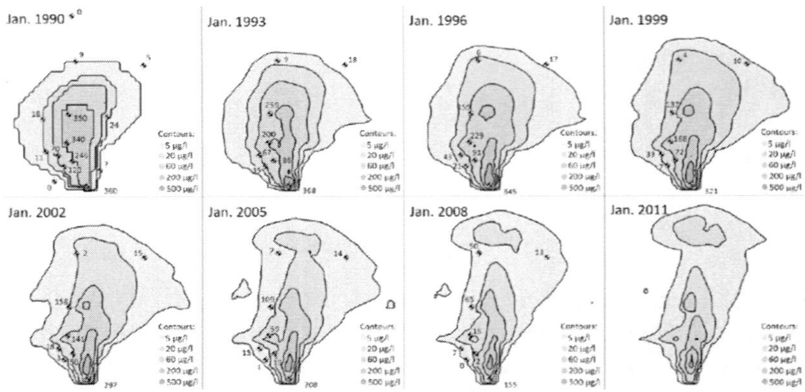

Figure 13. The migration and extent of TCE plume over two decades. Initial TCE contours were based on available data at January 1990, and simulated TCE contours and observed TCE data (unit: µg/l) were at eight monitoring wells at a depth of -30.48 m a.s.l between January 1993 and January 2011 in the focus area (4.6 × 4.8 km, see Fig 1). No observation data was available for 2011. https://doi.org/10.1371/ journal. pone. 0138954.g013

There are notable decreasing trends in the values and areas of TCE contamination on the temporal scale, as supported by the recent measured data. The total mass of TCE in Vega Alta aquifer was estimated as 5889.4 kg (equivalent to TCE volume of 4.033 m^3) in January 1990. Although the annual TCE influx (7.035 kg) was contributing to the subsurface plume, the total contamination reduced to estimated total mass of 2501.4 kg in January 2011. Besides sorption and natural attenuation, this is partially because public water supply wells (before closure) and industrial and agricultural pumping wells withdrew the aquifer within the extent of TCE plume in the focus area. The TCE removal rate is found to be higher at shallower layers where vertical hydraulic gradient is larger.

Freshwater resources within the northern karst systems of Puerto Rico are highly vulnerable to contamination. Especially the recent industrial and urban development and increased number of contaminated sites within the island lead to unintended contamination of the groundwater resources. The subsequent long-term contamination indicates a large capacity of the karst aquifers for storing and releasing contaminants which reflects potential lasting exposures.

4. CONCLUSIONS

This paper evaluates the application and adequacy of EPM approach to simulate groundwater flow and contaminant fate and transport processes within highly permeable karst environments, which is crucial for managing water resources management purposes. Where little to nothing is known about the occurrence and properties of conduit and fracture networks, approximating the complex nature of karst systems with equivalent hydraulic properties may sufficiently reproduce the aquifer behavior at regional and intermediate scales. Locating those networks and identifying the associated hydrogeological parameters are impossible without large scale exploration efforts. Using limited field data, the developed MODFLOW and MT3D models in this study adequately simulated the spatial distribution and high-resolution temporal variation of groundwater levels and the decreasing trend of TCE plume over 21 years, respectively, for Vega Alta karst aquifer. Withdrawals from a large well field, as the main regional discharge feature, are evaluated to significantly impact the water table dynamics and the dominant direction of plume migration. The occurrence of preferential flow influences the flow conditions and leads to irregular spreading of TCE plume in the aquifer. The main limitations of equivalent porous media approach in karst aquifers are failing to capture groundwater hydrodynamics on a local scale, and to simulate transient turbulent flow through conduit network and its interaction with the matrix.

ACKNOWLEDGMENTS

We thank Christine Gordon for her editorial help on the paper. We also thank the editor and the anonymous reviewers in advance for their valuable comments.

REFERENCES

1. Ford DC, Williams PW (2007) Karst hydrogeology and geomorphology. John Wiley & Sons 576.
2. Butscher C, Auckenthaler A, Scheidler S, Huggenberger P (2011) Validation of a numerical indicator of microbial contamination for karst springs. Ground Water 49(1):66–76. pmid:20180864
3. Chen Z, Goldscheider N (2014) Modeling spatially and temporally varied hydraulic behavior of a folded karst system with dominant conduit drainage at catchment scale, Hochifen-Gottesacker, Alps. Journal of Hydrology 514:41–52.
4. Liedl R, Sauter M, Huckinghaus D, Clemens T, Teutsch G (2003) Simulation of the development of karst aquifers using a coupled continuum pipe flow model. Water Resources Research 39(3):1057.
5. Scanlon BR, Mace RE, Barrett ME, Smith B (2003) Can we simulate regional groundwater flow in a karst system using equivalent porous media models? Case study, Barton Springs Edwards aquifer, USA. Journal of Hydrology 276:137–158.
6. Worthington SRH (2009) Diagnostic hydrogeologic characteristics of a karst aquifer (Kentucky, USA). Hydrogeology Journal 17:1665–1678.
7. Abusaada M, Sauter M (2013) Studying the Flow Dynamics of a Karst Aquifer System with an Equivalent Porous Medium Model. Groundwater 51:641–650.
8. Bodin J, Ackerer P, Boisson A, Bourbiaux B, Bruel D, Dreuzy JR de, et al. (2012) Predictive modelling of hydraulic head responses to dipole flow experiments in a fractured/ karstified limestone aquifer: Insights from a comparison of five modelling approaches to real-field experiments. Journal of Hydrology 454–455.
9. Ghasemizadeh R, Hellweger F, Butscher C, Padilla I, Vesper D, Field M, et al.(2012) Review: Groundwater flow and transport modeling of karst aquifers, with particular reference to the north coast limestone aquifer system of Puerto Rico. Hydrogeology Journal 20:1441–1461. pmid:23645996
10. Panagopoulos G (2012) Application of MODFLOW for simulating groundwater flow in the Trifilia karst aquifer, Greece. Environmental Earth Sciences 67:1877–1889.
11. Anderson MP, Woessner WW (1992) Applied groundwater modeling: simulation of flow and advective transport. Academic Press, New York, 381p.
12. Long JCS, Remer JS, Wilson CR, Witherspoon PA (1982) Porous Media Equivalents for Networks of Discontinuous Fractures. Water Resources Research 18(3):645–658.
13. Pankow JF, Johnson RL, Hewetson JP, Cherry JA (1986) An evaluation of contaminant migration patterns at two waste disposal sites on fractured porous media in terms of the equivalent porous medium (EPM) model. J. Contaminant Hydrology 1(1):65–76.
14. Hartmann A, Goldscheider N, Wagener T, Lange J, Weiler M (2014) Karst water resources in a changing world: Review of hydrological modeling approaches. Reviews of Geophysics, 52(3):218–242.
15. Putnam LD, Long AJ (2009) Numerical groundwater-flow model of the Minnelusa and Madison hydrogeologic units in the Rapid City area, South Dakota. U.S. Geological Survey Scientific Investigations Report 2009–5205.
16. Saller SP, Ronayne MJ, Long AJ (2013) Comparison of a karst groundwater model with and without discrete conduit flow. Hydrogeology Journal 21(7):1555–1566.
17. Quinn J, Tomasko DA (2000) Numerical Approach to Simulating Mixed Flow in Karst Aquifers. Groundwater Flow and Contaminant Transport in Carbonate Aquifers 147–156.
18. Quinn J, Tomasko D, Kuiper JA (2006) Modeling complex flow in a karst aquifer. Sedimentary Geology 184(3):343–351.
19. Lindgren RJ, Dutton AR, Hovorka SD, Worthington SRH, Painter S (2004) Conceptualization and simulation of the Edwards Aquifer, San Antonio Region, Texas. U.S. Geological Survey Scientific Investigations Report 2004–5277.
20. McDonald MG, Harbaugh AW (1988) A modular three-dimensional finite-difference ground-water flow model. U.S. Geological Survey Report 83–875.

21. Zheng C, Wang PP (1999) MT3DMS: A Modular Three-Dimensional Multispecies Transport Model for Simulation of Advection, Dispersion, and Chemical Reactions of Contaminants in Groundwater Systems; Documentation and User's Guide. U.S. Army Engineer Research and Development Center Contract Report SERDP-99-1; Vicksburg, Mississippi.
22. Lugo AE, Castro LM, Vale A, del Mar Lopez T, Prieto EH, Martino AG (2001) Puerto Rican karst—A vital resource. U.S. Department of Agriculture and Forest Service General Technical Report WO-65.
23. Molina-Rivera WL, Gómez-Gómez F (2008) Estimated water use in Puerto Rico, 2005: 454 U.S. Geological Survey Open-File Report 2008–1286.
24. Renken RA, Ward WC, Gill IP, Gómez GF, Rodriguez JM (2002) Geology and Hydrogeology of the Caribbean Islands Aquifer System of the Commonwealth of Puerto Rico and the US Virgin Islands. U.S. Geological Survey Professional Paper 1419:1–139.
25. Ginés A, Ginés J (2007) Eogenetic karst, glacioeustatic cave pools and anchialine environments on Mallorca Island: a discussion of coastal speleogenesis. International Journal of Speleology 36(2):57–67.
26. Padilla I (2013) Fate of DNAPLs in eogenetic karst aquifers: the case of northern Puerto Rico. 2013 GSA Annual Meeting, Denver CO.
27. Anaya A, Padilla I, Macchiavelli R, Vesper D, Meeker J, Alshawabkeh A (2014) Estimating preferential flow in karstic aquifers using statistical mixed models. Groundwater 52(4):584–596.
28. Gulley JD, Martin JB, Moore PJ, Brown A, Spellman PD, Ezell J (2015) Heterogeneous distributions of CO2 may be more important for dissolution and karstification in coastal eogenetic limestone than mixing dissolution. Earth Surface Processes and Landforms.
29. Budd DA, Vacher HL (2004) Matrix permeability of the confined Floridan Aquifer, Florida, USA. Hydrogeology Journal 12:531–549.
30. Monroe WH (1963) Geology of the Vega Alta Quadrangle, Puerto Rico, U.S. Geological Survey Geologic Quadrangle Map GQ-191.
31. Monroe WH (1980) Geology of the middle Tertiary formations of Puerto Rico. U.S. Geological Survey Professional Paper 953.
32. Giusti EV, Bennett GD (1976) Water resources of the North Coast limestone area. U.S. Geological Survey Water-Resources Investigations Report 42–75.
33. Giusti EV (1978) Hydrogeology of the karst of Puerto Rico. U.S. Geological Survey professional paper 1012. 68 p.
34. Sepúlveda N (1996) Three-dimensional ground-water-flow model of the water-table aquifer in Vega Alta, Puerto Rico. U.S. Geological Survey. Water-Resources Investigations Report 95–4184.
35. Rodriguez JM (1995) Hydrology of the North Coast limestone aquifer system in Puerto Rico. U.S. Geological Survey Water Resources Report 94–4249.
36. Rodriguez JM, Hartley JL (1994) Geologic and Hydrologic Data Collected at Test Holes NC-6 and NC-11, Hatillo and Isabela, Northwestern Puerto Rico. U.S. Department of the Interior, U.S. Geological Survey 93–465.
37. Rodriguez JM, Richards RT (2000) Detection of conduit-controlled ground-water flow in northwestern Puerto Rico using aerial photograph interpretation and geophysical methods. U.S. Geological Survey Water Resource Investigation Report 00–4147.
38. US Environmental Protection Agency (1997) Superfund Record of Decision (EPA Region 2): Vega Alta Public Supply Wells, Operable Unit 2, Vega Alta PR. OCLC Number: 48664598, Stock Number: PB97-963821; EPA ID: PRD980763775.
39. Rodriguez MJ (1997) Characterization of springflow in the North Coast limestone of Puerto Rico using physical, chemical, and stable isotopic methods. U.S. Geological Survey Water Resources Investigations Report 97–4122.
40. Veve TD, Taggart BE (1996) Atlas of groundwater resources in Puerto Rico and the US Virgin Islands. U.S. Geological Survey Water Resource Investigation Report 94–4198.
41. Gómez GF, Torres-Sierra H (1988) Hydrology and effects of development on the water-table aquifer in the Vega Alta quadrangle. U.S. Geological Survey Water-Resources Investigations Report 87–4105.

42. Guzmán RS, García R, Avilés A (1986) Reconnaissance of Volatile Synthetic Organic Chemicals at Public Water Supply Wells throughout Puerto Rico, November 1984–May, 1985. U.S. Geological Survey. Open-File Report 86–63.
43. Zack A, Rodriguez-Alonso T, Romas-Mas A (1987) Puerto Rico Groundwater Quality. U.S. Geological Survey. Open File Report 87–0749.
44. Guzmán RS, Quiñones MF (1984) Ground-water quality at selected sites throughout Puerto Rico, September 1982-July 1983. U.S. Geological Survey. Open-File Rep 84–05.
45. Schlumberger (2011) Visual MODFLOW 2011.1 User's Manual. Schlumberger Water Services (SWS). Kitchener, ON, Canada.
46. Sepúlveda N (1999) Ground-water flow, solute transport, and simulation of remedial alternatives for the water-table aquifer in Vega Alta, Puerto Rico. U.S. Geological Survey. Water-Resources Investigations Report 97–4170.
47. Gómez-Gómez F (2008) Estimation of the Change in Freshwater Volume in the North Coast Limestone Upper Aquifer of Puerto Rico in the Río Grande de Manatí-Río de la Plata Area between 1960 and 1990 and Implications on Public-Supply Water Availability. U.S. Geological Survey. Scientific Investigations Report 2007–5194.
48. Torres-Sierra H (1985) Potentiometric surface of the upper limestone aquifer in the Dorado-Vega Alta area, north-central Puerto Rico, February 1983. U.S. Geological Survey Water-Resources Investigations Report 85–4268.
49. Halihan T, Mace RE, Sharp JM (2000) Flow in the San Antonio segment of the Edwards aquifer: matrix, fractures, or conduits? Groundwater flow and contaminant transport in carbonate aquifers Wicks and Sasowsky (eds), AA Balkema, Rotterdam, Netherlands 129–146.
50. Romanov D, Gabrovšek F, Dreybrodt W (2003) Dam sites in soluble rocks: a model of increasing leakage by dissolutional widening of fractures beneath a dam. Engineering Geology 70(1):17–35.
51. Halihan T, Sharp JM, Mace RE (1999) Interpreting flow using permeability at multiple scales. Karst Modeling: Proceedings of the symposium Charlottesville, VA. Palmer, Palmer, and Sasowsky (eds). Special Publication 5, Karst Waters Institute 82–96.
52. Mehran M, Olsen RL, Rector BM (1987) Distribution Coefficient of Trichloroethylene in Soil Water Systems. Ground Water 25:275–282.
53. Cherry GS (2001) Simulation of flow in the upper North Coast limestone aquifer, Manatí-Vega Baja area, Puerto Rico. U.S. Geological Survey Water Resources Investigation Report 00–4266.
54. Larsen MC (2000) Analysis of 20th century rainfall and streamflow to characterize drought and water resources in Puerto Rico. Physical Geography 21(6):494–521.
55. Chuanmao J (1982) Variation of the groundwater regime under the effects of human activities and its artificial control Improvements of Methods of Long Term Prediction of Variations in Groundwater Resources and Regimes Due to Human Activity. IAHS Publ No. 136.
56. Yu X, Ghasemizadeh R, Padilla I, Irizarry C, Kaeli D, Alshawabkeh A (2015) Spatiotemporal changes of CVOC concentrations in karst aquifers: Analysis of three decades of data from Puerto Rico. Science of the Total Environment 511:1–10. pmid:25522355
57. Yu X., Ghasemizadeh R., Padilla I., Meeker J. D., Cordero J. F., & Alshawabkeh A (2015) Sociodemographic patterns of household water-use costs in Puerto Rico. Science of the Total Environment 524:300–309. pmid:25897735

CHAPTER 16

Developing an Approach for a Three-Dimensional Hydrogeologic Framework to Find Water in Northern Ghana

Shannon Holt[1], Alexandra Lutz[1*], Bansaga Saga[2], David L. Berger[1], James M. Thomas[1], Braimah Apambire[1]

[1]*Desert Research Institute, Division of Hydrologic Sciences, Reno, USA*

[2]*Ghana Integrated Water, Sanitation, and Hygiene Program, World Vision Ghana, Savelugu, Ghana*

ABSTRACT

The purpose of this study was to develop an approach for constructing a three-dimensional hydrogeologic framework using borehole driller logs. The ultimate goal is to increase drilling success rates via a better understanding of the regional hydrogeologic framework in northern Ghana. Groundwater development has increased in northern Ghana, but drilling successful boreholes is difficult due to complex geology and limited aquifer characteristic information. An approach was developed to construct a three-dimensional hydrogeologic framework of the basin using 900 borehole logs from World Vision International's Ghana Integrated Water, Sanitation, and Hygiene Project, located in northern Ghana. The study's approach consists of: evaluating potential software programs; collecting borehole drilling logs; data QA/QC; data standardization and normalization; analysis for trends and correlations; and creation of a three-dimensional hydrogeologic framework and two-dimensional cross sections. This approach can be used and adapted by others working to provide groundwater in developing countries.

Keywords: Groundwater Management, Hydrogeology, Groundwater Resources, Hydrogeologic Framework, Ghana

1. INTRODUCTION

This study develops an approach for a three-dimensional (3D) hydrogeologic framework that will contribute to a better understanding of the regional hydrogeological setting in the administrative Northern Region of Ghana, West Africa (Figure 1). Demand for groundwater in the Northern Region is intensifying because of rapidly increasing population, socio-economic development, and unreliable/poor quality surface-water resources. Developing groundwater, however, is not straightforward in the Northern Region; complicated geological features cause difficulties for drilling successful boreholes and low drilling success rates are reported both anecdotally and in the literature [1]. Complicated hydrogeologic features also cause difficulties for describing aquifer characteristics—they are highly spatially variable and often cannot be described in detail due to financial and temporal constraints.

With insufficient hydrogeological information to guide geophysicists and drillers, success rates remain low in particular areas of northern Ghana. In order to improve drilling success rates, a better understanding of regional hydrogeology is needed. The purpose of this study is to develop an approach for constructing a 3D hydrogeologic framework using driller's borehole logs. The ultimate goal is increasing drilling success rates via a better understanding of the regional hydrogeologic framework. In turn, a better understanding of the hydrogeology contributes to sustainable groundwater development and management. This approach can be used and adapted by others working to provide groundwater, as there is a need for regional studies to contribute to management frameworks and trans boundary issues [2].

For over two decades, World Vision International's Ghana Integrated Water, Sanitation, and Hygiene (GI-WASH) Project has been working to provide safe drinking water to populations in rural areas of Ghana. During the past decade, work has focused on the Northern Region for reasons described above. Hundreds of boreholes have been drilled, and driller's logs record the sub-surface hydrogeological information at each location.

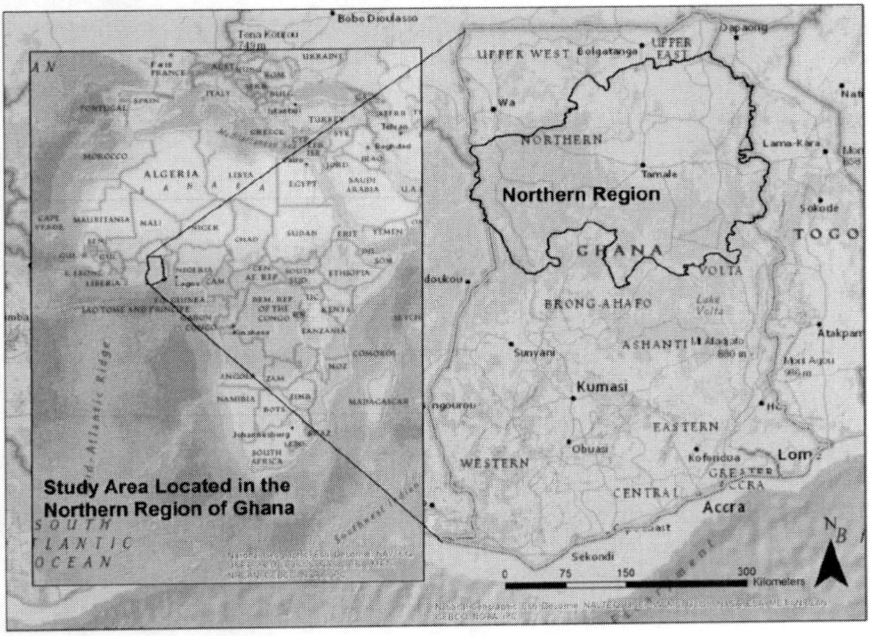

Figure 1. The administrative Northern Region of Ghana.

Information includes: well location, geology type at depth intervals, total depth, water level, season, and wet or dry borehole. Because these logs contain potentially valuable information, GI-WASH and the Desert Research Institute (DRI) collaborated to analyze these borehole logs in an effort to better understand the regional hydrogeology.

A regional hydrogeologic study typically consists of several steps and is conducted over a number of years. Other approaches taken on the African continent consist of regional groundwater flow models and transboundaryaquifer mapping [2] [3]. The United States Geological Survey (USGS) suggests a methodology consisting of: combining digital elevation models with geologic maps to provide points indicating surface exposures of geologic formations; creation of cross sections from borehole logs to define locations of hydrogeologic units; gridded surface interpolation for each of the hydrogeologic units; a map depicting the trace of faults; interpolation of surface and subsurface data using gridding algorithms; and applying appropriate stratigraphic principles [4]. Gathering of data is an initial and underlying procedure. Since this research focuses on a 3D framework that may contribute to a larger hydrogeological study, a modified approach of combined methods [4] - [6] is taken by GI-WASH and DRI. An outline of the modified approach is: evaluation of appropriate software; data collection; data QA/QC; data standardization; comparisons analysis; and aquifer calculations. More details are given in the approach section. Maps in Figure 1, Figure 2, Figure 5, and Figure 6 were created using ArcGIS® software by Esri. ArcGIS® and ArcMap™ are the intellectual property of ESRI and are used herein under license. For more information about ESRI®software, please visit www.esri.com.

2. STUDY AREA BACKGROUND

2.1. Groundwater Development and Management

Traditionally, surface water has been a primary water source in northern Ghana, but it is no longer viable due to water-borne pathogens, pollution, changes in rainfall patterns, and unavailability during long drought periods [7] [8]. This leads to groundwater development as a cost effective means for supplying rural communities with potable water [9]. Groundwater in Ghana generally has good chemical and microbial quality as compared to surface water, which makes it the preferred choice for safe drinking water [10].

Rural, subsistence family farms describe much of the setting of northern Ghana. The population in the Northern Region makes up approximately 20% of the total population of Ghana [7] [10] , which is about 5 million people based on a 2013 population estimate [11]. Sixty-eight percent of the total population of the Northern Region lives in rural communities. Rural communities are defined as less than 5000 inhabitants [12]. Approximately 70% of the population relies on subsistence farming as a primary source of income [8]. Across West Africa, food security, droughts, and famine are persistent problems. Groundwater is not only important as a source of potable water, but can also improve the socio-economic status of a community. In sub-Saharan Africa, poverty rose from 42% in 1981 to 47% in 2001, and was mainly attributed to issues of food security. In comparable developing countries in Asia, poverty levels decreased due to the development of groundwater for irrigation purposes, which increased food security [8].

Starting in the 1970s, the government of Ghana commissioned a formal policy that communities with fewer than 500 inhabitants be equipped with hand dug wells, and communities with up to 2000 inhabitants be supplied with boreholes fitted with hand pumps [12]. In 1998, approximately 52% of the rural population in Ghana depended on boreholes fitted with hand pumps or open wells [7]. In 2012, it was estimated that 81% of all rural Ghanaians had access to an improved water source [11]; it is noted that this number reflects the country as a whole and not northern Ghana in particular.

2.2. Study Area Hydrogeology

The Northern Region of Ghana is within the tropical continental or savannah zone, which experiences a single rainfall season and a long drought period, effectively separating climate into rainy and dry seasons. Rains begin in May, peak between July and August, and taper off by October; the remaining months experience dry, drought-like conditions. In the Savelugu-Nanton District, where the headquarters for GI-WASH is located, mean annual rainfall ranges between 1005 and 1150 mm [13]. Annual temperature is 18°C and humidity ranges from 65% to 85% during the rainy season, while temperature is 42°C and humidity is 20% during the dry season [9] [13]. Evapotranspiration rates in northern Ghana average 890 mm $year^{-1}$, but depends on season, location, and local vegetation [8].

As already discussed, there are complicated geological features in the Northern Region of Ghana. The Voltaian Sedimentary Basin, which underlies significant portions of the region, consists of sandstone, shale, mudstone, sandy and pebbly beds, shown in orange and red on Figure 2 [7] [12] [14]. More geological detail can be found in Yidana et al. [15]. Primary porosity of the Voltaian is limited due to high levels of consolidation and cementation. Thus, preferential groundwater flow paths are along the strike direction of fractures and weathered zones [7] [16] [17].

The fractures and weathered zones result in large variability in hydrogeologic conditions. For example, recharge of groundwater to the Voltaian can range from 2% to 4% of the annual rainfall, depending on the location [18]. Hydraulic conductivity has the potential to increase by several orders of magnitude over a short distance of a few meters [17]. Numerical modeling in a small sub-section of the Northern Region suggests hydraulic conductivity values for fractured sandstone range between 0.01 and 3 $m \cdot day^{-1}$and average transmissivity to be 13 $m^2 \cdot day^{-1}$ [7]. Across the region, transmissivity is reported to range from 1 to 30 $m^2 \cdot day^{-1}$ [19] and more specifically as 0.1 to 52.0 $m^2 \cdot day^{-1}$ and 0.2 to 16.0 $m^2 \cdot day^{-1}$ for sandstone and siltstone/mudstone aquifers, respectively [15].

The highly variable hydrogeologic conditions affect borehole drilling success rates. As much as possible, GI-WASH drillers target fractures and weathered zones of the Voltaian by using geophysical exploration, available information, and anecdotal knowledge. On average, boreholes reach a depth of 48 m and they rarely exceed 90 m [8]. In 1996, a 55% success rate for wet boreholes was reported [1]. Since then, anecdotal reports of drilling success rates vary between 50% and 80%, with the variance due to localized hydrogeological conditions [20].

Pumping yield can also be highly variable, ranging from 9.8 to 216 $m^3 \cdot day^{-1}$ (6.8 to 150 $L \cdot min^{-1}$) in the Voltaian [12]. In this study, yield is considered to be the maximum rate at which a borehole is pumped without changing the groundwater level during the test [21]. It is noted in this study that GI-WASH pump test methods are modified out of necessity to increase efficiency in the provision of water. Thus, several assumptions are made for representing the aquifer in this study. These assumptions and their limitations are discussed later. At any borehole location, GI-WASH strives for a minimum yield of 18.7 $m^3 \cdot day^{-1}$ (13 $L \cdot min^{-1}$) in order to be considered "successful" and develop into a production well to provide adequate water supply to a community of approximately 400 people [22]. In this study, boreholes producing less than the minimum yield are considered dry wells.

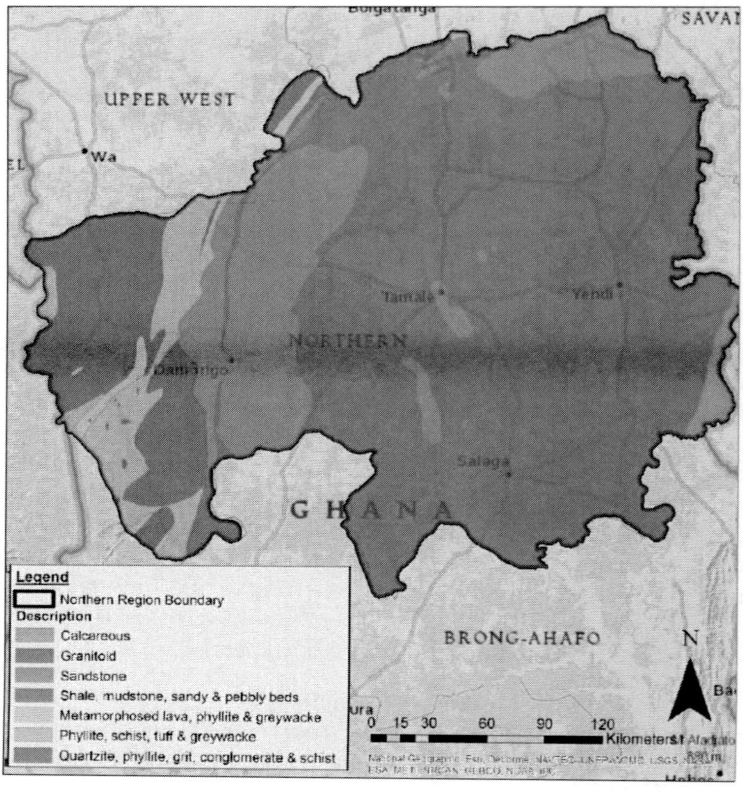

Figure 2. Geologic composition of the Northern Region of Ghana.

3. METHODS AND APPROACH

The following section describes in greater detail the approach used for constructing a three-dimensional hydrogeologic framework using existing GI-WASH driller's borehole logs. As already described, the approach is modified from several sources [4] - [6]. The modified approach consists of: evaluation of potential software programs; data collection; data QA/QC; data standardization; comparison analysis; aquifer calculation; and construction of cross sections. Figure 3 shows an outline of the approach. It is noted that construction of the cross sections is intended to show that the approach is feasible. Eventually, with more data, we intend to conduct a more rigorous analysis of the hydrogeology.

3.1. Software Evaluation

Multiple software packages exist for managing, interpreting, and visualizing hydrogeologic data derived from driller's logs. The software package Rockworks16 [23] was chosen because it meets several requirements including: a user-friendly interface; easy to import and export data; capability to interact with ArcGIS and GoogleEarth; a detailed visual output for 3D maps and 2D cross sections; and the ability to add additional information such as water-quality data. Several limitations, however, became apparent when using this software and the study area's data. These limitations are addressed in the discussion section below.

3.2. Data Collection

Data collection consisted of scanning a total of 916 paper borehole drilling logs, recorded from 2002-2010, from GI-WASH during June and July of 2013. Information from the logs was then manually entered into a Microsoft Excel spreadsheet that matched the input parameters for Rockworks16 Borehole Manager Software [23]. Importing an excel sheet directly into the software is the best way to enter a large amount of data quickly. The input parameters include: location; lithology; stratigraphy; I-data, which are numeric values for samples at depth intervals; I-text which is non-numeric information over depth intervals; T-data which are numeric values sampled at depth intervals and time intervals; P-data which are numeric values sampled at single depth points; P-test, which is non-numeric information sampled at a single depth; fractures; water level data; and observed colors down the borehole. These data were input into the Rockworks16 Borehole Manager for each borehole log.

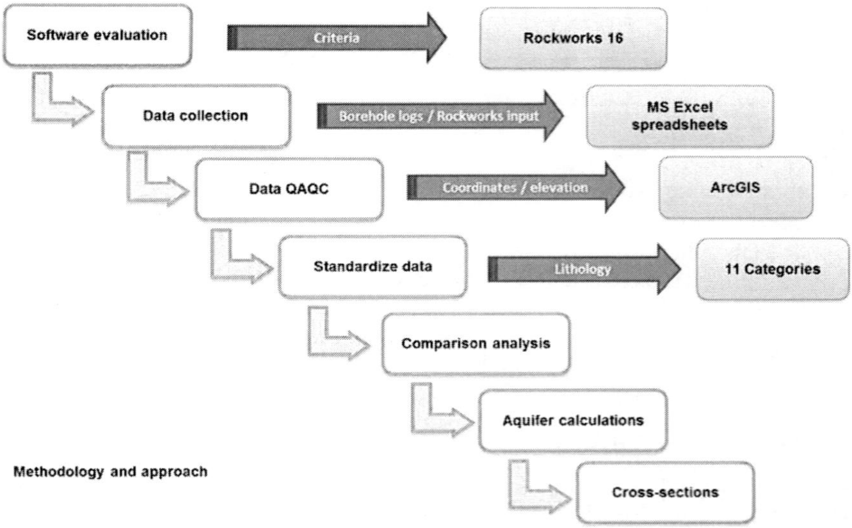

Figure 3. Flow chart of the methodologies and approach taken in this study.

3.3. Data QA/QC

Data QA/QC consisted of validating location (coordinates) and elevation information. The location of each well recorded in the borehole log was converted to a shapefile and overlain on the district map of the Northern Region using the ArcMap portion of Esri's ArcGIS. Each borehole was validated to ensure it resided in the correct district reported on the driller's log. When this information was incomplete, it was manually assigned a location in coordination with knowledge from GI-WASH staff and by using Ghana Survey Topography maps and ArcMap. Logs without coordinates, or coordinates that could not be corrected, were removed from the dataset. Missing elevation information was also updated using ArcMap. Digital elevation models were not used for this process since they are only freely available at 90 m resolution, which is too course for this study. Instead, boreholes were plotted over Ghana Geological Survey elevation contour maps of the country. Missing elevation data were read from the contour map. In cases where the borehole did not place on a contour line, the closest contour line's elevation was used. In areas where multiple known elevations were reported, an average was taken.

Upon completion of QA/QC, 879 of the 916 borehole logs were considered to have accurate coordinates and elevations. Assessing coordinate locations and elevation data was the only metric put into place to evaluate the quality of borehole logs before input into RockWorks16. Detailed information varied between logs, but due to the large study area it was decided that as many logs as possible be used to create cross sections in RockWork16 to minimize the interpolation of the regional hydrogeologic framework. Limitations due to interpolation processes are discussed later. Borehole location and elevation are critical for RockWorks16, as the software cannot manage borehole data lacking this information.

3.4. Data Standardization

Data standardization took place during data entry. Information from the borehole logs was entered into Microsoft Excel spreadsheets based on input parameters for the Borehole Manager RockWorks16 software and modified methodology [6] [24]. Diverse expertise of the drilling crews can result in variability of the level of detail described for soils and lithology formations, and the possibility of several different descriptions given for a single unit [24]. A standardization process developed by Ross et al. [5] groups attributes with similar lithology and Russell et al. [6] refined the methodology by rationalizing material descriptions to a limited number of groups that capture their main characteristics. In the rationalization process, descriptions that appear infrequently are removed or re-characterized, combination descriptions are simplified, and the remaining descriptions are reassigned to the appropriate groups. Because Russell et al. [6] used similar drilling methodology as performed in northern Ghana, methodology for standardizing lithology characteristics recorded by GI-WASH followed his example. In the case where descriptions are ambiguous, consultation with the drillers or a literature review took place to correctly characterize the description.

When reviewing the 879 GI-WASH borehole logs as per the methodology described [6] [24] , eleven groups emerged (Table 1). A literature review indicated that groundwater is most commonly found in fractures, therefore, lithology categories were focused around fracture characteristics as a dominant attribute. A fractured lithology was determined by the borehole log's description as either moderately or highly fractured. Slightly or not fractured were grouped into non-fractured lithologies. Sandstone, siltstone, mudstone, and shale were prevalent lithologies, which agrees with descriptions reported in the literature for the Voltaian [12] [14]. A relatively small percentage of borehole logs that describe areas of granite and other minerals found in granite (about 7%) were grouped into a single granite category.

In addition to location and lithology types, attributes from the logs also entered into the spreadsheets included: dominant lithology at depth intervals in the borehole; stratigraphic intervals such as confining layers and aquifer intervals; flow rate at intervals and final flow rate (yield); static and dynamic water levels; dry well; and well construction. The dominant lithology represented at the aquifer interval was recorded for each log. The aquifer interval refers to the interval between the depth to water and the final depth of the borehole. Depth to water is recorded on each log during the borehole drilling process. In all cases where the borehole intersects water during the drilling process the yield is measured at depth intervals from the first depth to water recording to the end of the borehole. If the aquifer interval spanned multiple lithology descriptions the most frequent lithologic unit and most frequent degree of fracturing was assigned as the dominant lithology. The dominant lithology for the aquifer interval was then used for comparison analysis. Of the 879 reported boreholes with logs, 368 were considered to be wet and were developed as water supply wells.

Table 1. Eleven dominant lithology types developed for this study.

1	Topsoil		
2	Sandstone	7	Fractured Sandstone
3	Mudstone	8	Fractured Mudstone
4	Siltstone	9	Fractured Siltstone
5	Shale	10	Fractured Shale
6	Granite	11	Fractured Granite

3.5. Comparison Analysis

Comparison analysis was done to evaluate well yield and aquifer lithology in order to begin to better characterize aquifers in the study area and determine favorable lithologic units for developing groundwater. For this comparison to be valid, each borehole must be constructed with the same well casing diameter and the method of measuring yield must be standardized. Fortunately, GI-WASH boreholes have the same well casing diameters and standardized yield measurements. During the drilling process, depth to water is recorded and yield is measured until the final depth of the borehole is reached, and a final yield is reported.

Yield values ranged from 7.2 to 424 $m^3 \cdot day^{-1}$ (5 to 295 $L \cdot min^{-1}$) with a median of 33.1 $m^3 \cdot day^{-1}$(22 $L \cdot min^{-1}$). For this study, "aquifer interval length" was used to normalize the yield for each well, for easier visualization of the data. The average length of the aquifer penetrated by the boreholes is 13 m. Minimum and maximum lengths are 3 and 46 m, respectively. Yield was divided by the aquifer interval length to obtain a yield value per meter of aquifer interval.

Boxplots were employed to evaluate yield and lithology. A boxplot graphically displays the distribution of the data set [23]. In summary, the mid line is the median of the data and the upper and lower quartile lines display where the bulk of the data points reside. The upper quartile line indicates 75% of the data is equal to or less than this line, and the lower quartile line indicates 25% of the data is equal to or less than this line. The whiskers or straight lines extending from the box represent the maximum and minimum data values, while asterisks display outliers.

Figure 4 shows the distribution of fractured and non-fractured lithologies (x-axis) and value of normalized yield per meter of aquifer interval (y-axis). Comparison of the distribution of fractured versus non-fractured lithologies shows that fractured granite, sandstone, and siltstone have higher upper quartile lines, indicating 75% of the data points have a higher yield than that of their non-fractured counterparts. These fractured lithologies also have comparatively higher median values. The approximate median value for all three lithology types is 5 $m^3 \cdot day^{-1}$ and the upper quarter line values are about 15 $m^3 \cdot day^{-1}$. Two wells had final yield values of 1242 (863 $L \cdot min^{-1}$) and 1080 $m^3 \cdot day^{-1}$ (750 $L \cdot min^{-1}$), which were categorized as outliers and removed from the dataset for easier visualization of the data. Three wells had yield values between 100 and 200 $m^3 \cdot day^{-1}$, and these were removed from the plot for easier visualization of the data. Of the 368 wet wells, 299 were drilled into fractures, which implies that 81% of the wet wells have screened intervals in a fractured rock.

3.6. Aquifer Calculations

Aquifer calculations were conducted on the 140 of the 879 boreholes with logs that contained aquifer test data and static and pumping water levels needed to calculate specific capacity and transmissivity. Most pumping tests were performed for two hours, but a few were between one and six hours. In instances where no time duration between the static and pumping water level was recorded a time period of two hours was assumed, which is the standard protocol for the drilling crews [22].

Transmissivity was calculated using an equation developed by Yidana et al. for Voltaian aquifers [16]:

$$T = 0.752SC^{1.076}$$

where: T is the transmissivity calculated in $m^2 \cdot day^{-1}$and SC is the specific capacity. Transmissivity ranges from less than 1 to 1753 $m^2 \cdot day^{-1}$, averaging 28 $m^2 \cdot day^{-1}$ and a median value of 2.2 $m^2 \cdot day^{-1}$. Seven wells with yields exceeding 300 $L \cdot min^{-1}$ were removed from the dataset, as yields exceeding this value are possible errors, resulting in an average transmissivity value of 12 $m^2 \cdot day^{-1}$, median value of 1.8 $m^2 \cdot day^{-1}$.

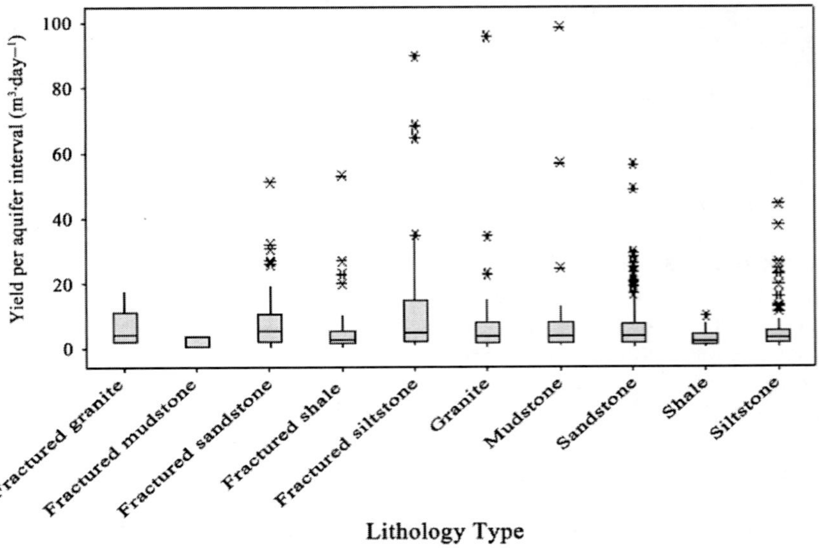

Figure 4. Boxplot distribution of fractured and non-fractured lithologies (x-axis) and values of normalized yield per meter of aquifer interval (y-axis).

3.7. Cross Section Development

The interpolation technique used in RockWorks16 is lithoblending, which horizontally extends the lithology from the control point (borehole log) until it hits a voxel (a three-dimensional volumetric cell) that has already been assigned a value as another control point (another borehole log). A more detailed description of lithoblending can be found in the RockWorks16 Training Manual [23]. In short, lithoblending must interpolate outliers or large gaps in lithology that exist throughout the domain area in order to easily map two-dimensional cross-sections. The 3D model was generated from the eleven lithology categories and interpolated across the domain space (Figure 5). The output dimensions of the interpolated solid model match those of the input boreholes, and a vertical exaggeration of 100 is applied to more easily view the model.

Smaller two-dimensional cross sections were developed from the 3D model to provide more sub-surface detail. First, the entire domain space is cut in four east-to-west directions and four north-to-south directions (Figure 6, upper). Then, smaller cross sections were cut in areas where large clusters of unsuccessful (dry or low-yielding) wells occur, or in areas where high yielding wells occur (Figure 6, lower). All water levels were plotted on the 2D cross sections using the Water Level Table in RockWorks16. Water levels are shown as a thick blue horizontal lines and should not be interpreted as the aquifer or the depth of the aquifer; rather, it is an interpreted aquifer profile, the limitations of which are discussed later. In this study, the upper part of the thick blue water level line represents depth to water during drilling, while the lower line represents the well bottom. Limited spatial and temporal data necessitated the assumption that the aquifer reported in the drilling logs is continuous and confined and depth to water recorded during the drilling process was at the same time of year for each well to eliminate variation in water table levels from the wet and dry season.

Figure 7 shows the four west to east cross sections developed using borehole logs near each profile line. Cross Section 4 has the most interpolation from the west-to-east direction due to limited and poorly spaced boreholes. Figure 8 shows the four north-to-south cross sections developed using borehole logs near each profile line. The majority of boreholes are located in the northern portion of the study domain compared to sparse boreholes in the southern portion of the domain. For this reason, a large amount of error and interpolation is expected on the southern portions of the north-south cross sections.

Figure 9 displays cross sections in areas where clusters of unsuccessful dry or higher yielding wells are found. Cross sections A, B, C, E, F, I, M, and N are drawn in areas where significantly more unsuccessful (dry or low yielding) than successful (higher yielding) wells were drilled. Cross sections D, G, H, and Jare drawn in areas where significantly more successful (higher yielding) wells were drilled. Higher yielding wells are found in the Zabzugu-Tatale district located along the eastern border of Ghana. These cross sections, along with spatial distributions visually displayed in any GIS mapping software, can provide insight into areas of preferential drilling. Limitations of the assumptions made above and potential misrepresentation in the 2D cross sections will be reviewed in the following section.

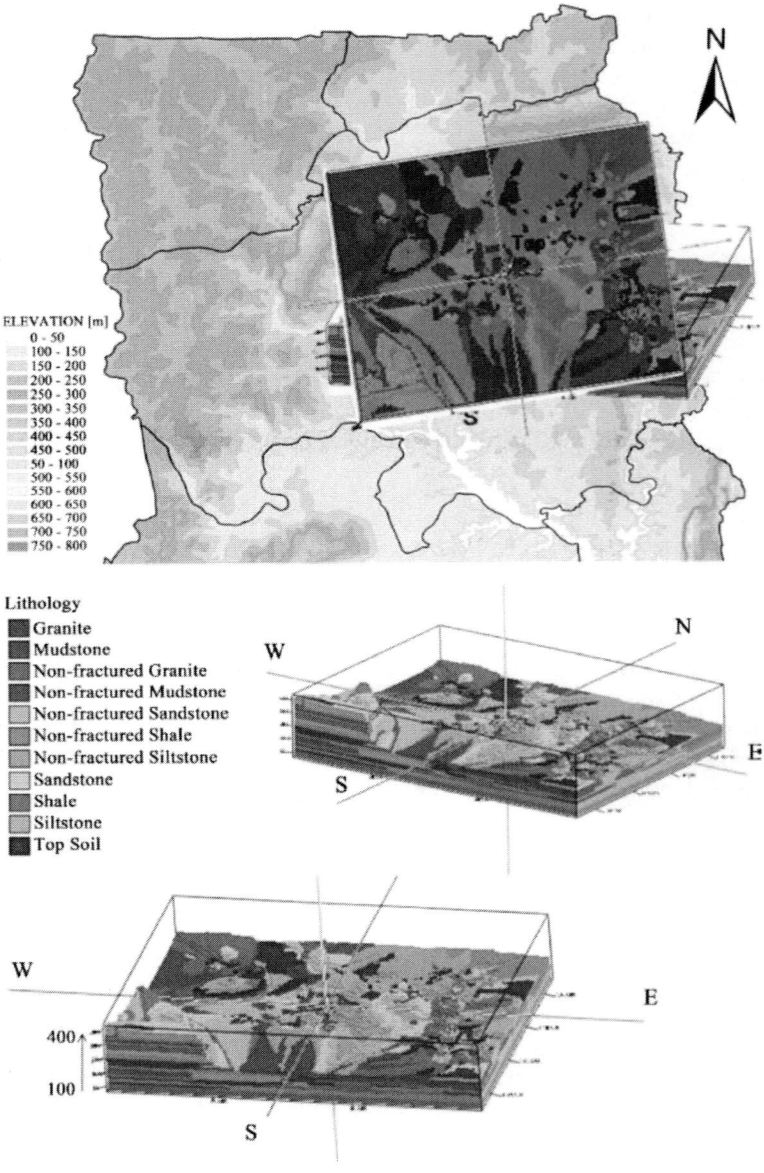

Figure 5. Left: three-dimensional model generated from lithology categories interpolated across domain space and overlaid on map of Ghana. Right: model and cross-section from southerly and south-easterly directions.

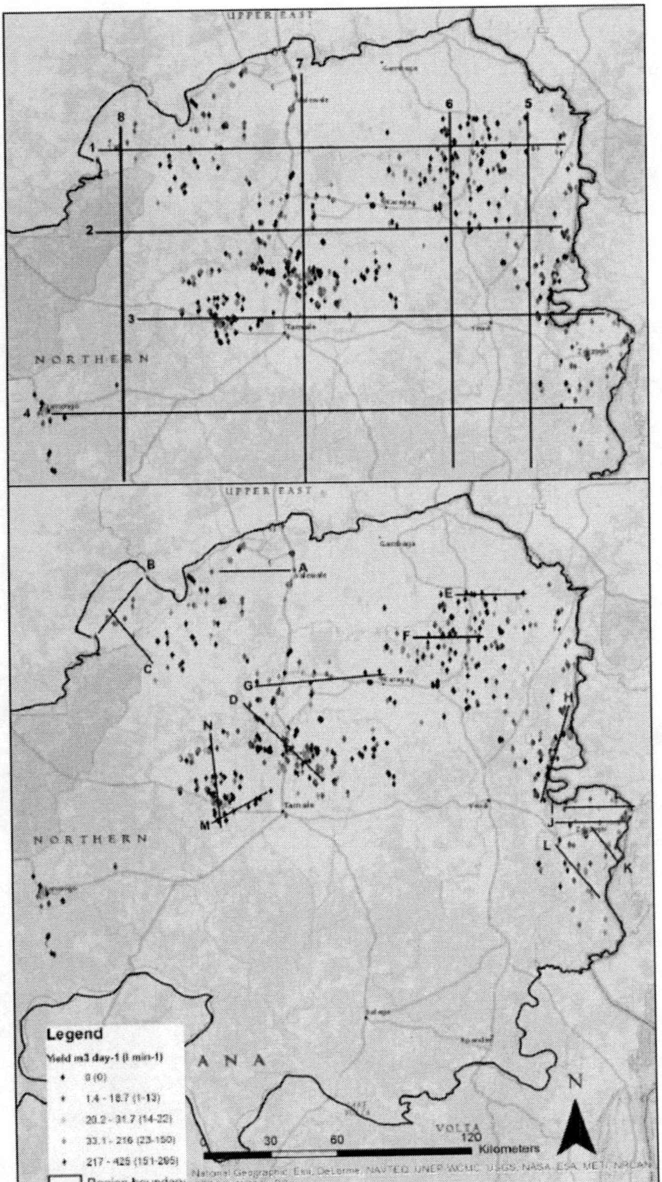

Figure 6. Upper: the domain is cut into four east-to-west (1 - 4) and north-to-south (5 - 8) directional cross sections. Lower: cross sections A, B, C, E, F, I, M, and N are drawn in areas where larger groups of unsuccessful (dry or low yielding) wells were drilled. Cross sections D, G, H, and J are drawn in areas where successful (higher yielding) wells were drilled.

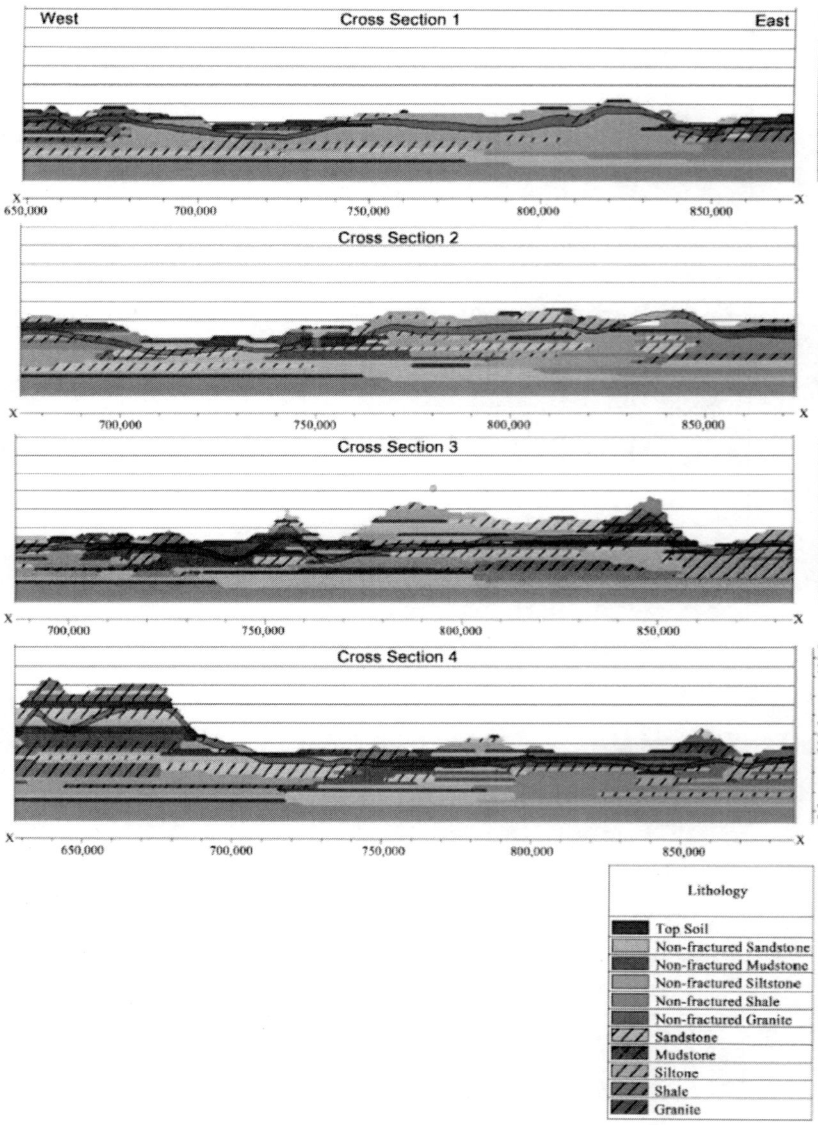

Figure 7. Four cross section profiles drawn in the west-to-east direction domain for an overall horizontal representation of the study domain. The blue line running through the profile is the interpolated aquifer profile. The x-axis is the distance along the lithology profile in meters. The vertical tick marks on the y-axis are elevation reported as meters above sea level [23].

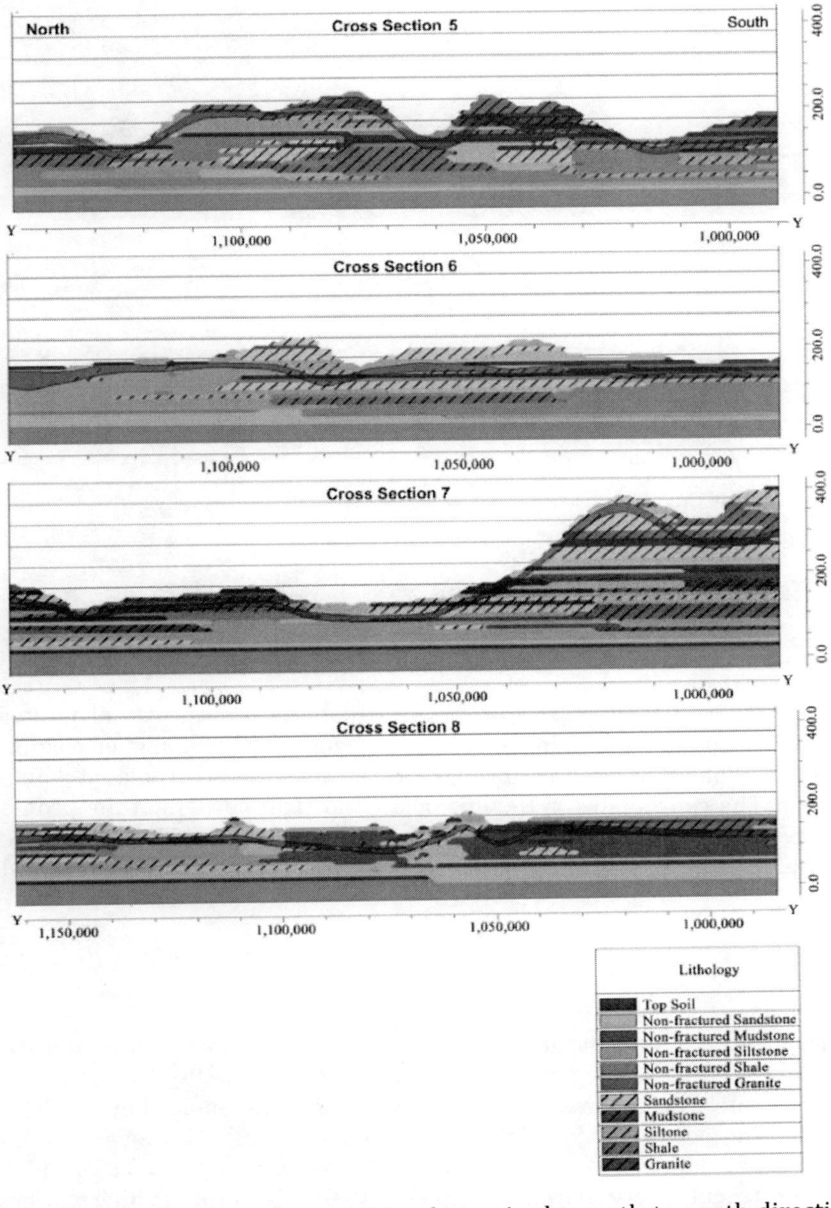

Figure 8. Four cross section profiles drawn in the north-to-south direction domain for an overall vertical representation of the study domain. The blue line running through the profile is the interpolated aquifer profile. The y-axis is the distance along the lithology profile in meters. The vertical tick marks on the y-axis are elevation reported as meters above sea level [23].

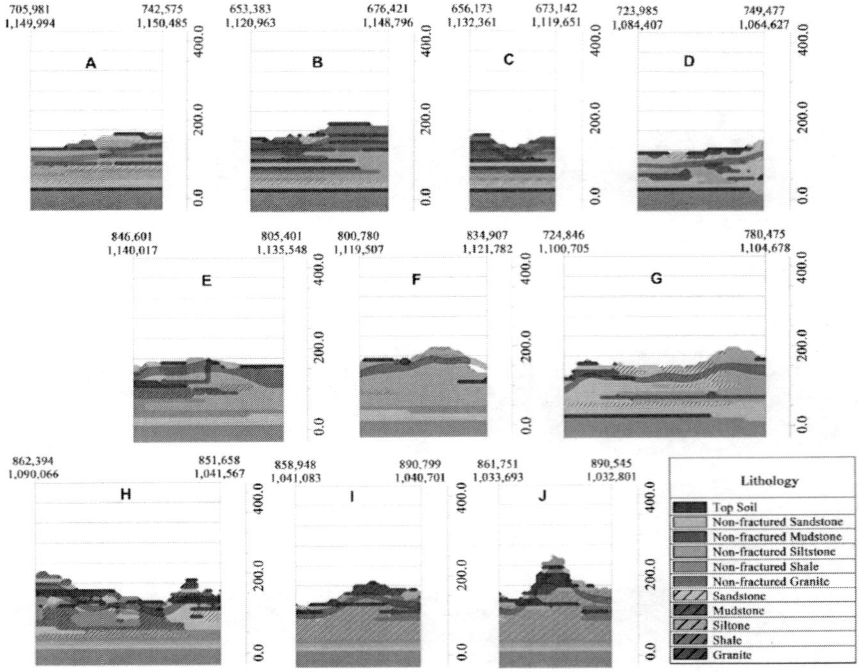

Figure 9. Smaller cross section profiles that represent areas of interest including clusters of dry, low yielding, and clusters of wet, higher yielding, wells. The blue line running through the profile is the interpolated aquifer profile. The coordinates in the upper left and right corners of the profiles are reported in UTM Zone 30, x-easting and y-northing. The vertical tick marks on the y-axis is the elevation reported in meters above sea level [23].

4. DISCUSSION

During the data standardization step, it was noted that 368 of the 879 boreholes with well logs were developed into wet wells. This is a 42% success rate, as compared with the 55% success rate reported by Sander et al. [1]. The data used to calculate a 55% success rate, however, are not known. The borehole logs used in this study span from 2002 to 2010 and do not represent every borehole drilled by GI-WASH in northern areas, therefore, the 42% success rate may not be representative of the overall drilling success rate.

The average transmissivity value of 28 $m^2 \cdot day^{-1}$ calculated in this study is within the higher range of transmissivity values reported in the literature. It is possible that several larger values are over-representing the average in this study. Removing seven wells with yields exceeding 300 L·min^{-1}results in an average transmissivity value of 12 m^{2}·day^{-1}, which is well within the reported regional range of 1 to 30 m^{2}·day^{-1} [15] and agrees with the 13 m^{2}·day^{-1} generated from the numerical model [7]. Grouping the study's remaining wells into sandstone and siltstone/mudstone gives average transmissivity values of 21 and 4.3 m^{2}·day^{-1}, respectively. These values are also well within the reported ranges of 0.1 to 52.0 m^{2}·day^{-1} for sandstone aquifers and 0.2 to 16.0 m^{2}·day^{-1} for siltstone/mudstone aquifers.

The boxplot figures of lithology and yield reveal higher median and upper quartile values for fractured sandstone, siltstone, and granite and non-fractured granite, suggesting these lithology types are more favorable for producing higher well yields. This statement is consistent with anecdotal knowledge and the literature that indicate fractured sandstone areas are more favorable for groundwater development. The Voltaian Sedimentary Basin is approximately 3000 to 4000 m thick and the average depth of each borehole is approximately 48 m [8]. Thus, a detailed representation of the entire basin formation is not reasonable in these relatively shallow cross sections due to limited data with depth. In order to better characterize the basin and make decisions based on the preferential depth to drill, experimental boreholes drilled more deeply would be necessary.

LIMITATIONS

Several limitations are apparent when using the Rockworks software with the study area's data. First is the limitation of scale. For this project the study area spans the majority of the Northern Region of Ghana but only 879 borehole logs available for this large region and these boreholes are generally clustered in small areas. When including all borehole locations in the 3D map and 2D cross sections, wells appear to be on top of each other and are illegible. This issue may be eliminated by decreasing the size of the study area or dividing the area into smaller areas for plotting purposes, however for this particular project that was not an option available for the larger scale analysis.

The next limitation is the spatial distribution of data and the interpolation method. Boreholes are clustered in certain areas and scattered in other areas throughout the domain space. In cases where boreholes are clustered, relatively less interpolation occurs across the sub-surface lithology. Conversely, where boreholes are scattered, relatively more interpolation occurs across the sub-surface lithology. This assumes sub-surface lithological conditions to be unchanging across long distances, such as Cross Section 4 (Figure 7), which is about 200 km. It is also a limitation that the

2D cross sections are developed from the 3D interpolated map, instead of going through a separate interpolation process at the smaller cross section scale. This issue may be eliminated by decreasing the size of the study area, but for the purpose of this project that was not an option.

Another limitation is that depth to water in each borehole is plotted using the aquifer profile for each cross section. Water levels are shown as a thick blue horizontal lines and should not be interpreted as the aquifer or the depth of the aquifer; rather, it is an interpreted aquifer profile. This displays a general thickness and location of the aquifer that each well is completed in. As already discussed, several assumptions are made when representing the aquifer in this manner. The first is that there is one confined and continuous aquifer throughout the basin. In some cases, boreholes are not drilled into fractures and yet yield water. Based on this observation we do not rule out the assumption that a non-continuous or confined aquifer exists. The second assumption is that all boreholes are drilled in the same season. Large variation in water table depth is present in this area due to heavy monsoonal rains followed by a long drought period [21] [24]. Some wells were drilled during times of high water table levels and others in the dry season when the water table has dropped; seasonal variation in water table level is not represented in this study and should be examined more closely in future work.

Finally, the aquifer profile has several limitations in the visual output and may compromise the accuracy of aquifer locations. The depth to water is interpolated across the lithology profile and connected to the next depth to water profile in the closest well. In some cases, the aquifer appears to be above ground since the data set includes dry wells and the aquifer profile may be connecting to the next wet well at a significant distance away. The 2D lithology and aquifer profiles are constructed from a 3D map, so the swath from which the 2D profile is created may consist of data above or below the profile line giving the illusion that the aquifer profile is above ground. This issue may be eliminated by decreasing the size of the study area, but for the purpose of this project that was not an option. Future work should focus on smaller areas of the region.

5. CONCLUSIONS

The purpose of this study was to develop an approach for constructing a three-dimensional hydrogeologic framework using existing borehole driller's logs. A modified approach to develop a three-dimensional hydrogeologic framework was developed by GI-WASH and DRI in combination with several methodologies [4] -[6]. The approach developed in this study consists of: evaluation of appropriate software; data collection;

data QA/QC; data standardization; comparisons analysis; and aquifer characteristic calculations.

Rockworks16 was chosen as the software program to develop a three-dimensional hydrogeologic framework for the study area. The program used information from World Vision's northern Ghana GI-WASH driller's logs as input to develop a 3D hydrogeologic framework for the study area. Some limitations of this approach include scale, spatial distribution of data and interpolation of these data, and location and seasonality of the depth to the water table. Data collection shows the importance of detailed information captured in the driller's log and transcription of this information to an electronic spreadsheet format. Data QAQC indicates that not all driller's logs contain accurate location and elevation information, so adjustments were required. Discussion is taking place between DRI and GI-WASH as to altering the format of the driller's logs to capture more detailed information whilst maintaining efficient field operations.

The standardization, comparison analysis, and aquifer characteristics calculations begin to reveal more information as to the local hydrogeology, especially with respect to well yield, transmissivity, and identifying more favorable lithologies to target for shallow groundwater development. This information, along with the 3D framework and 2D cross-sections is anticipated to contribute to a larger regional hydrogeological study in the administrative Northern Region of Ghana, West Africa. The framework and cross sections are intended to show that the approach is feasible, and make a case for compiling more data. Eventually, with more data, we intend to conduct a more rigorous analysis of the hydrogeology (including geophysics data and lineaments analysis). Ultimately, the results from this study should form the foundation for a more robust study of hydrogeology in this region to ensure sufficient, sustainable groundwater development, and management for all inhabitants in the region.

ACKNOWLEDGEMENTS

The authors wish to thank the numerous field staff at World Vision International's Ghana Integrated Water, Sanitation and Hygiene Project and the various communities in which the study is taking place. Funding for the project was provided by the Conrad N. Hilton Foundation. The authors also wish to thank three reviewers for their valuable suggestions and comments.

REFERENCES

1. Sander, P., Chesley, M. and Minor, T. (1996) Groundwater Assessment Using Remote Sensing and GIS in a Rural Groundwater Project in Ghana: Lessons Learned. Hydrogeology Journal, 4, 40-49. http://dx.doi.org/10.1007/s100400050086
2. Belcher, W., Faunt, C. and D'Agnese, F. (2002) Three-Dimensional Hydrogeologic Framework Model for Use with a Steady-State Numerical Ground-Water Flow Model of the Death Valley Regional Flow System, Nevada and California. Water-Resources Investigations Report 01-4254. USGS.
3. Ross, M., Parent, M. and Lefebvre, R. (2003) 3D Geologic Framework Models for Regional Hydrogeology and Land-Use Management: A Case Study from a Quaternary Basin of Southwestern Quebec, Canada. Hydrogeology Journal, 13, 690-707. http://dx.doi.org/10.1007/s10040-004-0365-x
4. Russell, H., Brennand, T., Logan, C. and Sharpe, D. (1998) Standardization and Assessment of Geological Descriptions from Water Well Records, Greater Toronto and Oak Ridges Moraine Areas, Southern Ontario. Geological Survey of Canada, Ottawa, 89-102.
5. Lutz, A., Thomas, J., Pohll, G. and McKay, A. (2007) Groundwater Resource Sustainability in the Nabogo Basin of Ghana. Journal of African Earth Sciences, 49, 61-70. http://dx.doi.org/10.1016/j.jafrearsci.2007.06.004
6. Akudago, J., Chegbeleh, L., Nishigaki, M. and Nanedo, N. (2009) Estimating the Storage and Maximum Annual Yield of the Aquifer System in Northern Ghana and the Potential for Irrigation. Electronic Journal of the International Association for Environmental Hydrology, 17, 1-13.
7. Anku, Y.S., Banoeng-Yakubo, B., Asiedu, D.K. and Yidana, S. (2008) Water Quality Analysis of Groundwater in Crystalline Basement Rocks, Northern Ghana. Journal of Environmental Geology, 58, 989-997 http://dx.doi.org/10.1007/s00254-008-1578-4
8. Obuobie, E., Diekkrueger, B., Agyekum, W. and Agodzo, S. (2012) Groundwater Level Monitoring and Recharge Estimation in the White Volta River Basin of Ghana. Journal of African Earth Sciences, 71-72, 80-86 http://dx.doi.org/10.1016/j.jafrearsci.2012.06.005
9. World Bank Data. http://data.worldbank.org/country/ghana (accessed on 10 October 2014)
10. Dapaah-Siakwan, S. and Gyau-Boakye, P. (2000) Hydrogeologic Framework and Borehole Yields in Ghana. Hydrogeology Journal, 8, 405-416. http://dx.doi.org/10.1007/PL00010976
11. Tay, C. (2012) Hydrochemistry of Groundwater in Savelugu-Nanton District, Northern Ghana. Journal of Environmental Earth Sciences, 67, 2077-2087 http://dx.doi.org/10.1007/s12665-012 1647-6
12. Acheampong, S. and Hess, J. (2000) Origin of Shallow Groundwater System in the Southern Voltaian Sedimentary Basin of Ghana: An Isotopic Approach. Journal of Hydrology, 233, 37-53 http://dx.doi.org/10.1016/S0022-1694(00)00221-3
13. Boadu. F., Gyamfi, J. and Owusu, E. (2005) Case History: Determining Subsurface Fracture Characteristics from Azimuthal Resistivity Surveys: A Case Study at Nsawam, Ghana. Journal of Geophysics, 70, 35-42.http://dx.doi.org/10.1190/1.2073888
14. Yidana, S., Yiran, G., Sakyi, P., Nude, P. and Banoeng-Yakubo, B. (2011) Groundwater Evolution in the Voltaian Basin, Ghana—An Application of Multivariate Statistical Analysesto Hydrochemical Data. Natural Science, 3, 837-854. http://dx. doi.org/10.4236/ns.2011.310109
15. Teeuw, R. (1995) Groundwater Exploration Using Remote Sensing and a Low-Cost Geographical Information System. Hydrogeology Journal, 3, 21-30. http://dx.doi.org/10.1007/s100400050057
16. Yidana, S., Banoeng-Yakubo, B., Aliou, A. and Akabzaa, T. (2012) Groundwater Quality in Some Voltaian and Birimian Aquifers in Northern Ghana—An Application of Multivariate Statistical Methods and Geographic Information Systems. Hydrological Sciences Journal, 57, 1168-1183.http://dx.doi.org/10.1080/02626667.2012.693612

17. Kwei, C. (1997) Evaluation of Groundwater Potential in the Northern Region of Ghana. Comwasan Consult for Canadian International Development Agency.
18. Nejoh, J. (2013) GI-WASH, Personal Communication.
19. Harvey, P. (2004) Borehole Sustainability in Rural Africa: An Analysis of Routine Field Data. In: Godfrey, S., Ed., People-Centred Approaches to Water and Environmental Sanitation. Proceedings of the 30th WEDC International Conference, Vientiane, WEDC, Loughborough, 339-346.
20. Edusei, S. (2013) GI-WASH, Personal Communication.
21. RockWorks16 Training Manual (2013) Solid Model Algorithms. RockWare, Golden.
22. Dumedah, G. and Schuurman, N. (2008) Minimizing the Effects of Inaccurate Sediment Description in Borehole Data Using Rough Sets and Transition Probability. Journal of Geographic Systems, 10, 291-315.http://dx.doi.org/10.1007/s10109-008-0066-4
23. Helsel, D. and Hirsch, R. (2002) Statistical Methods in Water Resources. USGS, Book 4, Hydrologic Analysis and Interpretation, 524 p.
24. Lutz, A., Minyila, S., Saga, B., Diarra, S., Apambire, B. and Thomas, J. (2015) Fluctuation of Groundwater Levels and Recharge Patterns in Northern Ghana. Climate, 3, 1-15.http://dx.doi.org/10.3390/cli3010001

CHAPTER 17

Identification of Hydraulic Parameters of Wadi El Natrun Pliocene Aquifer Using Artificial Neural Network

Khalaf S*, Ahmed AO, Abdalla MG and El Masry AA

Irrigation and Hydraulics Department, Faculty of Engineering, El-Mansoura University, Egypt

ABSTRACT

Many techniques, approaches and tools were used in this Research to achieve the Methodology. Using artificial neural network to simultaneous hydraulic parameters is one of these techniques. Transmissivity and storativity consider the most important parameter in each aquifer due to the reality of their effect on the aquifer properties. In this research, it is assumed that the transmissivity (T) and the storativity (S), represented by coordinates (X), (Y), hydraulic head (H), and observation times (t). These variables were chosen depending on the literature review. In the present study, the hydraulic head values at each cell (H) and the location of the cells (x, y) are considered as input parameters for finding the unknown parameters. The transmissivity (T) and storativity values (S) at cells are assumed and used in the finite difference method (Forward model) in order to find the value of hydraulic head at that cell. The hydraulic head values were used in the artificial neural networks (Inverse model) to estimate transmissivity (T) and storativity values (S) for Wadi El Natrun Depression. The study is based on coupling of forward model and inverse model. In general, the parameter estimation process consists of identifying a model that would reverse a complex forward relation.

Keywords: Parameter identification; Simultaneous; Inverse model; Neural networks; Groundwater flow; Forward model; Wadi El-Natrun

1. INTRODUCTION

Soft computing techniques analogous to biological nervous system are called as artificial neural networks (ANNs). The reason for wider acceptability of this technique can be attributed to its capability to develop computing tools, which may partially capture amazingly faster and complicated information-processing ability of the brain. Groundwater is an important source of water for drinking, irrigation, and industrial uses. It is also a major source for domestic water requirement. Ground water hydrology examines problems like prediction of groundwater head, distribution of transmissivity, storativity specific yield and estimation of parameters. Involved processes are nonlinear, complex, multivariate with variables having spatial and temporal variability. These are expressed by complex partial differential equations, which are normally solved with considerable approximations using complex numerical models. Precise conceptualization is nearly impossible for groundwater problems as physics of the system cannot be fully understood from the surface. ANNs can be used in groundwater hydrology since they do not require governing equations and their concomitant assumptions.

Strength of ANNs lies in mapping non-linear system data, which are capable of extracting the relation between the input and output of a process without adequate knowledge of the underlying principles. As the computational burden is primarily managed by replacing the numerical model with a surrogate simulator artificial neural network the technique of ANN is quite appropriate for groundwater problems.

Estimation of aquifer parameters demands large time, financial and manpower resources which are always scarce [1]. Therefore, the efforts are on increase to assess these by non-field methods using the technique of ANNs. The determination of aquifer parameters (also termed as inverse problem) has always been a challenge because of its ill-posedness [2-7]. Inversion of the trained feed forward neural network is done to estimate the transmissivity field for synthetic problem [5]. Most of the papers used synthetic or published data to assess parameters in confined aquifer [4-6,8] for the reason of nonavailability of sufficient number of patterns. Model sensitivity in terms of number of nodes in hidden layers is carried out by Shigdi et al. [5] using correlation coefficient, Y-intercept, and slope. An approach with a combination of ANNs and type curves based on Papadopoulos and Cooper analytical solution was used by Balkhair [4] whereas ANN and This solution coupling was performed by Lin GF et al. [9]. It was found that Levenberg-Marquardt achieves faster convergence than backpropagation algorithm [7]. To solve the problem of range of aquifer parameters to be used for training [4] carried out training in macro and micro scales. Macro scale is used for very wide and very narrow range whereas micro scale is for middle range of parameters. Lohani et al. [10] presents an efficient and stable artificial neural network (ANN) model for predicting groundwater level in south-east Punjab, India. Lohani et al. [11] used neural network configuration for predicting groundwater level in Amritsar and Gurdaspur districts of Punjab, India is identified. For predicting the model efficiency and accuracy, different types of network architectures and training algorithms are investigated and compared.

The determination of aquifer parameters (also termed as inverse problem) has always been a challenge because of its ill-posedness [2-7,12,13]. ANNs is solving many complex real world-predicting problems. ANNs have been applied to predicting groundwater levels [14,15], precipitation and runoff modeling, and aquifer parameter estimations [3,5,16-19].

The main objectives of this study may be summarized as follows:

(1) to develop an ANN model for solving the groundwater inverse problem by providing an ANN model to solve complex nonlinear relationships.

(2) to test the performance of the proposed ANN model.

(3) to simultaneously determine the transmissivity and storativity field of an aquifer by using limited observed head values.

1.1 Study area (wadi el natrun)

The study area lies at El Behera governorate, Western of Nile Delta, Egypt between Longitudes 30° 00' and 30° 33' E and Latitude 30° 20' and 30° 30' N. It parallels to (Cairo-Alex. Desert Highway) Km 90 to Km 110. The study area covers about 2016 Km^2 (**FIGURE 1**).

The study area comprises a total area of about 770 km^2. From the hydrological cross section, which pass through Wadi El Natrun area in West-East direction. Pliocene aquifer and from electrical sounding, this aquifer unit is mainly formed of alternating sand and clay and occasionally capped by thin layer of limestone. The Pliocene aquifer is considered as multi-layers' aquifer under confined to semiconfined condition. Most of these waters bearing layers belongs to the Upper and Middle Pliocene. In this study, the information of 111 pumping well and 14 observation well were collected as well as other parameters such as hydraulic conductivity. **TABLE 1** shows the values of hydraulic conductivity, transmissivity, and storativity of Pliocene at Wadi El Natrun area of different authors. According to the analysis of Pliocene aquifer in the area of study, we can notice that, the hydraulic conductivity (K) values ranged between 5 m/day at Wadi El Natrun area to 40 m/day in the east of Wadi El Natrun. The transmissivity (T) values ranged between 500 m2/day to 1600 m2/day in different zones of the Depression.

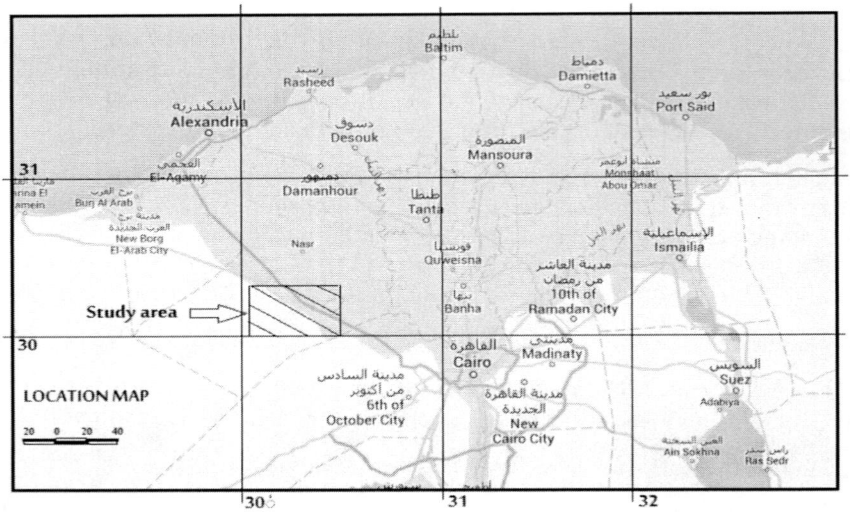

Figure 1: Location map of the study area.

Table 1: Hydraulic parameters of the Pliocene aquifer.

Author	Location of the test zone	Hydraulic parameters		
		K	T	S
Ahmed [27]	North of Wadi El Natrun	-	1043.9	-
Shata et al.[28]	Northeast Wadi El Natrun	11	-	-
Desert research [28]	Northeast Wadi El Natrun	31.9	-	-

General Petroleum [29]	East Wadi El Natrun	26	2600	3.9*10-3
Ahmed [27]	2 km from El hamaralake	-	794	-
Saad [30]	East side of Wadi El Natrun	52.9	1291.7	3.95*10-3
RIGW [31]	Wadi El Natrun area	9.8	500-327	1.7*10-2
Mostafa [32]	East Wadi El Natrun	47	943	7*10-4
Saad [30]	East Wadi El Natrun	38.9	695.5	1.35*10-3
Ahmed [27]	South Wadi El Natrun	-	1660	-
El-Sheikh [33]	BeniSalama	7.29	838	8.5*10-4
Sayed	East Wadi El Natrun (Jacob)	-	3618	0.10466
El Sayed	East Wadi El Natrun (Recovery)	-	2745	-

1.2 Geologic setting

The study area is occupied sedimentary rocks belonging to the Tertiary and Quaternary Eras. The sedimentary succession comprises several water-bearing formations, which are particularly influenced by structural features and thus affect the groundwater occurrences. However, the surface deposits dominating the area are studied through the geological map which shown in **FIGURE 2**. In the study area, Late Tertiary and Quaternary succession were studied by many authors. The subsurface geology of the study area is studied from the well logs as from the Wadi El Natrun deep well. The Pliocene aquifer is local in extent, covers the entire Wadi El Natrun area, and is discontinuously covered by Quaternary deposits of the Pleistocene Aquifer. As a result, the Pliocene aquifer is considered to be partially confined [20]. The Pliocene Aquifer's thickness ranges from 150 to 300 m thick [20].

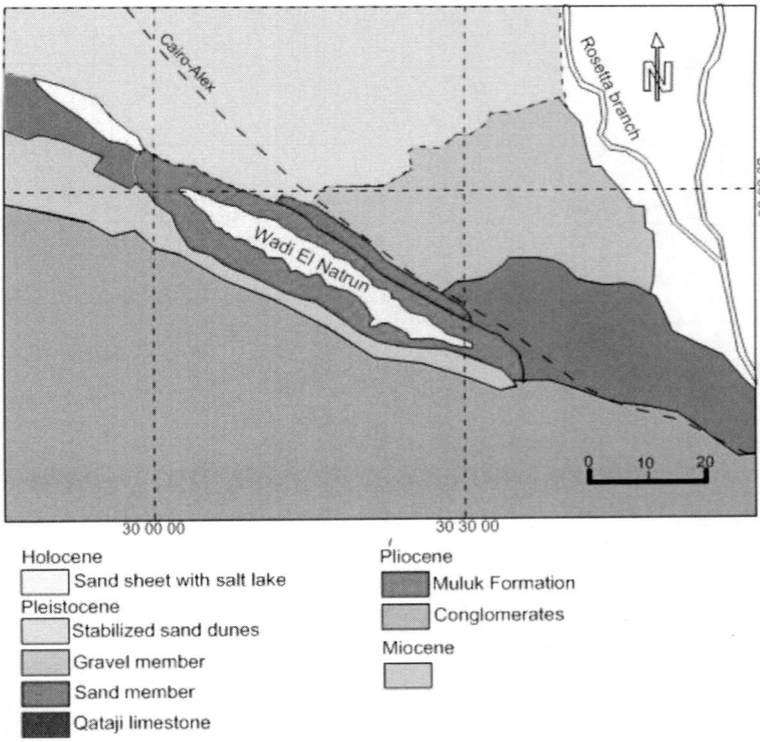

Figure 2: Geological map of Wadi El Natrun and its vicinities (Modified after Abu Zeid, 1984).

1.3 Hydrologic setting

Hydrogeologic setting in the Wadi El Natrun area is complex, and there is still significant uncertainty as to the precise flow regime and hydraulic connections between the aquifers. Nonetheless, multiple studies have observed both local and regional groundwater flow to be concentrated toward the base of the valley where it discharges to eleven saline lakes [20].

1.4 Govering equation

A general form of the governing equation, which describes the three dimensional movement of groundwater flow of constant density through the porous media is [21]:

$$\frac{\partial}{\partial x}\left[Kx\frac{\partial h}{\partial x}\right]+\frac{\partial}{\partial y}\left[Ky\frac{\partial h}{\partial y}\right]+\frac{\partial}{\partial z}\left[Kz\frac{\partial h}{\partial z}\right]-w=Ss\frac{\partial h}{\partial t} \tag{1}$$

Where: Kx, Ky, Kz are values of hydraulic conductivity along the x, y and z coordinate axes (L/t); h: is the potentiometric head (L); w: is the volumetric flux per unit volume and represents sources and/or sinks of water per unit time (t^{-1}); Ss: is the specific storage of the porous material (L^{-1}); and t: is time (t). The first part of this problem was run to get a steady state solution that takes the form:

$$\frac{\partial}{\partial x}\left[Kx\frac{\partial h}{\partial x}\right]+\frac{\partial}{\partial y}\left[Ky\frac{\partial h}{\partial y}\right]+\frac{\partial}{\partial z}\left[Kz\frac{\partial h}{\partial z}\right]-w=0.0 \tag{2}$$

From the steady state solution, the hydraulic conductivity for model aquifers can be found. Then the equation is solved for transient case in order to solve for storage coefficient.

Karahan et al. identified transmissivity distribution of a twodimensional aquifer system under steady-state flow conditions. In that study, randomly selected Cartesian coordinates and error-free hydraulic heads are used as input and corresponding transmissivity values are used as output in the ANN model [18].

Shigidi et al. determined aquifer parameters by using ANN. In order to obtain observed hydraulic heads, they used MODFLOW with a stochastically generated transmissivity distribution [5]. Their network architecture uses the transmissivity values as input and the hydraulic heads as output. The error residuals between computed and observed hydraulic heads are reduced by adjusting the transmissivity field, and this process is cycled until the convergence is met. On the other hand, Garcia et al. [17] solved the same problem in their previous study- i.e., Shigidi et al., by adding a noise term to the observed hydraulic heads and they obtained more precise results than their previous study [5].

Choose the appropriate architecture of network among the available networks based on the type of the data and the problem. After many trials, Multilayer Perceptron network (MLP) has been chosen because of its high capabilities to generalize well in problems plagued with significant heterogeneity and nonlinearity. The Multilayer Perception (MLP) trained with the back-propagation algorithm is perhaps the most popular network for hydrologic modeling [22]. The majority of the ANN applications in water resources engineering involve the employment of a conventional Feed Forward Back Propagation Method (FFBP).

The important feature of this network is its ability to self-adapt the weights of neurons in intermediate layers to learn the relationship between a set of patterns given as examples and their corresponding outputs. So that after having been trained it can apply the same relationship to new input vectors and produce appropriate outputs from inputs that the system has never seen before, a feature known as the generalizability of an ANN [1]. The artificial neural network with a multilayer back propagation network has been used successfully in several studies [5,14,23-26].

1.5 Conceptual model of wadi el natrun

On the light of the hydrogeologic properties of the Pliocene aquifer in Wadi El Natrun Depression (chapter five), a pictorial representation (conceptual model) of the water flow system is constructed to this aquifer. The constructed conceptual model depends on the following facts:

1) The aquifer of Wadi El Natrun consists of five Geoelectric layers appear as follows:

   - The first Geoelectric layer consist of dry sand, gravel and rock fragments, called the subsurface. The thickness of this layer is ranged between (0.9 m) and (3.8 m).
   - The second Geoelectric layer is formed from shaly, sand and presence of shale interbedded with sand and differential amounts of water. The thickness of this layer is ranged between (2.4 m) and (22 m).
   - The third geoelectric layer is made up of clayey, sand and gravel and underground water included in this layer, the thickness is ranged between (61.6 m) and (69 m).
   - The fourth geoelectric layer is composed of clayey, sand and gravel, and considered the continuation of the third geoelectric layer. It includes the same lithology. These values give indication about the presence of underground water in this layer, as shown as shown the thickness of this layer is ranged between (86.7 m) and (92.4 m).
   - -The fifth geoelectric layer is barrier separating it and from the fourth geoelectric layer.

   At the end, this section includes two water bearing layers, the third and the fourth geoelectric layers, that are saturated with water. These layers are considered the main aquifer of the study area.

2) The groundwater flow generally from NE to SW direction toward Wadi El Natrun Depression. The recharge from irrigation and the

evapotranspiration will be neglected. The main discharge source is the present flowing productive well (48 wells).

3) The Pliocene aquifer of Wadi El Natrun occurs under the confining conditions.

4) The base of the aquifer is not detected.

1.6 Model application

The model domain was selected to cover 770 kmM^2 (55 × 14 km). The model domain was discretized using 144 row and 224 columns square cells. This discrimination produces 32,250 cells in the model layer. The width of the cells along rows (in x-direction) is equal to the length along columns (in y-direction) 250 m as shown in **FIGURE 3**.

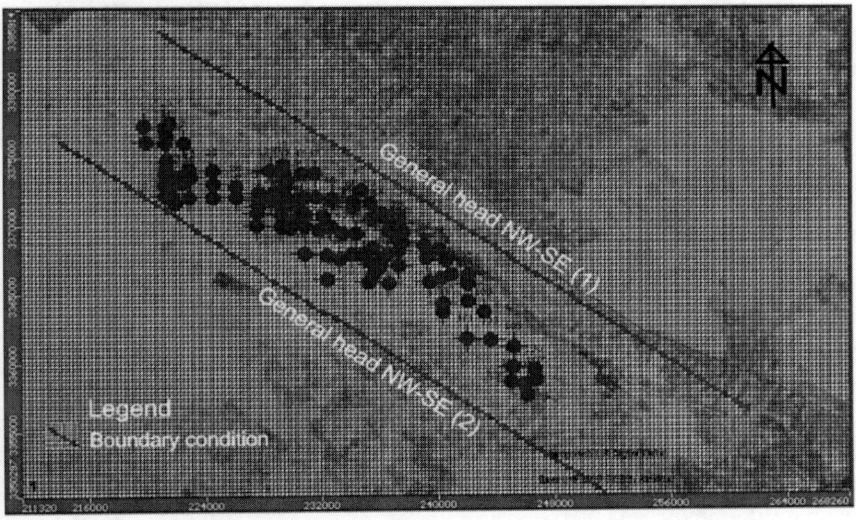

Figure 3: Flow model boundaries, grid, and pumping wells.

1.7 Estimating of hydraulic head

After finish the calibration of the MODFLOW model, and reached a good match between observed and calibrated head, as shown previous, the model will be ready to estimate the hydraulic distribution for all cells (**FIGURE 4**) which will use in the artificial neural network model. The values of hydraulic head in each cell as well as the coordinates of these cells can be exported as a .txt. File which will used later in the next chapter to predict the transmissivity and storativity using ANNs.

Figure 4: Head contour map of the model area.

1.8 Construction of ann model

Construction ANN model utilization FFBP Neural Networks (MLP) which built in MATLAB program version 7. Different ANN structure had been investigated to find optimum ANN model. The optimum neuron number in each hidden layer was also investigated.

In the ANN model architecture, the input layer includes 4 input (hydraulic head, Cartesian coordinates, and the time of observation), and the output layer include 2 outputs (transmissivity and storativity). In order to find the best number of hidden layers of ANN model many networks with different structure have been developed. These models had two hidden layers. The number of neuron in each hidden layer is also very important in the structure of ANN. The low number of neurons in each layer decreases the learning performance. In contrast, a high number of neurons increase the training time. Therefore, the optimum number of neurons in each layer is essential to build the ANN structure. In the present research, a different number of neurons (10-17) have been tested to find the best ANN structure. By comparing performance of developed ANN models the optimum number of neurons in each hidden layer for different structure (3 × 6 × 2), (3 × 10 × 2), (3 × 6 × 4 × 2), (3 × 8 × 4 × 2), (3 × 10 × 4 × 2), (3 × 12 × 6 × 2), (3 × 14 × 6 × 2), (3 × 14 × 8 × 2), (3 × 16 × 8 × 2) and (3 × 17 × 8 × 2). They had the same input and output layer, so they just differ with the hidden layers. The best results obtained from the network with two hidden layers with 17 and 8 hidden neurons, respectively.

For present study, FFBP (3 × 17 × 8 × 2) architecture is selected in the model. The logsig transfer function is applied in the first hidden layer and the purelin transfer function is applied in the second hidden layer (**FIGURE 5**).

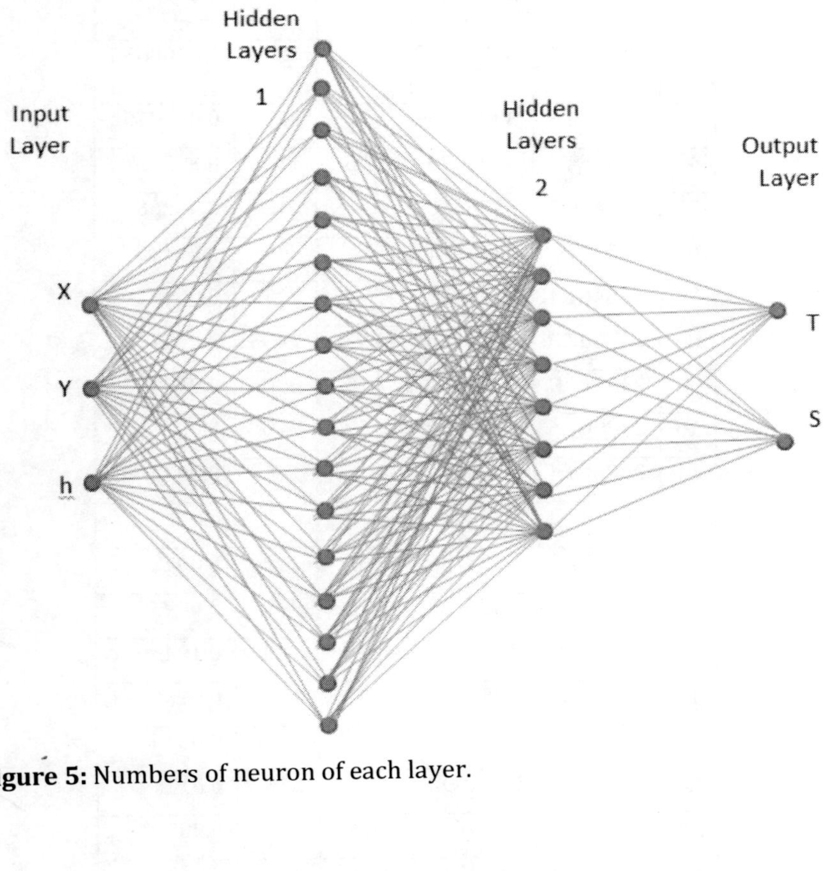

Figure 5: Numbers of neuron of each layer.

1.9 Construction data matrix of ann model

The matrixes of ANNs are different as compare with other matrixes, because that every parameter value should be arranged at one a row of this matrix. There are three matrixes need to be constructed in older to use in artificial neural networks model. Two matrixes for training the ANNs model, while the third matrix used as input parameters that we want to get its output (**TABLE 2**).

Table 2: Data for training ANN.

N	X	Y	h	T	S
1	219921.6	3378928	-10	1100	0.000155
2	219735.4	3378244	-11	731	0.0000554
3	221375.6	3373596	-18.25	955	0.0000152
4	224584.5	3374936	-12.5	1366	0.00288
5	227479.5	3375523	-8.5	613	0.00265
6	223747	3371761	-18	840	0.00135
7	228461.4	3370956	-16.25	827	0.00183
8	237002.8	3371180	-6.5	1435	0.00177
9	243511.9	3365666	-7.25	650	0.017
10	238908.6	3359950	-17.75	719	0.0075
11	248609.4	3360650	-8	1460	0.00181
12	250137.2	3359494	-6.75	870	0.0017
13	251607.1	3358735	-6.85	718	0.0012
14	252927.7	3357222	-7.25	595	0.0095
15	243511.9	3365666	-8.1	920	0.10466
16	238908.6	3359950	-13	1235	0.0166
17	248609.4	3360650	-7.5	630	0.0845
18	243887.9	3361092	-16	575	0.0124
19	247490.7	3356758	-16.5	535	0.0099
20	249098.6	3356194	-13.5	719	0.0075

1.10 Network training, validation, and testing

Generally, the default setting of artificial neural network divided date into three divisions, which are training, validation and testing with a proportion of 60%, 20% and 20% of data respectively. The training set the large proportion of data to learn pattern present in the data.

The conditions to stop training processes were set before the network is trained. Training was controlled by some of conditions as: the maximum number of iterations, target performance which specifies the tolerance between the neural network prediction and actual output, the maximum run time and the minimum allowed gradient and. The overall training of the ANN will involve the following processes; the input values of the first layer are weighted and passed on to the hidden layer; the neurons in the hidden layer will produce outputs by applying an activation function to the sum of the weighted input values; the resulting outputs are then weighted by the connections between the hidden and output layer.

The network technique enable the user to make good training by shows the result with network calculated by submitted a figure shows the training fitting. **FIGURE 6** shows good fitting which mean that the network gives a result data closed to the data that we had insert as output. The training coefficient R represent the fitting of observed invers calculated which is 0.98296, and this number is very closed to 1, so that mean we had good match.

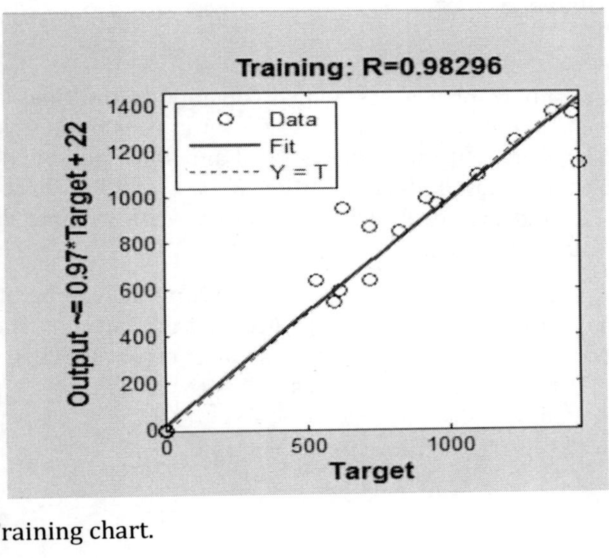

Figure 6: Training chart.

Validation is the simple check of the model which used 20% of data to ensure that the training is able to give accepted result and good match between observed and calculated data model is controlled using the same synaptic weights in the validation part. For validation R=0.898 consider well because that, there is a good fitting between both observed and validate data (**FIGURE 7**).

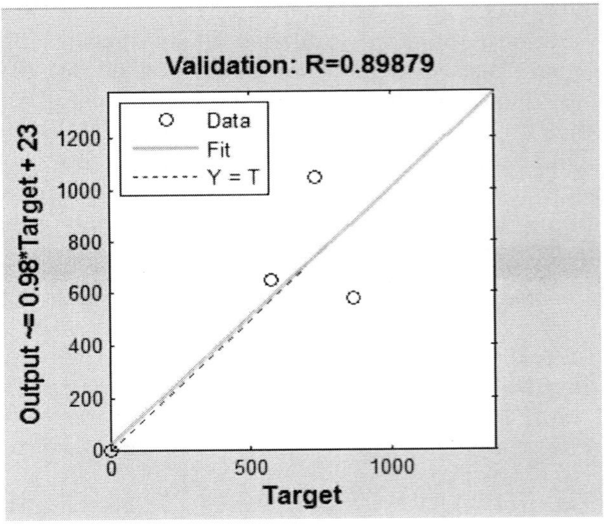

Figure 7: Validation chart.

Testing is an important process, which consider as the final check of the network to check the ability of this network to predict and gives accurate. Then we can decide that the model is good and show good match between observed and calculated data. **FIGURE 8** is a simple chart shows the coefficient (R) and the result of the fitting between validate data and the calculated data (**TABLE 3**).

PERFORMANCE: The performance of the ANN model can be quantified by statistical measure addressing the magnitude of the variable. The model can be validated in terms of root mean square error (RMSE), correlation coefficient (R) and scatter index (SI) as follows

$$RMSE = \sqrt{\frac{\sum_{i=1}^{N}(y_i - x_i)^2}{N}} \tag{3}$$

$$R = \frac{\sum_{i=1}^{N}(x_i - x^-) - (y_i - y^-)}{\sqrt{\sum_{i=1}^{N}(x_i - x^-)^2 \sum_{i=1}^{N}(y_i - y^-)^2}} \quad (4)$$

$$SI = \frac{RMSE}{x^-} \quad (5)$$

Where x_i is the observed values at the ith time step, y_i is the simulated values, N is the number of time increment and x^- and y^- are the mean values of observation and simulation, respectively.

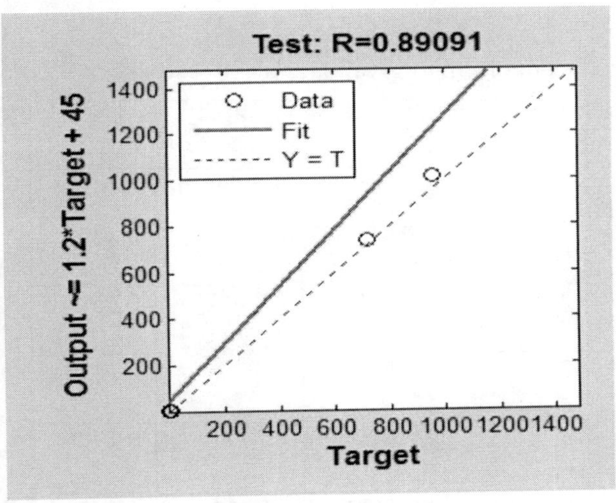

Figure 8: Chart shows the validation coefficient (R).

The desired results are generated in the output layer. The network achieves the desired learning by adjusting its interconnected weights continuously until there is a close match between the outputs from the neurons and the output from the training data. The difference between the predicted outputs and the original outputs is referred to as error. **FIGURE 9** shows the fitting and the coefficient for all training, validation, and testing.

SENSITIVITY ANALYSIS: Input variables are not, in general, independent-that is, there are interdependencies between variables. Sensitivity analysis rates variables according to the deterioration in modeling performance that occurs if that variable is no longer available to the model. In so doing, it assigns a single rating value to each variable. However, the interdependence between variables means that no scheme of single ratings per variable can ever reflect the subtlety of the true situation. Consider, for example, the case where two input variables encode the same information (they might even be copies of the same variable). A particular model might depend wholly on one, wholly on the other, or on some arbitrary combination of them. Then sensitivity analysis produces an arbitrary relative sensitivity to them. Moreover, if either is eliminated the model may compensate adequately because the other still provides the key information. It may therefore rate the variables as of low sensitivity, even though they might encode key information. Similarly, a variable that encodes relatively unimportant information, but is the only variable to do so, may have higher sensitivity than any number of variables that mutually encode more important information.

Table 3: Coefficient of training, validation and testing of ANN models.

Models No.	No. of coefficient		
	Training	Validation	Testing
1	0.6361	0.64	0.5443
2	0.7135	0.5443	0.6577
3	0.7974	0.6577	0.7411
4	0.8773	0.7411	0.7974
5	0.8812	0.7562	0.7934
6	0.91545	0.837	0.8253
7	0.9154	0.8423	0.8517
8	0.9626	0.8674	0.8742
9	0.9795	0.8734	0.8734
10	0.9995	0.87824	0.89091

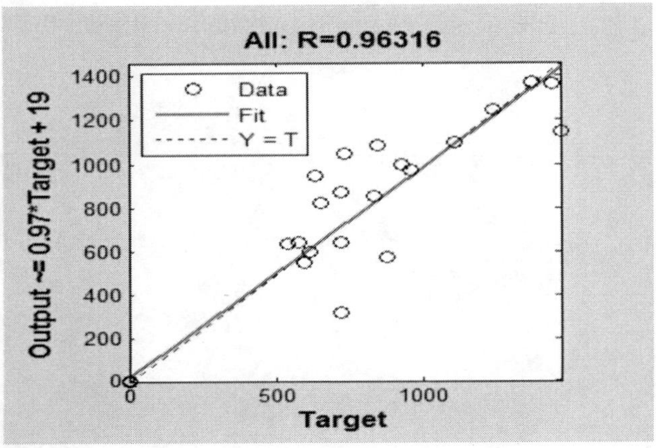

Figure 9: Fitting and the coefficient for training, validation, and testing.

2. DISCUSSION

From the previous section, it is shown that the proposed ANN model may be used to identify the transmissivity and storativity distributions of an aquifer in an inverse modeling framework. For real-world problems, there are some important issues that require further analysis. Note that the same input structure, network architecture, and solution parameters are used in this section. The contours of predicted vs. actual transmissivities and storativities for noise data conditions can be seen in **FIGURES 10 AND 11**, respectively. It is clearly seen from **FIGURES 10 AND 11** that the increase in the standard deviations may not significantly change the prediction performance of the proposed ANN model. This situation states that the proposed ANN model may have generalization ability. Therefore, it may be used in the real-world parameter estimation problems.

3. RESULTS OF ANN MODEL

After finish, the training by obtained good match between observed and predicted data with good validation and testing fitting then I saved the network in order to use it in prediction. The saving part of network mean saving of all weights that the network reached for obtaining good results. Then we use the same network to predict the distribution of both transmissivity and storativity. For that, we imported the new matrix of input, which include the distribution of head, which represented by the head (h) and the coordinates (*x* and *y*) which obtained from the MODFLOW model.

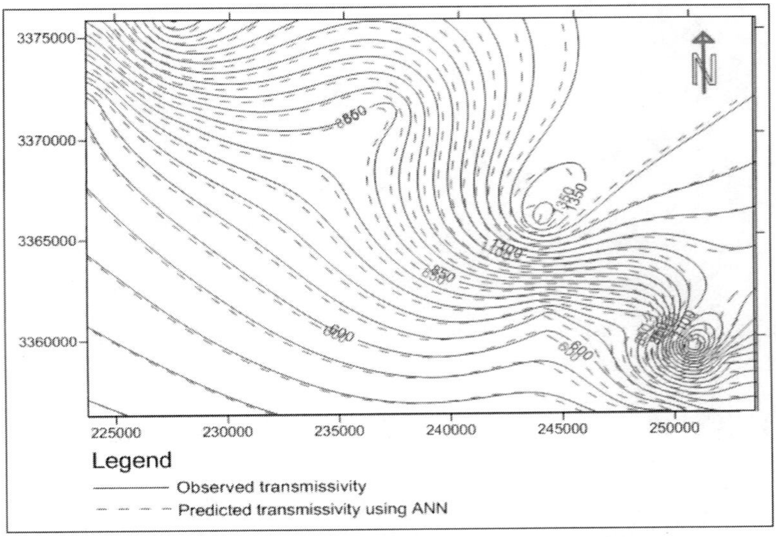

Figure 10: Actual and predicted distribution of transmissivity.

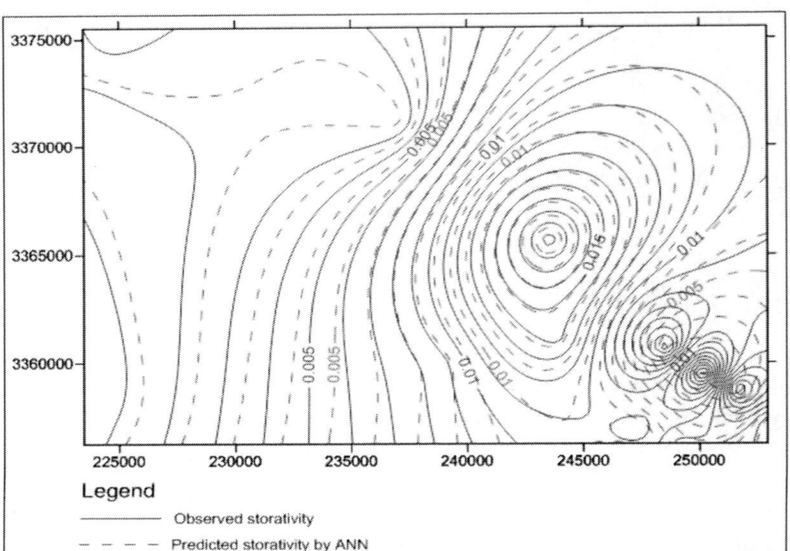

Figure 11: Actual and predicted distribution of storativity.

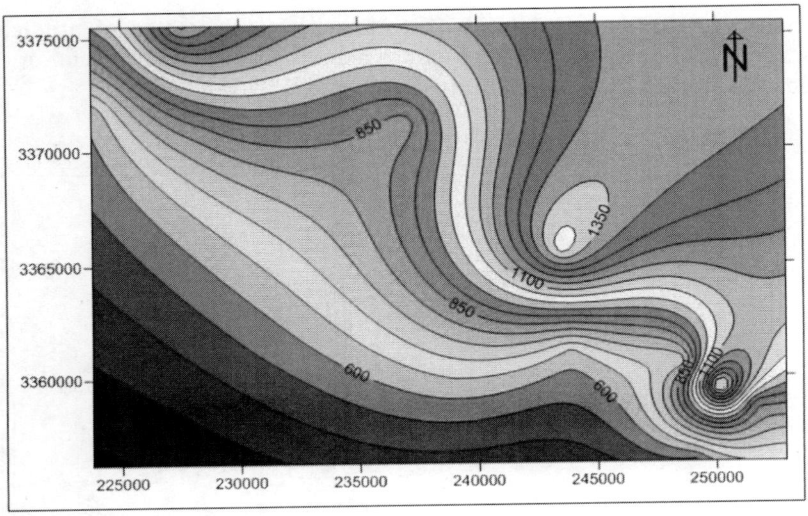

Figure 12: Distribution of predicted transmissivity using ANN.

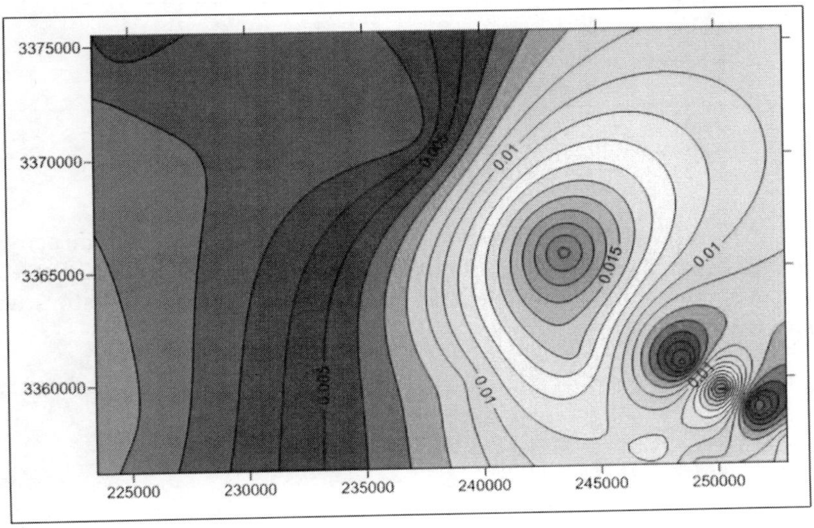

Figure 13: Distribution of predicted storativity using ANN.

As shown before the actual values of transmissivity ranged between 500 and 1600 m^2/day (**FIGURE 10**) and the actual values of storativity randed between 0.001 to 0.05. The values of predicted transmissivity are in the same range of the actual, also the distribution of the predicted transmisivity is closed to the actual one (**FIGURE 14**). Also, the values of predicted storativity are in the same range of the actual, also the distribution of the predicted storativity is closed to the actual one (**FIGURE 11**).

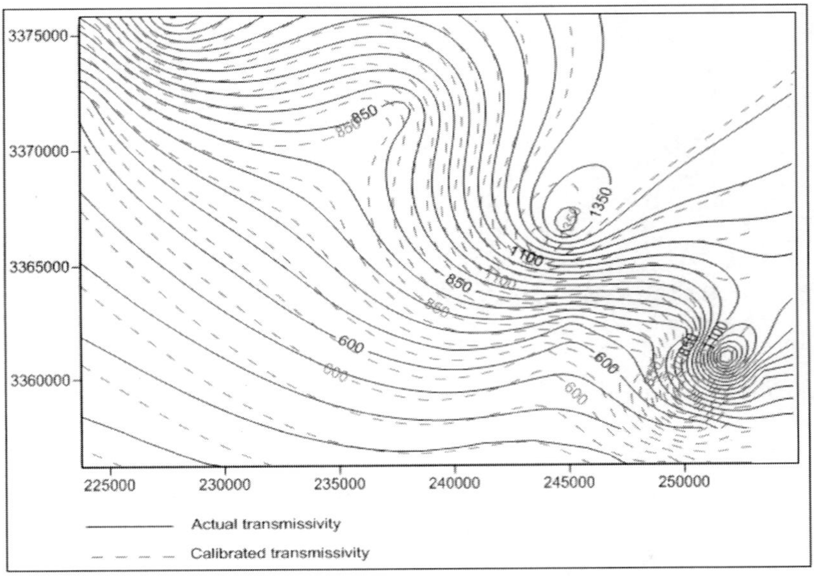

Figure 14: Actual and calibrated transmissivity.

By run the model to simulate the new data. The result appears directly in anew matrix include predicted transmissivity and storativity. **FIGURES 12 AND 13** shows the distribution of transmissivity and storativity respectively.

4. CONCLUSION

An ANN solution technique is proposed to simultaneously identify the transmissivity and storativity distributions of an aquifer system. A hypothetical aquifer is modeled on MODFLOW for transient flow conditions, and its results are used as observed values to train and validate the developed ANN model. The proposed ANN model requires Cartesian coordinates, sampled piezometric heads, and sampling times as input and associated transmissivities and storativities as output. The number of hidden layer neurons is determined using try and error. The synaptic weights of the ANN model are defaults. To prevent potential over-training problems, the available data set is divided into three parts (training, validation, and testing) and overtraining in the model is controlled using the same synaptic weights in the validation part. After obtaining the synaptic weights, transmissivity and storativity distributions are predicted head values for different simulation times. Results showed that predicted transmissivity and storativity distributions are very close to actual values. In addition, the performance of the ANN model is tested for noise data conditions in the observed head values. For this purpose, the available observed piezometric heads are corrupted with Gaussian noise of zero mean and different standard deviations. Results showed that there is no significant change in the prediction performance of the proposed ANN model and this model may be applied to real aquifer parameter estimation problems.

REFERENCES

1. Mehrotra P, Quaicoe JE, Venkatesan R (1997) Induction motor speed estimation using artificial neural networks. Canadian Conference on Electrical and Computer Engineering
2. Aziz ARA, Wong KV(1992)A neural network approach to the determination of aquifer parameters. Ground Water 30: 164-166.
3. Zio E (1997) Approaching the inverse problem of parameter estimation in groundwater models by means of artificial neural network. Prog Nuclear Energy 31: 303-315.
4. Balkhair KS(2002)Aquifer Parameters Determination for Large Diameter Wells Using Neural Network Approach. J Hydrology 265: 118-128.
5. Shigdi A, Gracia LA (2003)Parameter estimation in ground-water hydrology using artificial neural networks. J Comput Civil Eng17: 281-289.
6. Karahan H, Ayvaz MT(2004) Forecasting Parameters Using Artificial Neural Networks. J Porous Media 5: 43-49.
7. Li S, Liu Y (2005)Parameter Identification procedure in Groundwater Hydrology with Artificial Neural Networks ICIC 2005, Part II, LNCS 3645, 276-285.
8. Morshed J, KaluarachchiJJ (1998)Parameter estimation using artificial neural Network and genetic algorithm for free-product migration and recovery. Water Resour Res 34: 1101-1113.
9. Lin GF, Chen GR (2006)An improved Neural Network approach to the determination of aquifer parameters. J of Hydrology 316: 281-289.
10. Lohani AK, Krishan G (2015)Application of artificial neural network for groundwater level simulation in Amritsar and Gurdaspur districts of Punjab, India. J Earth Sci Climate Change 6: 274.
11. Lohani AK, Krishan G(2015)Groundwater level simulation using artificial neural network in southeast, Punjab, India. J GeologyGeophy 4:206.
12. Harrouni K,Ouazar D, Wrobel LC, Cheng AHD (1997) Groundwater parameter estimation by optimization and DRBEM. Eng Anal Boun Elem 19: 97-103
13. Maier HR, Dandy GC (1998)The Effect of Internal Parameters and Geometry on the Performance of Back-Propagation Neural Networks: An Empirical Study. Environmental Modeling & Software 13: 193-209.

14. Coppola E, Szidarovszky F, Poulton M,Charles E (2003) Artificial neural network approach for predicting transient water levels in a multilayered groundwater system under variable state, pumping, and climate conditions. J HydrolEng 8: 348-360.
15. Daliakopoulos IN, Coulibaly P, TsanisIK (2005) Groundwater level forecasting using artificial neural networks. J Hydrol 309: 229-240.
16. Lingireddy S (1998) Aquifer Parameter Estimation using genetic Algorithms and Neural networks. Civil EngrgEnvSyst 15:125-144.
17. Garcia LA, Shigidi A (2006) Using neural networks for parameter estimation in ground water. J Hydrol 318: 215-231.
18. Karahan H, Ayvaz MT (2006) Forecasting aquifer parameters using artificial neural networks. J Porous Media 9: 429-444.
19. Maier HR, Dandy GC (2000) Neural Networks for the Prediction and Forecasting of Water Resources Variables: A Review of Modeling Issues and Applications. Environmental Modeling & Software 15: 101-124.
20. Masoud AA, Atwia MG (2011) Spatio-temporal Characterization of the Pliocene Aquifer Conditions in Wadi El-Natrun Area, Egypt. Environmental Earth Sciences 62: 1361-1374.
21. Freeze RA, Cherry JA(1979) Groundwater: Englewood Cliffs, NJ, Prentice-Hall, p: 604.
22. ASCE Task Committee: on Application of Artificial Neural Networks in Hydrology I: Preliminary Concepts. J of Hydrologic Eng 5: 115-123.
23. CoppolaEA, RanaAJ, Poulton MM, Szidarovszky F, Uhl VW(2005)A Neural Network Model for Predicting Aquifer Water Level Elevations. Ground Water 43: 231-241.
24. Nayak PC, Rao YRS, SudheerKP(2006) Groundwater Level Forecasting in a shallow Aquifer Using Artificial Neural Network. Water Resour Manage 20: 77-90.
25. Karahan H, Ayvaz MT (2005) Groundwater parameter estimation by optimization and dual reciprocity finite differences method. J Porous Media 8: 211-223.
26. Karahan H, Ayvaz MT (2008) Simultaneous parameter identification of a heterogeneous aquifer system using artificial neural networks. Hydrogel J 16: 817-827.
27. Ahmed SA (1999) Hydrogeological and isotope assessment of groundwater in Wadi El-Natrun and Sadat City, Ain Shams University, Cairo, Egypt, p: 237.
28. Shata AA, El-Fayoumy IF (1970) Geomorphology, geology, and, the hydrogeology soil of Wadi El-Natrun- Maryut Agriculture Project Sympos. Hydrology of Nile Delta, UNISISCO 2: 385-396.
29. General Petroleum Company (GPC) (1977) Progress Report on geology and Stratigraphy of western Desert, Cairo, Egypt, p:88.
30. SaadKF (1962) Groundwater hydrology, in preliminary report on the geology, hydrogeology, and groundwater hydrology of Wadi El-Natrun and adjacent area, Part 3, Cairo, U.A.R., Desert Institute, The General Desert Development Organization, p:61.
31. RIGW/IWACO (1990) Hydrogeological inventory and development plan of groundwater Western Nile Delta region, Report, TN 77.01300-90-02.
32. Mostafa NE (1993) Hydrogeological and Hydro studies on Wadi El-Farigh Area, Western Desert, Egypt, M.Sc. Thesis, Faculty of Science, El-Minufiya University Shibin El-Kom, Egypt, p: 132.
33. El-Sheikh (2000) Hydrogeology of the area north and west of Wadi El-Natrun, M. Sc. Thesis Faculty of Science, MinufiyaUniv, Egypt, p: 151.

CHAPTER 18

Geological Control upon Groundwater Flow and Major Ion Chemistry with Influence on Basin Management in a Coastal Aquifer, South Australia

Nara Somaratne, Jacqueline Frizenschaf

South Australian Water Corporation, 250 Victoria Square, Adelaide, Australia

ABSTRACT

Estimation of natural recharge and potential for seawater intrusion are critical considerations for management of coastal freshwater aquifers. We show hydrochemical signatures of groundwater to identify the influence of geological control on chemical processes in a coastal groundwater system. We used dominant hydrochemical facies, salinity and magnesium ions to determine two main groundwater flow paths with different origins and ages. Mixing of groundwater with different origins and ages results in unreliable recharge estimates using chlorofluorocarbon (CFC) and chloride mass balance (CMB) methods, thus limiting available methods for recharge assessment. Interpretation of hydrochemical data suggests that calcium carbonate dissolution, ion exchange processes and mixing with sea aerosol in coastal zones are the main influencing factors on groundwater chemistry.

Restricted groundwater flows due to occurrence of a basement high at the southern side of the basin boundary influence the distance to the toe of the saline wedge. Thus, knowledge of geological control over groundwater systems forms an important part of characterising basins and contributes toward effective management of groundwater resources.

Keywords: Hydrochemistry; Groundwater Mixing; Groundwater Management; Recharge

1. INTRODUCTION

Spatial distribution of groundwater chemistry is primarily influenced by mineralogical composition of sediments in contact with groundwater [1]. The nature and distribution of aquifers and aquitards in a geologic system are controlled by the lithology, stratigraphy and structure of geologic formations. The main geological controls are the stratigraphy and structural features such as cleavages, fractures, faults, and solution features which are the geometrical properties of the geologic systems. The influence of geological control changes natural water chemistry and therefore knowledge of main hydrochemical processes that control groundwater chemistry is a prerequisite for effective basin management [2]. Several studies examined the influence of geological controls: the presence of faults [3,4]; geology and structure of bedrock [5,6]; fracture control [7] which influences groundwater flow and water chemistry; and hydrogeological study of groundwater flow diversion caused by anticlinal structures [8].

Among geological controls, the presence of faults and their influence on flow paths and water chemistry are widely reported and these are known to be effective flow barriers [3]. Green *et al.* [4] reported flows across regional scale faults and found that groundwater flow either enhanced or inhibited, depended on orientation of fractures with respect to the cross fault hydraulic gradient. Bensel *et al.* [9] and Meeder [10] related the impact of faults on near surface processes to surface features such as vegetation patterns and drainage networks; and thus, faults can have great influence on transport processes in the sub-surface and are important in water management.

Eaton et al. [7] noted that there are few studies on the hydrogeology of sedimentary rock aquitards, although they are important controls in regional ground water flow systems. Cherry *et al.* [11] highlighted that determination of aquitard integrity is an important aspect of groundwater management because it influences water quality. Thus, the presence or absence of an aquitard, which is not necessarily a confining unit, can have a significant impact on basin water quality. We examined effects of basement rock outcrop and discontinuity of the clay aquitard on groundwater salinity, water chemistry, influence on recharge estimation methods and extent of the saline wedge in a coastal aquifer in South Australia.

2. STUDY AREA

The Uley South basin is located in the Southern Eyre Peninsula in South Australia (Figure 1). Average annual rainfall and pan evaporation is 550 mm and 1550 mm, respectively. The basin is used for town water supply with current allocation of about 6.8×10^6 m^3 per year. The geology and hydrogeology of the basin is characterised based on: Segnit [12], Johns [13], Morton and Steel [14], Wilson [15], and Shepherd [16]. These works include development of a geological model from borehole information, geological surface mapping, delineation of saturated limestone boundaries and pumping tests with attention focussed on Uley Wanilla, Uley East and Uley South groundwater lenses.

The aquifer system in Uley South is comprised of Quaternary Bridgewater Formation limestone (QL aquifer), beneath this is the Upper Tertiary Clay (Uley Formation) unit with thickness of 5 - 25 m, which forms an aquitard between the Tertiary Sand (TS) of Wanilla Formation and the Quaternary aquifer systems [14]. Recent studies [17,18] show that there is a hydraulic connection between the Uley South unconfined QL aquifer and the Uley Wanilla and Big Swamp areas through the TS aquifer (**Figures 1** and **2**). The basin is topographically closed and the low-lying central portion contains numerous sinkholes. Runoff is highly ephemeral, occurs only after moderate to high intensity rainfall and persists only tens to hundreds of meters before entering a sinkhole [17, 19,20]. Surface drainage is poorly developed as is typical of semi-arid karstic terrains that support only thin soils combined to produce scant vegetation. The groundwater flow direction in Uley South is from north-east to southwest.

An important surface feature is the Big Swamp (**Figure 1**) with an underlying Tertiary Clay with the primary mechanism of discharge through evaporation. In wetter years, Big Swamp overflows and infiltrates into the Quaternary Limestone unconfined aquifer and leaks into the TS aquifer raising salinity up to 5200 mg/L near Big Swamp.

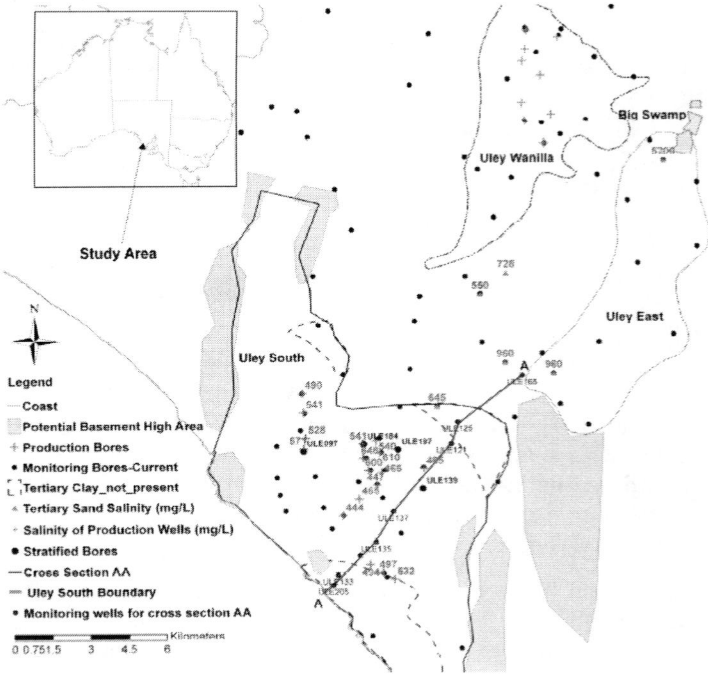

Figure 1. Location of the study area.

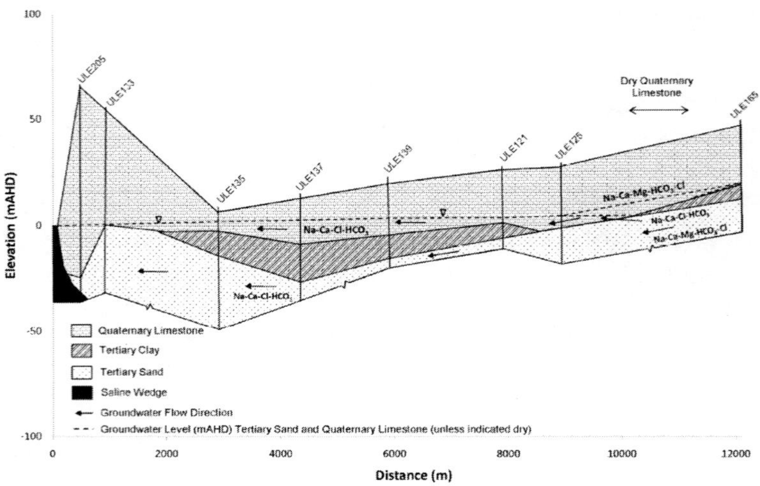

Figure 2. Cross section AA from the coast in the Uley South basin to Uley East basin’s western edge.

3. RESULTS AND DISCUSSION

The influence of geological control is divided into three sections: influence on QL aquifer salinity, hydrochemistry of QL groundwater, and the effect of extension of the saline wedge in TS aquifer. These are important factors in managing the Uley South basin.

3.1. Influence on Salinity

In Uley South, the occurrence of basement high at the basin boundary and discontinuity of aquitards and solution features constitute the most important controls upon salinity and water chemistry. Salinity measurements taken from the town water supply (TWS) wells indicate higher salinity in the north-central portion than elsewhere in the basin (**Figure 1**). Salinity distribution in the basin is influenced by two main geological controls: the occurrence of basement high in the southern half of the basin, and the area of Tertiary Clay absence in the central part (**Figures 1** and **2**).

Harrington *et al.* [17] and Somaratne [18] show that the high-salinity plume of groundwater in the TS aquifer originates from the Big Swamp area as a result of downward leakage of Big Swamp's surface water through the clay aquitard. As the plume moves down gradient along its main flow path, the western edge of the Uley East basin, it is diluted from 5200 mg/L salinity at Big Swamp to 960 mg/L at the southern boundary of the Uley East basin. The area where aquitard is absent in Uley South, the TS aquifer receives direct recharge, resulting in further dilution of salinity to 645 mg/L (**Figure 1**). As a result, pronounced salinity stratification is found in QL aquifer monitoring well ULE197, down-gradient to the Tertiary Clay absence area (**Figure 3**(**a**)). In the ULE197 well, salinity of the QL base is identical to that in TS aquifer monitoring well ULE109, located at the same site. Harrington *et al.* [17] indicated that residence time in Big Swamp and Uley Wanilla groundwater age is >45 years. Mixing of old waters originating from Uley Wanilla and Big Swamp with younger water (<20 years) in Uley South has important ramifications on applicable recharge estimation methods. When mixing occurs of different groundwater systems with different ages and different chlorides concentrations, this results in unreliable recharge estimates using chlorofluorocarbon [21,22] and chloride mass balance methods [23,24], thus limiting the methods available for recharge assessment.

In contrast, the southern half of the QL aquifer has no link to the Big Swamp area through either QL or TS aquifers due to occurrence of the basement high at the Uley South boundary (Figure 1). Therefore, QL water in the southern part of Uley South is primarily derived from direct rainfall recharge within Uley South. ULE139 is a QL aquifer monitoring well, which shows no evidence of salinity stratification (**Figure 3**(**b**)).

3.2. Influence on Hydrochemical Signatures

The combination of aquifer heterogeneity, low hydraulic gradient, recharge through solution features in addition to diffuse recharge, mixing with different water types and modification along flow paths results in Uley South being hydrochemically complex. Hydrochemistry was assessed principally by major ion chemistry within the Uley Wanilla, Uley East and Uley South lenses, and how these are related to evolution of hydrochemical processes in the Uley South basin.

Groundwater can be classified into "water types" on the basis of the absolute and relative abundance of major ions [25]. However, due to the predominance of Na and Cl ions in both QL and TS aquifers, the major ion differences present are inconclusive in distinguishing between QL and TS waters. For all QL and TS groundwaters, Ordens *et al.* [20] has shown that Cl/Br in both rainfall and groundwater samples are close to the Cl/Br ratio of seawater indicating groundwater chloride are marine derived implying that both QL and TS waters may have similar chemical histories.

Discrimination based on relative major ion abundance is more successful where recharging freshwater is of continental (Ca-HCO_3 type) rather than marine (Na-Cl type) origin as in Uley South. QL and TS groundwater are analysed for pH, electrical conductivity, anions (HCO^-_3,SO^{2-}_4 CL^-,) and cations (Ca^{2+}, Mg^{2+}, Na^+, K^+). Although the majority of water has no dominant ion composition, Uley Wanilla water contains distinctively high Mg ions (**Figure 4**) as observed by Harrington *et al.* [17]. High chloride in TS waters is derived from the Big Swamp area and the Na-Cl type water is diluted towards Na-Cl-HCO_3 waters at Uley South. This represents either waters that have chemically evolved along flow paths or that result from mixing between Na-HCO_3 waters and Na-Cl waters.

Classification of groundwater into water types on the basis of major ion composition and other selected characteristics provides a framework for interpretation of groundwater pathways [26]. Eighty-five groundwater samples were used for QL aquifer. This includes the data of Evans [19] and town water supply wells' water quality data. For the TS aquifer, twenty-seven water samples were taken from Evans [19].

Hydrochemical examination shows that carbonate dissolution and ion exchange are the two dominant chemical processes active in the QL aquifer. Water types are categorised into four types of groundwater, which represents a chemical evolution from Na-Mg-Cl-HCO_3 groundwater at Uley Wanilla lens to end member Na-Cl groundwater at Big Swamp (**Figure 5**).

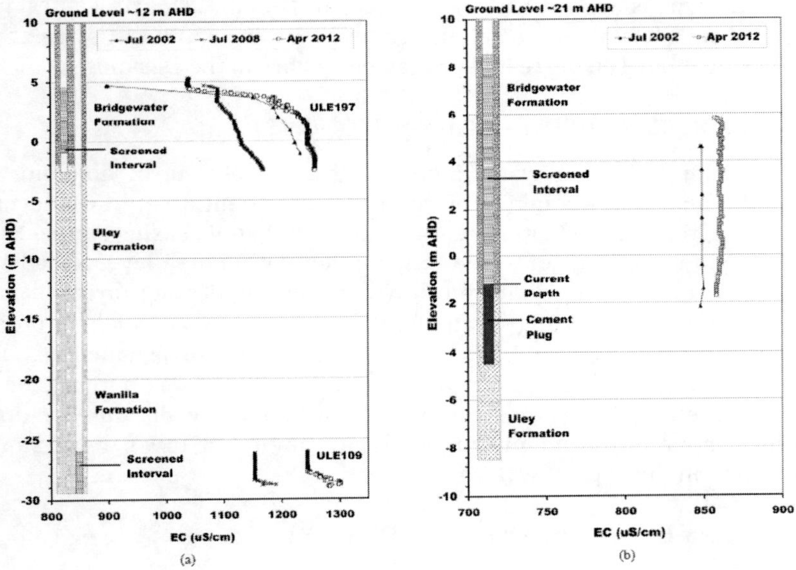

Figure 3. Salinity stratification of QL groundwater in the central part of the basin. (a) ULE109 and ULE197; (b) ULE139.

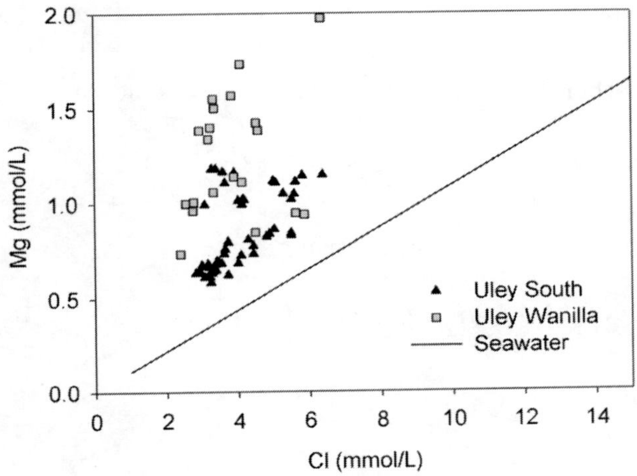

Figure 4. Cl vs Mg in Uley Wanilla and Uley South lenses.

3.2.1. Na-Cl or Na-Ca-Cl (1)

In the absence of evaporites, this type of water occurs in the Big Swamp due to enrichment of Na and Cl ions though evaporation. The end member of this water type is Na-Ca-Cl-HCO_3 or Na-Ca-HCO_3-Cl in the north-central part of Uley South basin, where salinity of town water supply wells are about 650 mg/L; about 100 mg/L higher than elsewhere in the QL aquifer.

3.2.2. Na-Mg-Ca-Cl-HCO_3/Na-Ca-Mg-HCO_3-Cl (2)

Na-Mg-Ca or Na-Ca-Mg water indicates that dissolution of dolomite (or dolomitic limestone) is taking place in the catchment area. In Quaternary Limestone, this type of water occurs in the Uley Wanilla basin (**Figure 6**). In the TS aquifer, this type of water occurs either within the Uley Wanila basin or directly down-gradient in the flow direction, indicating direct leakage (interconnection between two aquifers) from QL to TS aquifer. This type of water in the Uley South QL aquifer is found only in production bores directly down gradient to Tertiary Clay absence area as a result of QL aquifer receiving TS waters. Apart from production wells directly down gradient, none of the monitoring wells nor any other town water supply wells contain this type of water.

3.2.3. Ca-Na-HCO3-Cl/Ca-Na-Mg-HCO3-Cl (3)

This is the typical water type found in limestone aquifers. The Ca-Na-HCO_3-Cl water type represents recently recharged water with an abundance of dissolved CO_2; thus calcite dissolution is taking place. This water type occurs in the southern part of the central basin where there is no TS water mixing with Uley South QL aquifer water. Uley South basin being a karstic limestone aquifer, sink- holes collect runoff containing organics; this provides sufficient CO_2 for the following reactions to dissolve calcite.

The main mechanism for dissolution of calcite is as follows:

$$CaCO_3 + CO_2 + H_2O = Ca^{2+} + 2HCO_3^- \quad (1)$$

The above reaction includes the following steps:

$$CO_2 + H_2O = H^+ + HCO_3^- \quad (2)$$

$$CaCO_3 + H^+ = Ca^{2+} + HCO_3^- \quad (3)$$

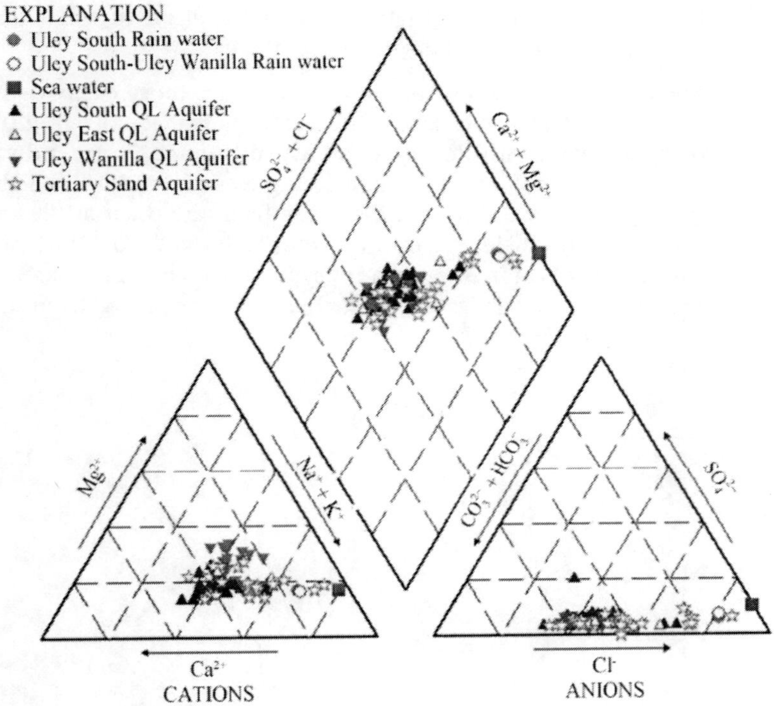

Figure 5. Piper diagram showing water types

3.2.4. Na-Ca-Cl-HCO3/Na-Ca-HCO3-Cl (4)

Na-Ca-HCO_3-Cl water types represent Ca-HCO_3 water affected by mixing with Na-Cl waters or that has undergone cation exchange processes. The most common occurrence of such water is in the central basin (**Figure 6**), where high salinity waters originating from Big Swamp enter the QL aquifer. In addition another possible reason for the occurrence of this water type is cation exchanges, along with sea aerosol in the coastal zone.

3.3. Influence on Coastal Salinity

Seawater intrusion is generally identified through chemical analysis of groundwater. No single analysis definitively identifies seawater intrusion [27]; however by examining various analysis methods, it is evident when fresh groundwater mixes with seawater. According to HydroMetric WRI [27], at low chloride concentrations, it is often difficult to identify incipient seawater intrusion. This is due to the natural variation in fresh water chemistry at chloride concentrations below 1000 mg/L.

Mixing trends between groundwater and seawater are more easily defined when chloride concentrations exceed 1000 mg/L, as is evident in **Figure 7** where mixing increases with depth. As shown in **Figure 7**, groundwater mixing takes place in coastal aquifer monitoring wells from Na-Ca-Cl-HCO_3 to Na-Cl water types at two different distances from the coast at the same elevation. On the northern side, TS groundwater is derived from downward leakage from the Uley Wanilla-Big Swamp system, which is approximately 22 km from the coastline. TS receives further direct recharge from areas where Tertiary Clay is absent.

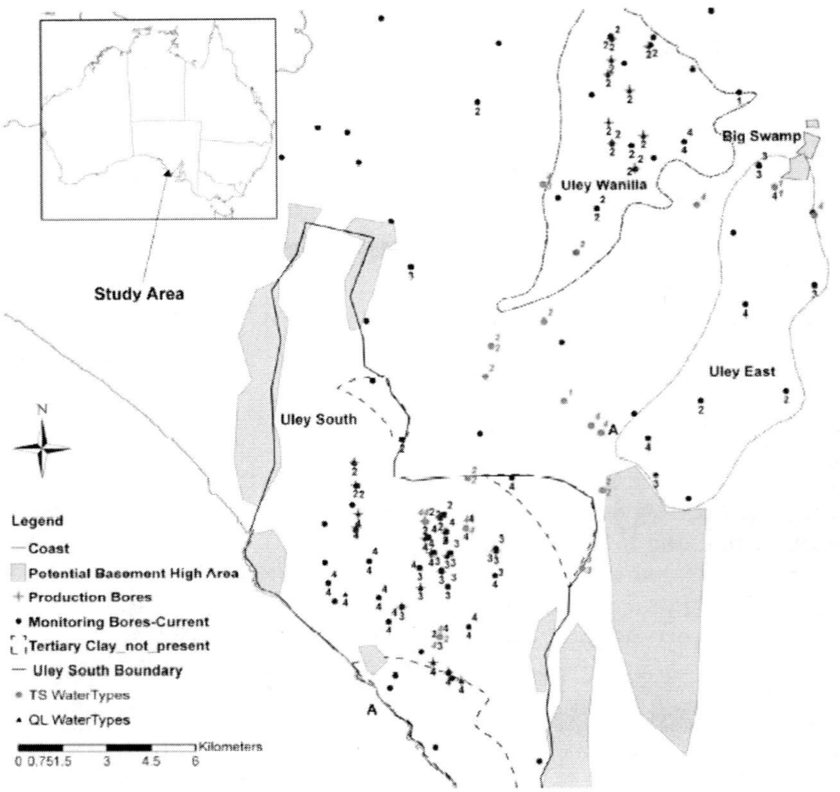

Figure 6. QL and TS aquifer water type distribution (1 = Na-Cl or Na-Ca-Cl; 2 = Na-Mg-Ca-Cl-HCO_3 or a-Ca-Mg-HCO_3-Cl or Ca-Na-Mg-HCO_3-Cl; 3 = Ca-Na-HCO_3-Cl or Ca-Na-Cl-HCO_3; 4 = Na-Ca-Cl-HCO_3 or Na-Ca-HCO_3-Cl).

Therefore, it is conceptualised that a large quantity of TS groundwater flows on the northern side with a higher hydraulic gradient (0.00458). This makes the distance to the toe of the saline wedge in TS smaller, less than 500 m on the northern side. In contrast, the TS aquifer on the southern side receives leakage from the QL aquifer within Uley South basin. The extent of the TS is limited only to the Uley South boundary, which is on average approximately 5 km from the coast, yielding an average hydraulic gradient of 0.00117. This results in comparatively small groundwater flows through TS along the southern side, creating a longer saline wedge for pressure balance at the same elevation. Distance to the toe of the saline wedge is about 900 m from the coastline on the southern side. In Uley South coastal aquifer monitoring wells, sea water is intercepted in Tertiary Sand aquifer in SLE069 at −23 m AHD and in ULE205 at −30 m AHD (**Figure 7(b)**) showing increasing salinity with depth. No QL aquifer coastal bores intercept saline waters indicating that the saline wedge in QL aquifer is less than 500 m.

4. CONCLUSIONS

We highlight the importance of integrating knowledge of hydrochemistry of groundwater sources into estimation of recharge and the potential for saline intrusion, both of which are essential components in assisting with effective management of groundwater resources. In particular, we analysed the groundwater chemistry from the Uley South coastal aquifer in South Australia, and examined the processes that control groundwater chemistry. The major outcome of this study is identification of two distinct groundwater sources due to geological controls and their influence on groundwater chemistry. The geologic controls that led to these two distinct sources of different origin and age, were evidenced in the presence of two features, basement rock highs and absence of an aquitard.

Evaluation of hydrochemical data suggests that calcium carbonate dissolution, ion exchange processes and mixing with sea aerosols in the coastal zone are the main factors influencing groundwater chemistry. When the two groundwater sources with different ages and different chloride concentrations mix, results in unreliable recharge estimates are likely to result in using CMB and CFC methods. In addition, the subsurface basement rock high influences the currently observed seawater wedge. The toe of the saline wedge has been observed between 900 m and 500 m inland. Thus knowledge of geological control in the groundwater system, as well as the hydrogeochemical makeup of groundwater source(s), forms important parts of the characterisation of groundwater basin, ultimately contributing to effective management of the groundwater resources.

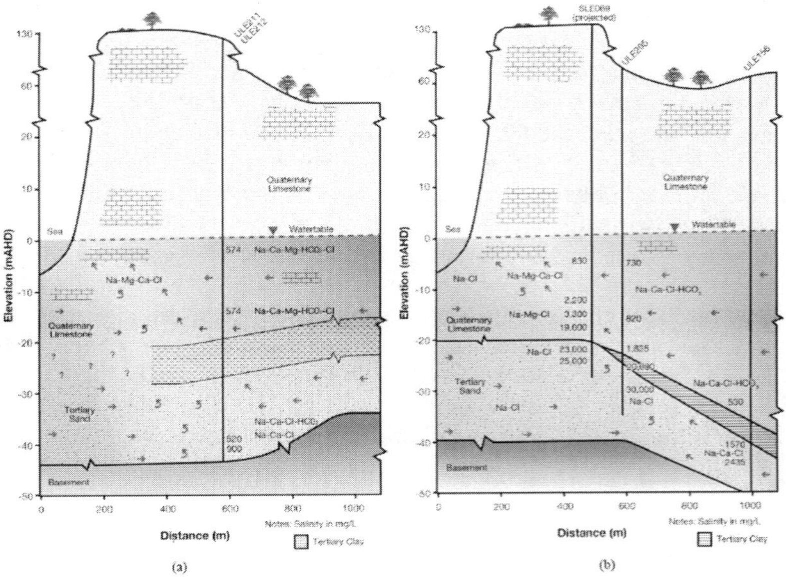

Figure 7. Uley South Saline Wedge. (a) Northern side; (b) Southern side.

ACKNOWLEDGEMENTS

The editor and reviewers are thanked for evaluating this paper.

REFERENCES

1. L. Lakshmanan, R. Kanan and M. Senthi Kumar, "Major Ion Chemistry and Identification of Hydrogeochemical Processes of Groundwater in a Part of Kancheepuram District, Tamil Nadu, India," Environmental Geosciences, Vol. 10, No. 4, 2003, pp. 157-166. http://dx.doi.org/10.1306/eg100403011
2. M. Lorite-Herrera and R. Jiménez-Espinosa, "Impact of Agricultural Activity and Geologic Controls on Groundwater Quality of the Alluvial Aquifer of the Guadalquivir River (Province of Jaén, Spain): A Case Study," Environmental Geology, Vol. 54, No. 7, 2008, pp. 1391-1402. http://dx.doi.org/10.1007/s00254-007-0920-6
3. R. M. Delinom, "Structural Geology Controls on Groundwater Flow: Lembang Fault Case Study, West Java, Indonesia," Hydrogeology Journal, Vol. 17, No. 4, 2009, pp 1011-1023. http://dx.doi.org/10.1007/s10040-009-0453-z
4. G. Green, E. Watts, D. Alcoe, A. Costar, and L. Mortimer, "Groundwater Flow across Regional Scale Fault," Department for Water, South Australia, Technical Report DFW 2010/15, 2010.
5. S. T. Finch, A. Mccoy and E. Melis, "Geologic Controls on Groundwater Flow in the Mimbres Basin, South Western New Mexico," New Mexico Geological Society Guide Book, 59th Field Conference, 2008, pp. 189-198.

6. J. K. Mulwa, S. J. Garciri, J. O. Barongo, N. OpiyoAkech and G. K. Kianji, "Geological and Structural Control Influence on Groundwater Distribution and Flow in Ngong Area, Kenya," African Journal of Science and Technology, Science and Engineering Series, Vol. 6, No. 1, 2005, pp. 105-115.
7. T. T. Eaton, M. P. Anderson and K. R. Bradbury, "Fracture Control of Ground Water Flow and Water Chemistry in a Rock Aquitard," Groundwater, Vol. 45, No. 5, 2007, pp. 601-615. http://dx.doi.org/10.1111/j.1745-6584.2007.00335.x
8. T. E. S. Odeh, M. Schirmer and G. Strauch, "Structural Control of Groundwater Flow Regimes and Groundwater Chemistry along the Lower Reaches of the Zerka River, West Jordan, Using Remote Sensing, GIS and Field Methods," Vol. 58, No. 8, 2009, pp. 1707-1810.
9. V. F. Bensel, R. T. Van Balen and J. J. De Vries, "The Impact of Faults on the Hydrogeological Conditions in the Roer Valley Rift System: An overview," The Netherlands Journal of Geosciences/Geologie en Mijnbouw, Vol. 82, No. 1, 2003, pp. 41-54.
10. J. Meeder, "Karst Control of the Biscayne Aquifer Groundwater Flow Patterns. The Biscayne and Other Eogenetic Karst Aquifers: Characterization, Modeling, and Management," Geological Society of America, Annual Meeting and Exposition, Paper 77-6, 2012.
11. J. A. Cherry, B. L. Parker, K. R. Bradbury, T. T. Eaton, M. G. Gotkowitz, D. J. Hart and M. A. Borchardt, "Role of Aquitards in the Protection of Aquifers from Contamination: A State of the Science Report," AWWA Research Foundation, Denver, 2004.
12. R. W. Segnit, "Final Report on the Uley Wanilla Fresh Water Basin, Eyre Peninsula," South Australian Department of Mines, 1942.
13. R. K. Johns, "Geology and Mineral Resources of Southern Eyre Peninsula," Geological Survey of South Australia, Bulletin 37, 1961.
14. W. Morton and T. M. Steel, "Eyre Peninsula Groundwater Study-Uley South Basin, Progress Report 1, Southern Eyre Peninsula," Mineral Resources Reviews, South Australia, 128, 1970, pp. 33-48.
15. C. C. Wilson, "Geology of the Quaternary Bridgewater Formation of Southwest and Central South Australia," PhD Thesis, Flinders University of South Australia (unpublished), 1991.
16. R. G. Shepherd, "Report on Underground Water Investigations-Sleaford Bay-Coffin Bay Area," Department of Mines (Unpublished Report), 1963.
17. N. Harrington, D. Zulfic and D. Wohling, "Uley Basin Groundwater Modelling Project. Vol. 1: Project Overview and Conceptual Model Development," DWLBC Report 2006/01, Government of South Australia, 2006.
18. N. Somaratne, "Hydrogeology of the Uley South Basin," SA Water Hydrogeological Research, SA Water/2013-1, 2013.
19. S. L. Evans, "Estimating Long-Term Recharge to Thin, Unconfined Carbonate Aquifers Using Conventional and Environmental Isotopes Techniques: Eyre Peninsula, South Australia," Masters Thesis, Flinders University of South Australia (unpublished), 1997.
20. C. M. Ordens, A. D. Werner, V. E. A. Post, J. L. Hutson, C. T. Simmons and B. M. Irvine, "Groundwater Recharge to a Sedimentary Aquifer in the Topographically Closed Uley South Basin, South Australia," Hydrogeology Journal, 2011.
21. G. S. Weissmann, Y. Zhang, E. M. LaBolle and G. E. Fogg, "Dispersion of Groundwater Age in an Alluvial Aquifer System," Water Resources Research, Vol. 38, No. 10, 2002, p. 1198. http://dx.doi.org/10.1029/2001WR000907
22. L. N. Plummer, E. Busenberg and L. F. Han, "Chapter 5: CFCs in Binary Mixtures of Young and Old Groundwater. Use of Chloroflurocarbons in Hydrology: A Guide Book," International Atomic Energy Agency, 2006.
23. W. W. Wood and W. E. Sanford, "Chemical and Isotopic Methods for Quantifying Groundwater Recharge in a Regional Semi-Arid Environment," Ground Water, Vol. 33, No. 3, 1995, pp. 458-468. http://dx.doi.org/10.1111/j.1745-6584.1995.tb00302.x
24. W. W. Wood, "Use and Misuse of the Chloride Mass Balance Method in Estimating Groundwater Recharge," Groundwater, Vol. 37, No. 1, 1999. http://dx.doi.org/10.1111/j.1745-6584.1999.tb00949.x

25. E. Custodio, "Hydrogeochemistry and Traces," In: E. Custodio and G. A. Brugeman, Eds., Groundwater Problems in Coastal Areas, Studies and Reports in Hydrology, Vol. 45, 1987, pp. 213-269.
26. P. Heisig, "Groundwater Resources of the Clifton Park Area, Saratoga County, New York," US Geological Survey, Water Resources Investigation Report 01-4104, 2002.
27. W. R. I. HydroMetrics, "Seawater Intrusion Analysis Report, Seaside Basin, Monterey County, California," 2011.

INDEX

W